A FIELD GUIDE TO THE

REPTILES AND AMPHIBIANS

OF BRITAIN AND EUROPE

Collins is an imprint of HarperCollinsPublishers Ltd.
77–85 Fulham Palace Road
London
W6 8JB

The Collins website address is: www.collins.co.uk

First edition published 1978
Second edition published 2002

2 4 6 8 10 9 7 5 3 1

02 04 06 08 07 05 03

ISBN 0 00 219964 5

Colour reproduction by Colourscan, Singapore
Printed and bound by Printing Express Ltd.

HxR

A FIELD GUIDE TO THE

REPTILES AND AMPHIBIANS

OF BRITAIN AND EUROPE

E. Nicholas Arnold

Illustrated by
Denys W. Ovenden

HarperCollins*Publishers*

PREFACE TO THE SECOND EDITION

The second edition of this book is rather different from the first which appeared 24 years ago in 1978. Some 201 species are included compared with just 125 in the original version and the number of amphibians alone have changed from 45 to 74 species, an increase of about 65%. This contrasts with the situation in, say, European birds and mammals where alterations have been quite trivial over the same period. Some of the increase results from a slight extension in the area covered by the book. The Canary islands and Madeira are now included as well as the small Greek islands that lie along the west coast of Asiatic Turkey, and this adds about 20 species. A couple of introductions whose fate was initially uncertain are now well established and one or two forms in neighbouring areas have been found to extend naturally into Europe as well. But there have also been totally new discoveries. These include the Majorcan Midwife Toad, a form first encountered as a fossil but found alive in 1980, the large Tenerife Speckled Lizard discovered in 1997 and the Gomera Giant Lizard first seen alive by scientists in 2000. The rest of the increase comes from study of well known forms that turn out to really consist of more than one species. The main source of these discoveries, especially among amphibians, has been the use of novel techniques involving studies of proteins and DNA sequence. Over the period between the two editions there has also been a huge increase in knowledge of the natural history of European reptiles and amphibians, something that has expanded almost exponentially, especially in the south of the continent. This book attempts to summarise much of this new information and provides a synopsis of knowledge about two of the most interesting animal groups in Europe.

ACKNOWLEDGEMENTS

I am extremely grateful to Denys Ovenden who illustrated both the first and second editions of this field guide, and to John A. Burton who also participated in the first edition.

I have also been greatly helped by past and present colleagues at the Natural History Museum, London, including Salvador Carranza, Colin McCarthy, Barry Clarke, Garth Underwood, James Harris, Mark Wilkinson and the staff of the Libraries. I should like to thank the Trustees of the Museum for access to these and to the research collections over the years. Richard Ranft of the British Library provided information on recordings of frog and toad calls. Other people who have freely shared their knowledge of European reptiles and amphibians or helped in other ways include, in no special order, Keith Corbett, Gabriel King, Elizabeth Platts, Angus Bellairs, Alistair Reid, Anthony Chaplin, Christopher Lever, Miranda Armour-Chelu, Edward Wade, Anthea Gentry, Herman in den Bosch, Marinus Hoogmoed, Raoul van Damme, Bieke van Hooydonk, Dirk Bauwens, Roger Bour, Alain Dubois, Ann Marie Ohler, Wolfgang Böhme, Konrad Klemmer, Ulrich Joger, Beat Schätti, Valentin Peréz Mellado, J. Mateo, Aurora Castilla, J. Castroviejo, Luis Felipe López-Jurado, Alfredo Salvador, Peter Hopkins, Claudia Corti, Benedetto Lanza, G. Lanfranco, Eustatis Valakos, Panagiota Maragou, Chloe Adamopoulou, Richard Clark, Tom Uzzell, Hans-Jurg Hotz, and Renate, Nicholas and Christopher Arnold. Katie Piper and Myles Archibald greatly facilitated the production of this book and showed great forebearance.

Finally, I owe a great debt to all those who have studied and written about European reptiles and amphibians over the past 150 years, especially George A. Boulenger.

CONTENTS

How to use this book   **9**

Introduction   **12**
The biology of reptiles and amphibians   13
Reptiles, amphibians and people   19
Reptiles and amphibians in the British Isles   24

Salamanders and Newts   **28**
Key   28
Typical salamanders and newts, *Salamandridae*,   30
    Key to Pond newts, *Triturus*   41
Cave salamanders, *Plethodontidae*   50
Olm, *Proteidae*   53

Frogs and Toads   **54**
Key   55
Fire-bellied toads, Painted frogs and Midwife toads, *Discoglossidae*   59
Spadefoots, *Pelobatidae*   68
Parsley frogs, *Pelodytidae*   71
Typical toads, *Bufonidae*   73
Tree frogs, *Hylidae*   77
Typical frogs, *Ranidae*   80

Tortoises, Terrapins and Sea turtles   **98**
Key   98
Tortoises, *Testudinidae*   100
Terrapins, *Emydidae* and *Bataguridae*   103
Sea turtles, *Dermochelyidae* and *Cheloniidae*   106

Lizards   **111**
Key   113
Agamas, *Agamidae*   118
Chameleons, *Chamaeleonidae*   119
Geckos, *Gekkota*   120
Lacertid lizards, *Lacertidae*   126
Green lizards, *Lacerta*   134
    Key to Green lizards, *Lacerta*   135
Small lacertas, *Lacerta* and *Podarcis*   142
    Area 1: North, western and central Europe   144
    Area 2: Iberia and the adjoining France   147
    Area 3: Balearic Islands   155
    Area 4: Corsica and Sardinia   158
    Area 5: Italy, Sicily and Malta   161
    Area 6: East Adriatic coastal area   165
    Area 7: South-eastern Europe   171

Canary Island lizards, *Gallotia* 180
Skinks, *Scincidae* 185
Slow worms and Glass lizard, *Anguidae* 191
Worm lizards, *Amphisbaenidae* 194

Snakes **195**
Key 196
Worm snakes, *Typhlopidae* 200
Sand boas, *Boidae* 201
Typical snakes, *Colubridae* (Montpellier Snake, Whip snakes,
Dwarf Snake, Rat snakes, Water snakes, Smooth snakes,
False Smooth Snake and Cat Snake) 202
Vipers, *Viperidae* 226
Key to Vipers, *Vipera* 227

Identification of Amphibian Eggs **238**
Key 239

Identification of Amphibian Larvae **242**
Key to Salamander and Newt Larvae 244
Key to Frog and Toad larvae 248

Internal characters **253**

Glossary **255**

Further Reading **260**

Distribution maps **264**

Index **286**

COLOUR PLATES

1. SALAMANDERS 1
2. SALAMANDERS 2
3. NEWTS and SALAMANDERS 3
4. LARGE NEWTS
5. SMALLER NEWTS 1
6. SMALLER NEWTS 2
7. PAINTED FROGS
8. PARSLEY FROGS, MIDWIFE TOADS
9. SPADEFOOTS
10. TYPICAL TOADS
11. FIRE-BELLIED TOADS, TREE FROGS
12. BROWN FROGS 1
13. BROWN FROGS 2
14. WATER FROGS
15. TORTOISES
16. TERRAPINS
17. SEA TURTLES 1
18. SEA TURTLES 2
19. AGAMA and CHAMELEONS
20. GECKOS
21. MISCELLANEOUS LACERTIDS
22. ALGYROIDES
23. GREEN LIZARDS 1: Ocellated and Schreiber's Green Lizard
24. GREEN LIZARDS 2: Green and Balkan Green lizards
25. GREEN LIZARDS 3: Sand Lizard
26. GREEN LIZARDS 4: Juveniles
27. SMALL LACERTAS 1: Viviparous Lizard and Common Wall Lizard
28. SMALL LACERTAS 2a: Iberian peninsula and adjoining France
29. SMALL LACERTAS 2b: Iberian peninsula and Madeira
30. SMALL LACERTAS 3: Balearic Islands
31. SMALL LACERTAS 4: Corsica and Sardinia; Malta

32. SMALL LACERTAS 5: Italy and Sicily

33. SMALL LACERTAS 6: East Adriatic coastal area

34. SMALL LACERTAS 7a: South-eastern Europe, mainland

35. SMALL LACERTAS 7b: Southern Greece (Peloponnese)

36. SMALL LACERTAS 7c: South Balkans and Aegean Islands

37. CANARY ISLAND LIZARDS 1

38. CANARY ISLAND LIZARDS 2

39. SKINKS 1

40. SKINKS 2

41. LIMBLESS LIZARDS

42. WORM SNAKE, SAND BOA, MONTPELLIER SNAKE and HORSESHOE WHIP SNAKE

43. WHIP SNAKES

44. WHIP, DWARF and RAT SNAKES

45. RAT SNAKES

46. WATER SNAKES

47. SMOOTH SNAKES, FALSE SMOOTH SNAKE and CAT SNAKE

48. VIPERS 1

49. VIPERS 2

HOW TO USE THIS BOOK

Identification

Novices will find identification of reptiles and amphibians easier if they first read the Introduction (p.12) and the preliminary remarks at the beginning of the sections on each main animal group:

Salamanders and Newts	p. 28
Frogs and Toads	p. 54
Tortoises, Terrapins and Turtles	p. 98
Lizards	p. 111
Snakes	p. 195

Then the following approach is recommended:

★ Look quickly through the plates and try to match your animal's appearance as closely as possible to one of the species shown, bearing in mind that many species are rather variable.

★ Once this is done, check with the text and maps that you are within the range of the species concerned. This is very important. Apart from sea turtles, most reptiles and amphibians do not move about much and are only rarely found outside their normal ranges.

★ If the range fits, then compare your animal with the description in the **Identification** and **Variation** texts for the species. If it does not agree with these, look at the forms listed under **Similar Species**.

★ Should the above method fail, and the animal can be examined at close quarters, try using the keys. This is rather laborious but often results in more certain identification. Main keys are: Salamanders and Newts, p. 28; Frogs and Toads, p. 55; Tortoises, Terrapins and Turtles, p. 98; Lizards, p. 113; Snakes, p. 196. There are also subsidiary keys in the text to particular groups such as Pond newts and Vipers.

Keys are made up of sections each containing two or more contrasted statements, as in the following example:

1.	Fore-feet with four toes	2
	Fore-feet with three toes	7
2.	Toes stubby	3
	Toes not stubby	5

Starting at section 1, select the statement that agrees most closely with your animal. When you have done this, go to the section of the key indicated immediately after the statement, and so on until you arrive at a positive identification. For instance, in the example above, if your animal has four toes on the fore-feet, proceed to section 2. Having arrived there, check toe shape and if it is not stubby go on to section 5.

The descriptive texts

For easy reference, the text describing each species has been divided into sections as follows.

Names. Both an English vernacular name and a scientific name are given for each species. In most languages including English, only the more obvious kinds of reptile and amphibian have widely accepted names and a number of species are often covered by a single term. The precise vernacular names given in books are often recent inventions and alternatives frequently exist. However, vernaculars will probably stabilise over time. The scientific name consists of two words, the first (always written with a capital letter) is the *genus*, the second the *species* name. In most cases species with the same genus name are rather similar and closely related. The scientific names are used by reptile and amphibian enthusiasts throughout the world, irrespective of their native language. They often simplify communication and are well worth learning.

Range. Ranges of species are described in the text and shown on the maps. An indication of distribution outside Europe is also given.

Identification. Descriptions are often quite short for easily recognised species but longer for the more difficult ones. On occasions technical terms have to be used; these are explained in the Glossary (p. 255) or at the beginning of the relevant main sections, just before the keys. Sizes are given in centimetres (2.54cm = 1 inch) or millimetres (25.4mm = 1 inch). In most cases total length is given, but for lacertid lizards, where the tail is usually long but often broken, the basic measurement is distance from snout to vent. For tortoises and terrapins the length of the upper shell (carapace) is given: this is a straight line measurement, not over the curve of the shell as is usual in sea turtles.

Variation. Individual and regional variation are briefly described here and, where appropriate, subspecies described. (See p. 14).

Similar Species. The species most likely to cause confusion, particularly in the field, are listed. Distinguishing features are also often mentioned but in other cases these may be discussed under **Identification**.

Voice. This is only described for frogs and toads. Some other reptiles and amphibians are capable of producing usually weak grunts, squeaks and hisses but these are rarely of use as identification characters. (See also p. 17).

Habits. As much information as possible about habits is given, with some emphasis on habitat. This is because the place where an animal is found is often characteristic of the species concerned, and as such is a useful clue to identity. Specific information about food is given for forms with specialised diets. Basic facts about reproduction are also listed, although the amount of information available is quite varied. In general within a species, size of eggs and often their number increases with the size of the mother. Egg sizes are given in many cases and often a fairly wide size range is indicated. This is not only because eggs vary intrinsically in their dimensions, in many cases they also swell after being laid.

Venom. Discussion is limited to snakes with venom and distinct fangs (enlarged tubular or grooved teeth that inject the venom), the only species in Europe that might be dangerous to people. Some other snakes appear to produce saliva with toxic properties and many amphibians have poisonous skin secretions (p. 18), but these are only marginally harmful to human beings in normal circumstances.

Other names. Vernacular and scientific names of many reptiles and amphibians are not stable and alternatives may exist. Those that have been used recently are listed here, but there are usually reasons why the names at the head of each species text are preferable. The first part of a scientific name, the genus, is sometimes changed to reflect new ideas of relationships of species. For instance, the Moroccan Rock Lizard has been called *Lacerta perspicillata* for many years but is now sometimes named as *Podarcis perspicillata* to reflect its apparent relationship to Wall lizards, or *Teira perspicillata* to emphasise its supposed relationships elsewhere. Such changes are sometimes unavoidable but they are inevitably confusing for non-specialists (and many specialists too). Fortunately, in many cases including this one, it is reasonable to retain the old name in the interest of easier communication and, where possible, this course has been taken elsewhere.

Family introductions. To avoid frequent repetition in the species texts, many facts common to several different animals are placed in the introductions to families or other major groups. These include the general characteristics of the members of the family or group and often some information on their habits and breeding.

Illustrations. The colour plates have been drawn from life whenever possible and should give a fairly good idea of how animals look in the field. Individuals of most species occurring in Europe are shown in colour. Where they are markedly different, both sexes and juveniles are often illustrated.

Because many reptile and amphibian species are very variable it is not possible to illustrate more than a selection of these differences. Where a choice had to be made, the more usually encountered variants have been shown; thus common patterns are illustrated in preference to rare ones and mainland and large-island forms rather than ones confined to isolated rocks and islets. Further details about variation are given in the texts. Often, the animals illustrated are not typical of any particular locality and, where possible, are 'average', specimens.

Scale. In most cases, the animals in each plate are drawn to the same scale, and unless otherwise stated are approximately life-sized.

INTRODUCTION

The main area covered by this book is shown in the map below. It takes in all of Europe west of a line running just east of Moscow. All east Atlantic and Mediterranean islands that are politically part of European countries are included, ranging from Madeira and the Canary islands in the west to the Greek islands along the coast of Asiatic Turkey in the east, although Asiatic Turkey itself is not included. In this book, Europe, is used for the area defined above, and does not include areas further east formally included in that continent. The maps on pp. 22–23 show the Atlantic and Greek islands in some detail.

The biology of reptiles and amphibians

Reptiles (including tortoises, terrapins, turtles, lizards, and snakes) can be easily recognised by their typically dry, scaly skin, whereas that of amphibians (including frogs, toads, newts and salamanders) is usually moist and lacks obvious scales. Both reptiles and amphibians evolved earlier than mammals and birds. They differ most obviously from these in lacking both hair and feathers and also function on a quite different energy budget. Birds and mammals usually maintain a constant high temperature, for instance about 37°C in people. This gives them many advantages and allows them to be almost constantly active in a wide range of conditions, but it often means that a great deal of internal heat has to be produced, which requires a high food intake. Reptiles and amphibians, on the other hand, have variable temperatures and either live close to the temperature of the air or water that surrounds them or get heat by basking in the sun or making contact with sun-heated surfaces. As a result of such basking, some lizards run at even higher temperatures than people when fully active on hot days. In contrast, in very cold conditions, especially when sun is unavailable, activity for all reptiles and amphibians may become impossible, but this disadvantage is balanced by the fact that little or no internal heat needs to be produced and reptiles and amphibians can therefore manage on a very low food intake. Because of this, they do very well in situations where food is sparse or intermittent, which is why they survive in spite of competition from the more sophisticated birds and mammals.

Species

Species are made up of individual animals that can interbreed with each other and are distinguishable from individuals assigned to other species. In most cases members of species can be recognised by their appearance, but sometimes other characters are used, such as features of internal anatomy, voice, or molecular features. These involve distinctive body chemistry such as the precise kinds of proteins present or the detailed structure of parts of the DNA that form their genes. In many cases, different species have had long separate histories often running into millions of years and are expected to remain separate in the future, if they survive.

Members of one species do not usually hybridise with those of another, especially in the wild. This may be because the species concerned do not come into contact, but breeding may fail to take place even when they do, and this is usual for species coexisting in a particular geographical area. Even when such interbreeding occurs, it is usually not successful. For instance few or no offspring may be produced or their ability to breed or survive is very restricted. As with most generalisations, there are exceptions. For example, Yellow-bellied and Fire-bellied toads interbreed in some very restricted areas. Also, a few species have been produced by hybridisation. This is true of some relatives of the Caucasian Rock Lizard (*Lacerta saxicola*) outside our area. These consist of all-female populations that reproduce without intervention of males and arose from crosses between more normal sexual species. Some special forms of water frog, like the Edible Frog (*Rana* kl. *esculenta*) have also arisen by hybridisation and can usually only continue to exist if they interbreed with one of the parent species (see p. 88).

Variation within species

Within a species, not all animals are exactly alike. Even excluding variation due to sex and age, members of a population can sometimes look very different. Reptiles and amphibians often show great variability in colour and pattern, which can be a source of confusion when trying to identify them. Within a population, variation may be *continuous*, so that the animals could be arranged in a series showing gradual change; or *polymorphic* in which case all individuals can be assigned to one of two or more distinct groups (*morphs*), for example the Painted Frog (*Discoglossus pictus*) has distinct spotted and striped variants, and some small lacertid lizards have striped and unmarked morphs. Morphs are often mistaken for separate species.

Widespread species also often show regional variation. For instance, in most places, the Common Wall Lizard (*Podarcis muralis*) is rather small and brown with some dark markings, but in some areas it is larger and has a green back and heavy spotting. Many north-western Italian populations show these features especially the males. This sort of variation is often formally recognised by naming subspecies, which are indicated by adding a third word to the scientific name. The Common Wall Lizard becomes *Podarcis muralis muralis* in most areas and *Podarcis muralis nigriventris* in north-western Italy. Subspecies are geographical concepts, which means that all the animals in a particular population belong to the same subspecies, even though some of them lack its characteristic features. For example, some Wall lizards in north-west Italy are indistinguishable from ordinary brown *P. muralis muralis* but they would still be called *P. muralis nigriventris* like the green-backed animals among which they occur.

In the past it was very fashionable to name subspecies and many of them have been described. This is especially so for species that are found in groups of islands with numerous isolated populations that show at least average differences. Some 24 subspecies of the Ibiza Wall Lizard (*Podarcis pityusensis*) are recognised and about 48 of the Italian Wall Lizard (*Podarcis sicula*). On the European mainland, subspecies often lack clear-cut boundaries and there may often be large intergrade areas between them where populations cannot really be assigned to one or to another. It is also possible for one set of features, such as colouring, to suggest a particular pattern of subspecies while another set, such as scaling, suggest another.

These kinds of problems persuaded many biologists to abandon the use of subspecies, regarding them as an unnecessary complication. However, they continue to be used for European reptiles and amphibians and the recent development of molecular systematics, where differences in proteins and DNA are used to study populations, shows that subspecies are very varied in their significance. In some cases they do seem to be based on trivial differences in anatomy and colour that are not always consistent, but at the other extreme, some turn out to be groups of populations that have had long separate histories and deserve to be regarded as species in their own right. The recent increase of the number of species recognised in Europe is partly due to such discoveries and there are probably more to be made.

Because they are still quite widely used, attract attention to at least superficial differences, and sometimes turn out to be very distinct entities, many subspecies are listed in the **Variation** section of the species texts, although their validity often requires further research.

Rare variations

Very occasionally animals that are strikingly different from other members of their species are found. For instance, individuals sometimes turn up which completely lack pigment (albinos) or which have very odd colour patterns. These, together with two-headed terrapins and snakes and six-legged frogs, often receive a lot of attention and are of some pathological interest. But such animals are very uncommon and do not usually survive long in nature. They are therefore usually not included in the species texts.

Food

Most European reptiles and amphibians feed mainly on live animals. The principal exceptions are tortoises, certain turtles and the tadpoles of some frogs and toads, all of which eat mostly plants. Animal food is very frequently swallowed whole after, at most, a rather cursory chewing which helps to subdue it. The kind of prey varies: most reptiles and amphibians take a wide range of edible animals that they can over-power but snakes tend to specialise and often eat mainly one or a few particular types. For instance Smooth snakes (*Coronella*) usually eat lizards, while Water snakes (*Natrix*) take amphibians and fish. Snakes eat relatively large animals for their size and often have specialised ways of reducing the activity of their prey: Smooth snakes and Rat snakes (*Elaphe*) hold their victim with one or more coils of their body; other species, such as vipers, inject venom which rapidly kills prey. Most European reptiles pick up food directly with their jaws but chameleons, many frogs and toads, and sala-manders and newts have tongues which are sticky and can be flicked forwards to catch food.

Ecological niches

Different species of reptiles and amphibians can coexist in the same area because they occupy different ecological niches in which they often exploit different resources. They may eat different kinds of food or sizes of prey, be active at differ-ent times of day or night, or differ in which parts of the environment they occupy. Food differences are common between snake species in the same area. Most other reptiles and amphibians are not very specialised in their diet and so avoid compet-ing in other ways. Thus, large species of lizards eat bigger prey than smaller ones and, while most lizards forage by day, geckos hunt after dark. Differences in habitat are often very marked in closely related species. For example, in southern Bosnia, Marsh Frogs (*Rana ridibunda*) are common in lakes and slow and weedy rivers, Balkan Stream Frogs (*Rana graeca*) near springs and along cool streams, Agile Frogs (*Rana dalmatina*) in open woods and in neighbouring moist meadows, and the Common Frog (*Rana temporaria*) is the dominant species high in the mountains. Lizards show analogous patterns of separation in the same area. Sand Lizards (*Lacerta agilis*) are found on the ground in dry pastures being replaced by Viviparous Lizards (*Lacerta vivipara*) in moist places along streams and drainage ditches. In stonier areas, the Dalmatian Wall Lizard (*Podarcis melisellensis*) lives close to the ground while the Viviparous Lizard (*Lacerta vivipara*) climbs more extensively on boulders and low walls preferring slightly moister places. On rock outcrops the Sharp-snouted Rock Lizard (*Lacerta oxycephala*) climbs expertly in bare sunny places but is replaced in shadier, more wooded areas by the Mosor Rock Lizard (*Lacerta mosorensis*). The Green Lizard (*Lacerta viridis*) overlaps especially with the two Wall lizards in bushy places but takes larger prey and is more strongly associated with vegetation.

Times of activity

As reptiles and amphibians depend on external heat for their activity, they cannot remain active when the temperature is very low, and in colder areas they must hibernate. The winter months are passed in a state of torpor in some safe, usually frost-free refuge such as a hole in the ground or a deep rock crevice. Some amphibians and terrapins hibernate under water. The time of inactivity varies with local conditions: in the far North and at high altitudes it can be up to two thirds of the year, while in southern countries, some species may not really hibernate at all. Because of this regional variability, times of hibernation are not given in the species texts. There are great differences in cold sensitivity between species: some amphibians are active in near-freezing conditions, while many reptiles require a minimum temperature of 15°C before they show themselves. Most species in southern Europe become less active in summer. This is particularly noticeable in aquatic forms which suspend activity and retreat into holes or wet mud when the water in which they live dries up. Other species tend to appear only for short periods and are usually more wary than in the spring.

Most reptiles and many amphibians are diurnal, but others are only active in the evening or at night. The diurnal ones are often most apparent in the morning and may retreat around the middle of the day to reappear in the late afternoon. This pattern is most obvious in hot weather. In cooler conditions, peak activity is often in the middle of the day.

Breeding

Most European reptiles produce eggs. In the case of tortoises, some terrapins, and geckos, these are hard-shelled and often laid in quite dry places, while in other species they have soft, flexible shells and are usually deposited in moist sand, earth or dead vegetation. The young that hatch from these eggs are small versions of their parents, differing only in size, proportions and sometimes colouring. They are essentially independent and usually receive no care from their parents. In a few species eggs are retained in the body of the mother who eventually gives birth to fully developed young or eggs which hatch almost immediately. This happens in most vipers, the Sand Boa (*Eryx jaculus*), the Smooth Snake (*Coronella austriaca*), Slow worms (*Anguis*), some skinks (*Chalcides*) and most populations of the Viviparous Lizard (*Lacerta vivipara*). Most amphibians produce eggs with a gelatinous coating. They are usually laid in water where they develop into animals quite unlike their parents. These aquatic larvae (better known as tadpoles in the case of frogs and toads) spend a considerable time feeding and then change quite abruptly (metamorphose) into miniatures of their parents. Amphibian larvae are more fully discussed on p. 244. A minority of salamanders give birth directly to larvae or fully-formed babies, or their eggs are laid in the normal way but hatch into metamorphosed young.

Fertilisation in reptiles is internal, mating sometimes taking place with relatively few preliminaries, although rival males may fight, or, more often, display to each other and to any females within range. Copulation may be prolonged and males often guard their mates to prevent them immediately seeking new partners, so giving the first male's sperm a better chance of fertilising her eggs. Male lizards and snakes also leave a mating plug in the female. which again increases their chances of being the only fertiliser. In spite of these strategies, females do often seek, and find, additional partners and a single clutch of eggs may sometimes have more than one father. Fertilisation is also internal in European newts and salamanders, but not in our frogs and toads. In these animals, males embrace the females from above but the eggs are only fertilised as they are laid.

Voice

Newts, salamanders and most reptiles are virtually silent or their voice is limited to faint grunts, squeaks or hisses. Geckos, one or two other lizards and some tortoises produce more distinct calls but it is only frogs and toads that are really highly vocal. In the breeding season, the males produce a distinct, and often loud, mating call which attracts females to the breeding area. The call of each species is usually highly characteristic, which enables females to present themselves to the right males. This is very important in ponds where more than one species breeds at the same time and where females are likely to be grasped by the wrong males (in many cases, male frogs and toads are not very particular about their partners and will often grab females of other species and in some cases even fish or pieces of wood). Calling is expensive in terms of energy and the increased risk of attracting a predator rather than a mate. A minority of males in some species avoid these problems by not calling but stop close to another male where they may be able to intercept a female attracted by his voice.

In addition to the mating call, males may have a release call, which is produced when they are inadvertently grasped by another male. As its name implies, this call usually results in the caller being freed. Some forms have a number of other vocalisations and some, such as the Typical frogs (*Rana*) may scream when injured or captured.

Mating calls are mainly limited to the breeding season which is usually in the spring in northern Europe and in winter and spring in more southern areas. However, a few species may breed in the summer as well, such as discoglossids, Parsley frogs and the Natterjack Toad (*Bufo calamita*). Sporadic and isolated calling may occur at other times, especially in the autumn, and Water frogs and Fire-bellied toads may call throughout the warmer months of the year, even when not actively breeding.

Calls are often amplified by air-filled resonators called vocal sacs. These are soft, bag-like structures that lead off the mouth cavity and are inflated with air when calling takes place. In many species, there is a single vocal sac under the floor of the mouth. It may be relatively small and the throat-skin covering it unmodified, in which case the vocal sac is termed internal. External sacs are usually larger and the skin covering them is thin, wrinkled and elastic, so that when the sac is inflated, it often forms a large, translucent balloon underneath the throat. Water Frogs differ from other European species in having an external vocal sac at each corner of the mouth.

Frog and toad calls are often good field characters and are usually very easy to distinguish, once they have been heard. The easiest way to become familiar with calls is by listening to recordings of them (see p. 264). The often rather guttural sounds produced by frogs and toads can be difficult to put into words, but an attempt has been made to describe them in the species texts, either by transliteration or by comparing them with more familiar sounds. It should be borne in mind that there is often variation in the call of a species, especially in speed, which usually varies with temperature, and in pitch, which can alter with the size of the calling individual.

Natural enemies and defence

Most reptiles and amphibians are small, helpless and potentially edible. They are consequently eaten by many larger animals. Many snakes eat them and so do a variety of birds including shrikes, hawks, owls, crows, storks and herons, and small mammals such as rats, hedgehogs, stoats, weasels, foxes and badgers. Most of the warm-blooded predators take reptiles and amphibians as part of a varied diet and the only reptile specialist in Europe that is not a reptile itself is the Short-toed Eagle, which eats a large proportion of snakes.

Methods of combating predators are varied. Many species have markings that make them inconspicuous in their natural habitats and they often make use of secure retreats which their enemies cannot easily enter: for instance, many rock lizards are very flattened so that they can go deep into narrow crevices. Other species depend on speed and agility to escape predators.

If cornered, many reptiles will bite, and this is most effective in the larger species, especially the venomous snakes. Many other forms supplement their limited defensive abilities with elaborate bluffs. Frogs and toads often make themselves appear larger by adopting special postures and inflating with air, and Spadefoots *(Pelobates)* will scream and jump towards their attacker. Some non-poisonous snakes, like the Viperine Snake *(Natrix maura)* may have viper-like markings and when threatened will increase their resemblance to a venomous snake by flattening the head, hissing and striking, although often with the mouth closed. Many lizards and two salamanders will shed (autotomise) the tail when this is grasped, allowing them to escape and leaving the predator with a distraction (see Autotomy in the Glossary, p. 255).

Many amphibians have noxious skin secretions which deter predators. In some cases the species concerned have black and yellow or black and red markings which have a warning function. If a predator has already suffered from attacking a brightly coloured, noxious amphibian it may associate the markings with its previous experience and be wary of a similar animal. In some instances the bright markings are always visible, for example on the Fire Salamander *(Salamandra salamandra)*, while in others they are usually hidden but are exposed by special postures in time of danger, as in Fire-bellied Toads *(Bombina)* and the Spectacled Salamander *(Salamandrina terdigitata)*.

Skin shedding

Amphibians and reptiles shed the whole outer layer of their skin at intervals, in contrast to mammals and birds which shed it more or less continuously in small flakes. Amphibian and especially snake and lizard skin often comes off as a continuous transparent layer. In snakes the skin is robust and frequently shed whole, so it may be found in the field. Characteristically it is turned inside out as the snake escapes from it. The skin faithfully represents the surface of the snake and even the convex window over the eye remains in place. This preservation of detail means that snake skins can often be identified to species by looking at the arrangement of the scales which is still represented in the shed skin. Shedding the skin in this way, which is often called sloughing (pronounced 'sluffing'), permits growth and may help remove dirt adhering to the outside of the animal. Amphibians and soft-skinned lizards like geckos often eat the shed skin, recouping some of the costs of replacing it.

Reptiles, amphibians and people

As a group, reptiles and amphibians have had relatively little material effect on people although they have sometimes figured in their diet. Tortoises, terrapins, turtles and their eggs, frogs and even snakes and lizards are, or have been, regularly eaten in many places, including parts of Europe. Reptiles and amphibians have also featured in religious rites and mythology. A snake cult existed in ancient Greece, associated with medicine and the god Asclepius, a connection celebrated in the vernacular name of one of the better-known European species, the Aesculapian Snake (*Elaphe longissima*). Snakes are still used today in Christian worship on particular saints', days in a small area of central Italy and on the Greek island of Cephalonia. In his play *The Frogs*, Aristophanes famously included the song of one of the more noisy local species, the Greek Marsh Frog (*Rana balcanica*). He rendered its call as 'Brekekekek, coax, coax', letting us know that in the 5th Century BC this frog sounded much like it does now. Another Greek playwright, Aeschylus, had less luck with reptiles. The Delphic Oracle predicted he would be killed by the fall of a house. Living in an area prone to earthquakes he sensibly took to an outdoor life but died when a bird dropped a tortoise on his head. This is not entirely far-fetched, as Lammergeier vultures regularly drop objects such as bones, and perhaps tortoises, from a height, breaking them open so the birds can eat the contents. Amphibians can also be worrying to the superstitious. In parts of the Balkans there is a saying that, if a frog sees your teeth your mother will die, presumably making gloomy unsmiling children a blessing. This legend also exists as far away as Pakistan although it is very difficult to see how it could have arisen.

The reciprocal effect of people on reptiles and amphibians has been considerable. Apart from the still widespread practice of killing any snake-like animal encountered, there are many recent developments which are harmful to reptiles and amphibians. Pollution and draining of wetlands and their conversion to agriculture has undoubtedly reduced the numbers of many amphibians and endangered whole species. Spraying crops with pesticides often appears to have an adverse effect, either directly or by destroying their food; run-off of pesticides into pools can have a devastating result on the breeding success of amphibians. This also is true of the habit of running road drains into nearby ponds, polluting them with oil-residues from the road surface. Heavy traffic accounts for large numbers of amphibians when migrating to their breeding ponds, and also for a wide variety of reptiles, especially tortoises. Road surfaces also have a fatal attraction to many nocturnal snakes and lizards which move on to them after dark because they retain heat. In north-west Europe, the destruction of heathland areas, removal of hedgerows and filling in of ponds have contributed to a sharp decline in several species.

Collection of animals for the pet-trade is another factor that can reduce natural populations. In many cases, the effect of modest collecting is probably slight, but in areas where a species has already been restricted to a few isolated populations by other factors, it can be important in extermination, at least on a local level. Conspicuous, easily caught animals, such as tortoises are most likely to suffer greatly from collecting. There are now laws restricting such activities in most European countries. Few European species will be completely exterminated by human activities, at least in the near future. But there are some that are seriously endangered and others that are likely to be reduced from their present abundance. One of the charms of southern Europe is the large numbers of reptiles and amphibians found there. It would be a pity if this were to change.

Not all human activities have been harmful to reptiles and amphibians. The ranges of some species have been increased by human transportation, either intended or accidental. Isolated colonies of the Italian Wall Lizard (*Podarcis sicula*) in several Mediterranean countries are almost certainly the result of introduction by people. This species has also been introduced into the United States and the Turkish Gecko (*Hemidactylus turcicus*) is now also found in the warmer parts of North America. Part felling of extensive woodland and its replacement by cultivation has undoubtedly reduced the numbers of some forms. At the same time it has benefited those species that survive best in disturbed and more open habitats. The stone-piles and dry-stone walls that now characterise so many Mediterranean areas provide a host of refuges for reptiles and their prey and allow some species to exist in concentrations that would have been very rare in natural conditions. Construction of garden ponds and of open irrigation cisterns in dry areas has also benefited many kinds of amphibians.

Watching reptiles and amphibians

To see reptiles and amphibians in reasonable numbers, it is necessary to choose time and place carefully. In general, these animals are much easier to see and more abundant in southern countries, where they may be encountered almost anywhere. In the North, a high proportion of forms are restricted to a few specialised habitats, and are often rather secretive. Some idea of the sort of places that repay investigation is given in the **Habits** section in the species texts.

Frogs, toads and newts breed communally in ponds and lakes and are much easier to see during the mating season (spring for many species and areas) than at other times. Season is also important for other groups. In northern countries all forms hibernate during the winter months but are generally conspicuous in the spring, and remain quite active for the rest of the summer and autumn. In the Mediterranean area on the other hand, at least some species are active through most of the winter. Spring is again the season of greatest activity, but with the heat of summer many forms retreat and at most are seen only briefly or in small numbers. For most diurnal species, sunny weather is best.

More reptiles and amphibians will be seen by walking over suitable country slowly and quietly. Surroundings should be scanned with care, for many species are well camouflaged and depend on inactivity to avoid recognition. Any movement or rustling in vegetation may repay investigation. Breeding frogs and toads can often be located by their call, which may carry for considerable distances. Nocturnal and secretive species can be found by looking for them after dark with a broad-beamed torch. For nocturnal amphibians, the best nights are those preceded by rain.

Many species will allow an observer to get very close to them if movements are slow (although the distance at which an animal retreats often varies with the season, time of day and many other factors). On the whole, terrapins, the larger lizards and snakes tend to be more wary than other forms. For observing the more active species, binoculars are useful. High magnification is not particularly important (6x to 8x binoculars are probably best), but near-point of focusing and light-gathering power are. As many reptiles and amphibians allow a relatively close approach, it is essential to have binoculars that are still focusable at short range. Light-gathering power can be roughly assessed by dividing the diameter in millimetres of the front (object) lens by the magnification. Lens diameter is usually marked on the body of binoculars just after the magnification, e.g. 6 x 30, which would give a result of 5. Any result above 4 is quite satisfactory.

It is often possible to make a photographic record of reptiles and amphibians seen, and this may sometimes be helpful in identifying difficult animals. Unlike birds and mammals, these animals are relatively easy to photograph since many of them allow close approach, often remain still for long periods and bask in bright sun.

Collecting data

An essential requirement for any field study is a well-kept record. A lot of interesting information on behaviour and habits can be collected, even by an absolute beginner, provided that information is recorded in a thorough manner. At first, note-keeping may seem superfluous, it is only when a record has been kept for some time that it becomes obvious how much would have been forgotten without it. For all observations, date, locality, time of day and weather should be noted, preferably with a brief description of habitat. Only put in firm identifications if you are sure of them. In cases where there is doubt, it is much better to note that this is so.

Preserving dead animals. Interesting animals found dead can be preserved, although it is important to check that such collection is not illegal; the kind of animal concerned, or all animals in the area where it was found, may be protected by law. Preservation enables doubtful identifications to be more carefully checked and may provide firm evidence of range extensions. Ideally, dead animals should be placed in about 70% alcohol (= 140° above proof spirit). Larger reptiles should be liberally injected with alcohol first, or the body opened with scissors, and this is also beneficial for preserving smaller ones. In emergencies, quite good results can be obtained using distilled drinking alcohol, although this tends to be rather weak. The stronger grades of brandy, gin, anis, slivovica, tuica, rakija, schnapps, or ouzo all work quite well, at least in the short term. A 10 per cent formalin solution (commercial formaldehyde solution diluted with about nine times its volume of water) also used to be recommended for preservation but formalin is now regarded as a potentially cancer-producing agent. Most snakes can be identified from the head and neck alone, so these are worth preserving even when the rest of the body is badly damaged.

Handling reptiles and amphibians

In many instances it is helpful to examine reptiles and amphibians closely when confirming identifications, but again it is important to check it is legal to handle the kind of animal concerned in the area where you are. Remember too that these animals are delicate and even slight injury can seriously reduce the chances of their survival when released. Many species can be most easily examined by placing them in a transparent plastic bag, which should contain a little water if amphibians are to be put in it. Many amphibians produce noxious skin secretions that are harmful to other animals, so the same bag should not be used for reptiles or even other amphibian species. As plastic bags are not permeable to air, animals should not be kept in them very long.

If reptiles and amphibians have to be restrained more closely, this should be done gently. Most injuries result from holding them too tightly, often because the holder is worried about being bitten. In fact, European amphibians scarcely bite at all, and although lizards and the smaller non-venomous snakes may bite, they are often incapable of breaking the skin. It is only the larger snakes that can draw blood in any but minute quantities, and provided they are held correctly this can be avoided. Of course, venomous snakes should not be handled at all.

If amphibians are held, this should be done with a wet hand to avoid rubbing off the layer of protective mucus that covers them. Most amphibians are sensitive to excessive heat and the prolonged warmth can harm them, so they should not be held too long. Newts and salamanders should not be gripped closely, indeed this is usually unnecessary as most of them are relatively slow moving. Frogs and toads often need some restraint, and it is easiest to hold them for short periods with the thumb and forefinger round their waist and their legs lightly gripped by the remaining fingers and palm of the hand. Lizards are most easily held with the thumb and forefinger just behind the head, and the body in the lightly closed fist; the close contact appears to make the lizard feel secure and less inclined to struggle. Lizards should never be picked up by their tail as, in many cases, it will break off. Non-venomous snakes should also be held just behind the head, but the body needs to be supported to save putting too great a strain on the neck. Do not pull snakes out of holes or crevices by their tails; it can injure them seriously.

The Greek and Atlantic islands

Greek islands covered by this book (except Corfu). The broken line marks the southern and eastern limits of the area considered.

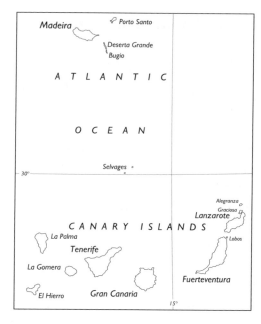

East Atlantic islands covered by this book. Madeira is about 700km west of the coast of Morocco and Fuertaventura about 100km from it.

Reptiles and Amphibians in the British Isles

Relatively few kinds of reptiles and amphibians are native to the British Isles. In Britain, there are only six amphibians and six reptiles that are certainly native, and in Ireland only three amphibians and one reptile are definitely known to be long established. A number of other species have been introduced, either accidentally or intentionally, and a few of these have survived and bred in England and Wales.

Species occurring naturally in the British Isles

Northern Crested Newt (*Triturus cristatus*, p. 43) Widespread in Britain, although less abundant in the north and west.

Common Newt (*Triturus vulgaris*, p. 46). Widespread in Britain, although less common in the north and west. Also present in Ireland.

Palmate Newt (*Triturus helveticus*, p. 48). Widespread in Britain, although rare in the east Midlands and East Anglia.

Common Toad (*Bufo bufo*, p. 73). Widespread in Britain.

Natterjack Toad (*Bufo calamita*, p. 74). A very restricted natural distribution in north-west England and adjoining Scotland from Merseyside to the Solway Firth, in coastal Lincolnshire and Norfolk, and at one site in Hampshire. Also present in Kerry, south-west Ireland. Originally also in North Wales, Dorset, Surrey, West Sussex, Kent, Suffolk and Bedfordshire. Reintroduced into North Wales, Dorset and Bedfordshire.

Common Frog (*Rana temporaria*, p. 80). Widespread in Britain and Ireland.

Pool Frog (*Rana lessonae*, p. 90). Possibly native in East Anglia, but now probably extinct. Introductions have also been made in Britain.

Viviparous Lizard (*Lacerta vivipara*, p. 144). Widespread in Britain and Ireland.

Sand Lizard (*Lacerta agilis*, p. 140). A very restricted distribution in southern England in Dorset and Surrey; also in north-west England in coastal Merseyside. Originally present in coastal areas of Kent, East Sussex, the Wirral, North Wales, and perhaps Berkshire and around London. Recent small introductions have been made in the West Country, West Sussex, Wales and Scotland.

Slow Worm (*Anguis fragilis*, p. 192). Widespread in Britain. One small introduction in Ireland, on the Burren in County Clare.

Grass Snake (*Natrix natrix*, p. 216). England and Wales.

Smooth Snake (*Coronella austriaca*, p. 221). Very restricted in southern England, to small areas of Dorset, Hampshire and Surrey. Originally also in Wiltshire, Berkshire, East Sussex, and possibly Devon.

Adder (*Vipera berus*, p. 230). Widespread in Britain.

Leathery Turtle (*Dermochelys coriacea*, p. 110). Occurs regularly around the coasts of the British Isles in summer and a few animals are usually stranded or washed up dead on beaches each year.

Other turtles turn up occasionally, particularly the **Loggerhead Turtle** (*Caretta caretta*, p. 107) and less commonly **Kemp's Ridley** (*Lepidochelys kempii*, p. 107). **Green Turtle** (*Chelonia mydas*, p. 108) and **Hawksbill Turtle** (*Eretmochelys imbricata*, p. 109) have also been recorded but extremely rarely. The Green turtles at least may have been parts of commercial consignments which were regularly brought to Britain by ship in the past.

Introduced species in the British Isles

All kinds of species escape or are released deliberately and may be encountered in the British Isles. Most of them do not persist very long, but a few, mainly European, species survive quite well and may breed, occasionally for a long time. The most successful introductions in terms of persistence are listed below. Only those introduced species that have probably lived for at least 25 years in wild or semi-wild conditions are included.

Alpine Newt (*Triturus alpestris*, p. 44). Colonies have been reported from Kent, east Surrey, south-east London, Birmingham, Shropshire, Sunderland and central Scotland. The Surrey colony at least has persisted for eighty years.

Italian Crested Newt (*Triturus carnifex*, p. 44). Reported from Surrey and Birmingham.

Clawed Toad (*Xenopus laevis*, Pl. 7). First reported on the Isle of Wight where it may be extinct, but now found in Wales. This is a very aquatic southern African species and is a member of the family Pipidae which occurs in Africa and South America. It is a distinctive flattened frog, with a very smooth skin, eyes on top of the head, small forelimbs and very fully webbed feet.

Common Midwife Toad (*Alytes obstetricians*, p. 64). Isolated introductions into Bedfordshire, Yorkshire, Northamptonshire, Hampshire (Littleton), Devon and south-west London. The Bedford colony may have survived for a century.

Edible Frog (*Rana* kl. *esculenta*, p. 92). Many small introductions have been made as far apart as Surrey and Yorkshire, but frequently do not survive very long.

Marsh Frog (*Rana ridibunda*, p. 89). The most successful introduction. The first colony was started in Walland Marsh, Kent in 1935 and the frog is now found in several areas of Kent and in East Sussex. Other introductions are present in south-west and west London, although the last colony may in fact really be of the Levant Water Frog (*Rana bedriagae*, p. 97).

Common Wall Lizard (*Podarcis muralis*, p. 145). Isle of Wight, East Sussex, Dorset, Devon, Surrey and south-east London. Most colonies are of the typical form of this species, *P. m. muralis*, but some are of the north-west Italian subspecies, *P. m. nigriventris*, with a green back and heavy spotting on the belly, at least in some males.

Aesculapian Snake (*Elaphe longissima*, p. 214). Introduced about thirty years ago near Colwyn Bay in North Wales, where it still survives.

Grass Snake (*Natrix natrix*, p. 216). Southeast European individuals *(N. n. persa)* were introduced to a small area of Surrey and their genes may persist in the local population as characteristic striped individuals have been seen.

Other species have persisted for a while but probably do not form breeding colonies at the present time. They include the following.

Yellow-bellied Toad (*Bombina variegata*, p. 60)

Common Tree Frog (*Hyla arborea*, p. 77). At least one introduced colony in Hampshire, but this is now extinct.

American Bullfrog (*Rana catesbeiana*, p. 97). Recently bred near the border between Sussex and Kent, but efforts were made to exterminate it.

Green Lizard (*Lacerta viridis*, p. 138)

European Pond Terrapin (*Emys orbicularis*, p. 103)

Red-eared Terrapin (*Trachemys scripta elegans*, p. 105). Introduced into several localities, including canals in the London region. It survives well but is not believed to breed in the British Isles. However, terrapins are very long-lived and, if rare exceptionally hot summers occasionally permitted breeding and successful hatching of the eggs, they might persist without further introductions.

Other freshwater terrapins occasionally released into fresh waters in Britain include the **Snapping Turtle** (*Chelydra serpentina*) and the **Painted Terrapin** (*Chrysemys picta*) both from the south-east United States, and the **Chinese Soft-shelled Turtle** (*Pelodiscus sinensis*).

The Channel Islands

The reptiles and amphibians on these islands off the coast of Normandy, are typical of the neighbouring French mainland; not all of them are native in Britain or Ireland.

Common Newt (*Triturus vulgaris*, p. 46). Guernsey.

Palmate Newt (*Triturus helveticus*, p. 48). Jersey.

Common Toad (*Bufo bufo*, p. 73). Jersey, Guernsey.

Common Frog (*Rana temporaria*, p. 80). Guernsey, Alderney, Sark.

Agile Frog (*Rana dalmatina*, p. 83). Jersey.

Common Wall Lizard (*Podarcis muralis*, p. 145). Jersey.

Green Lizard (*Lacerta viridis*, p. 138). Jersey; introduced to Guernsey.

Slow Worm (*Anguis fragilis*, p. 192). Jersey, Guernsey, Herm, and Alderney.

Grass Snake (*Natrix natrix*, p. 216). Jersey. Animals here tend to be fairly uniform, often without an obvious pale collar.

SALAMANDERS AND NEWTS
(Tailed Amphibians, Caudata)

This group includes some 420 species, 29 of which occur in our area. These all have rather long bodies, soft often moist skin without scales, and well developed tails. The term salamander is used for almost all tailed amphibians, while newt is often applied to various semi-aquatic species. Some of the most important features for identifying newts and salamanders are shown below. Others include general proportions, skin texture and colouring. For identification of eggs and larvae, see p. 238 and p. 242.

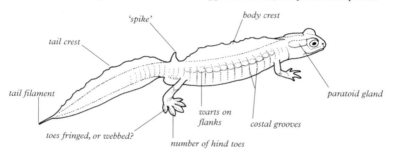

Some features to check when identifying salamanders and newts

Key to Salamanders and Newts

Pond newts (*Triturus*), which come out here in section 11, have their own key on p. 41.

1. East Adriatic area only. Usually in subterranean waters. Fore-feet with only three toes, hind feet with only two; usually very pale with pink feathery gills; eyes very small (vestigial in adults), body very elongate
   Olm, *Proteus anguinus* (p. 53. Pl. 3)
   Fore-feet with four toes, hind-feet with four or five. Eyes not very small   **2**

2. Sardinia, south-east France, north and central Italy only. Feet partly webbed, toes usually appear stubby; nasolabial groove visible with lens (Fig., below)
   Cave salamanders, *Speleomantes spp.* (p. 50, Pl. 2)
   Toes not partly webbed and stubby; no nasolabial groove   **3**

Cave Salamanders (*Speleomantes*)

3. Italy only. Four toes on hind feet. Small and dark with usually yellow or red
    mark on top of head; underside of tail bright red
                 Spectacled Salamander, *Salamandrina terdigitata* (p. 35, Pl. 2)
    Five toes on hind feet; colouring not as above   **4**

4. Paratoid glands large, raised and porous   **5**
    Paratoid glands not large and porous, sometimes absent   **9**

5. Uniform jet black (rarely brown), no bright spots   **6**
    Jet black, blackish or brown usually with lighter markings   **7**

6. Central and Eastern Alps, east Adriatic mountains. Tail tip pointed
                 Alpine Salamander, *Salamandra atra* (p. 33, Pl. 1)
    Alps of north-west Italy and south-east France only. Tail tip rounded in adults
                 Lanza's Alpine Salamander, *Salamandra lanzai* (p. 34, Pl.1)

7. North-east Italy only (Vicenza province above 1,300m). Usually with
    whitish, grey, yellowish or intense yellow markings which vary from
    restricted diffuse spotting to a broad irregular band along the back.
                 Alpine Salamander, *Salamandra atra aurorae* (p. 33, Pl. 1)
    Most areas. Light markings only rarely arranged in a single irregular band
    along the back   8

8. Not Corsica or south-east Aegean islands.
                 Fire Salamander, *Salamandra salamandra* (p. 31, Pl.1)
    Corsica only
                 Corsican Fire Salamander, *Salamandra corsica* (p. 33, Pl. 1)
    South-east Aegean islands. Usually brownish, often with lighter spots; belly
    pale, underside of tail orange-yellow. Males with a soft 'spike' on upper
    surface of tail base
                 Luschan's Salamander, *Mertensiella luschani* (p. 34, Pl. 3)

9. North-west Iberian peninsula only. Very slender. Tail of adults very long, up
    to 2.5 times length of body, and cylindrical. Eyes very prominent, vertical
    grooves present on sides of body, typically two copper or gold stripes on
    back
            Golden-striped Salamander, *Chioglossa lusitanica* (p. 35, Pl.2)
    Not very slender, tail rarely much longer than body and often distinctly
    flattened from side to side   **10**

10. Spain and Portugal only. Adults large (up to 30cm); head very flat with small
    eyes; usually a row of prominent warts on each flank; the ribs may
    protrude through these and can often be felt
            Sharp-ribbed Newt, *Pleurodeles waltl* (p. 36, Pl. 3)
    Not like above; adults not larger than 18cm, and usually much smaller   **11**

11. Pyrenees, Corsica and Sardinia only. Usually in or near cold and often
    running water. Often flattened salamanders; Pyrenean species has very
    rough skin   **12**

Not Corsica or Sardinia. Rarely in very cold running water. Not very
strongly flattened; skin never very rough in Pyrenees region

                    Pond newts, *Triturus spp.* (see separate key on p. 41)
12. Pyrenees only. No paratoid glands; skin very rough; males lack spurs on hind
legs

                    Pyrenean Brook Newt, *Euproctus asper* (p. 37, Pl. 3)
Corsica only. Paratoid glands fairly distinct; skin not very rough; no spots on
throat; males have spurs on hind legs

                    Corsican Brook Newt, *Euproctus montanus* (p. 38, Pl. 3)
Sardinia only. Paratoid glands often not very obvious; skin not very rough;
throat usually spotted, males have spurs on hind legs

                    Sardinian Brook Newt, *Euproctus platycephalus* (p. 39, Pl. 3)

Typical Salamanders and Newts
(Family *Salamandridae*)

All but seven of the European species of tailed amphibian belong to this family, which
contains about 56 species and is widely distributed in north-west Africa, Europe and
adjoining Asia, eastern Asia and western and eastern North America. In Europe a
number of quite distinct kinds are present. Some of these are terrestrial like; the Fire
Salamander group (*Salamandra*), Luschan's Salamander (*Mertensiella*), the Spectacled
Salamander (*Salamandrina*) and the Golden-striped Salamander (*Chioglossa*). Others
are more aquatic including the Sharp-ribbed Newt (*Pleurodeles*), Brook newts
(*Euproctus*) and Pond newts (*Triturus*), although they too are often partly terrestrial
even though they spend at least the breeding season in water (see p. 39 for a detailed
account of Pond newts). Only the Fire Salamander group and Pond newts are
widespread, the others being confined to fairly small areas in southern Europe.

All salamandrids are secretive, at least outside the breeding season, and are usually only encountered on land during the day by turning over stones and logs and
searching in crevices in moist places. Typically, they forage in the evening or at night,
especially when the weather is damp. All species are mainly invertebrate feeders, a
wide range of prey that can be captured by these mainly rather slow moving animals
being eaten.

European salamandrids vary considerably in their breeding behaviour, but in all
cases the male produces a compact package of spermatozoa, the spermatophore. In
Fire Salamanders and their relatives and in Luschan's and Golden-striped salamanders, the male carries the female on his back for some time before depositing the
spermatophore on the ground, after which he lowers her cloacal region on to it. The
courtship of the Sharp-ribbed Salamander is essentially like that of the Fire
Salamander, but takes place in water. In Brook newts the male clasps the female with
his tail and often with his jaws as well, and the spermatophore is deposited in or close
to her cloaca. In the Spectacled Salamander there is no contact but the partners
gyrate around each other for some time. The Pond newts differ from all other
European salamandrids in the elaborate breeding dress of the males and a complex
underwater courtship in which they do not grasp the females.

Most salamandrids lay eggs singly in water, attached to plants or stones according
to their habitat. However, the Spectacled Salamander and Sharp-ribbed Salamander
lay their eggs in clumps. Alpine and Luschan's Salamanders produce fully formed

young and Fire and Corsican Salamanders do this occasionally but most usually give birth to well developed aquatic larvae.

A larval stage that is markedly different from the adults is usual in amphibians and that of frogs and toads is familiar as round-bodied vegetarian tadpoles which change radically when they metamorphose into the adult body form. In contrast, larvae of salamanders and newts are entirely carnivorous and more like the adults in appearance, so they do not undergo such marked changes at metamorphosis. The front legs are developed before the hind ones and the gills are always external (see p. 242). The general shape of salamandrid larvae varies with habitat, species usually living in cold running water have lower crests on the tail and body and shorter gills than those living in ponds.

FIRE SALAMANDER *Salamandra salamandra*   Pl. 1.

Range West, central and south Europe. Similar animals occur in north-west Africa and parts of south-west Asia, but these are now treated as separate species. Map 1.

Identification Adults usually substantially less than 20cm including tail, but up to 25cm in some populations. Unmistakable; a large, often robust, rather short-tailed salamander with large paratoid glands. Virtually the only European species that usually has the upper surface marked with intense yellow, orange or reddish spots or stripes on a black ground colour (exceptionally the bright markings are reduced to a few small blotches. This characteristic pattern is present even in newly metamorphosed young. The underside may be entirely dark or spotted.

Variation There is much variation in body proportions and colouring (Fig., p. 32 and Pl. l). Size, shape and number of bright markings vary considerably even within populations and even entirely yellow or black animals rarely occur. Main recognised variants are as follows. Salamanders from the Balkans and eastern Europe to the Alps (*S. s. salamandra*) usually have a pattern of irregular orange or yellow spots. In the Pirin mountains of Bulgaria (*S. s. beschkovi*) animals have short extremities and sometimes a yellow stripe along the mid-back, and in Italy (*S. s. giglioli*) red spots may be present on the belly. Salamanders from France, the north-east Pyrenees area, Belgium and Holland (*S. s. terrestris*) have spots arranged in two lines along the back. This form integrades with *S. s. salamandra* in a wide area of central Germany, the Czech Republic and adjoining Poland. In the central and west Pyrenees and eastern Cantabrian mountains (*S. s. fastuosa*) the two lines of spots become continuous stripes and there is an additional stripe on each flank; similar more western populations from Asturias to North Galicia (*S. s. bernardezi*) are rather smaller. However, some populations in Asturias are quite different being brown with a yellowish head. Striped animals integrade with spotted ones in the more central areas of Spain (*S. s. bejarae*) where salamanders may sometimes have parts of their spots coloured reddish or brownish. High in the Sierra de Gredos mountains of central Spain animals may be small with a compressed tail and are mainly black (*S. s. almanzoris*). In north-west Spain (west and south Galicia) and in north and central Portugal (*S. s. galliaca*) spots are often dull and may include red pigment, while in extreme southern Portugal, in the Serra de Monchique (*S. s. crespoi*), salamanders are sometimes very big, up to even 25cm, with short limbs and many small jagged-edged spots. Animals in the Sierra Morena north of the Guadalquivir river (*S. s. morenica*) have small spots with a lot of red on the head and less on the body, tail and limbs. Finally, south of the Guadalquivir river (*S. s. longirostris*), salamanders lack red in their spots and characteristically have a spot over each eye and one over each parietal gland, as well as others on the body.

Salamanders from the Iberian peninsula, except the Pyrenees and Cantabrian areas, appear to have had a long separate history from those in other places.

Similar Species Unlikely to be confused with any other species, although very dark individuals can look rather like typical Alpine Salamanders (opposite), however, they usually have at least some small yellow spots. Range is also often useful in identifying these aberrant specimens. See also *S. atra aurorae* (opposite).

Habits Over much of range typically occurs in cool moist broad-leaved forests, especially at their edges, and more rarely in coniferous woods. In the Iberian peninsula also found in Mediterranean woodland, scrub, meadows, and rocky areas around mountain lakes. May live up to 2,500m in central Spain, but appears commonest under 800m in the Alps. Nearly always terrestrial in damp situations and rarely found very far from water. Exceptionally, may be aquatic at high altitudes. Strictly nocturnal and frequently active after rain. Very slow moving, rarely foraging more than a few metres from its daytime refuge. Otherwise often found when turning logs, dead bark and stones. Protected by abundant white noxious skin secretion which may be actively squirted from glands on the upper surface and irritates the mouth and eyes of predators; bright body colours warn of this defence.

About 6–8 months after mating, females most usually give birth to 8–70 larvae 2.5–3.5cm long, which are often deposited in cool clean, flowing water and at the edges of mountain lakes, although occasionally stagnant and weedy water is used. Larvae live mainly on the bottom and may overwinter, though often they metamorphose in about 2–4 months at a length of around 4–6.5cm. In parts of Spain very big larvae are sometimes produced and metamorphose in as little as two weeks. In contrast, the high altitude *S. s. almanzoris* produces small 2.5cm long larvae that metamorphose at around 9cm but only after a long interval. In *S. s. bernardezi* and *S. s. fastuosa*, females may give birth to just 2–8 fully metamorphosed salamanders 3–5cm long; this method of reproduction also occasionally occurs in *S. s. salamandra*. Developing babies may cannibalise other eggs or larvae within their mother, reducing the brood to 1–15. Sexual maturity is reached in about 2–4 years. Fire Salamanders may live 20 years in the wild and 50 in captivity.

Fire Salamander (*Salamandra salamandra*), some variations in pattern in south-west Europe.

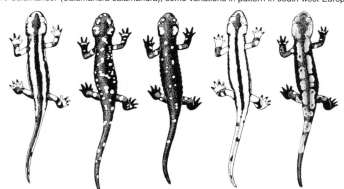

S.s. bernardezi   *S. s. almanzoris*   *S. s. almanzoris*   *S. s. fastuosa*   *S. s. terrestris*

CORSICAN FIRE SALAMANDER *Salamandra corsica*   Pl. 1.

Range Corsica. Map 2.

Identification Up to about 20cm. Very similar to some Fire Salamanders but paratoid glands tend to be relatively small, the body plump and the toes short. Spots may be yellow or orange. A species recognised as separate on the basis of its biochemistry and best identified by distribution.

Similar Species None within range.

Habits Found in a wide range of habitats, particularly forests of beech, sweet chestnut and pine but also in dry scrub, subalpine degraded habitats, humid gorges and even caves. Occurs from 50–1,750m but commonest from 500–1,300m.

Breeding similar to Fire Salamander. Usually gives birth to larvae, which may sometimes be deposited in even poorly oxygenated water. Fully metamorphosed young are occasionally produced.

ALPINE SALAMANDER *Salamandra atra*   Pl. 1.

Range Alps and neighbouring mountains in Switzerland, extreme southern Germany, Austria, northern Italy, Slovenia and north-west Croatia. Isolated populations occur in southern Bosnia (Hercegovina), on the north Albanian border and perhaps in between. Map 3.

Identification Adults up to 15cm including tail, but usually smaller. A moderately robust salamander with large paratoid glands and a pointed tail tip. Generally similar to Fire Salamander but usually distinguishable by uniform, all-black colouring and sometimes more 'ribbed' appearance.

Variation Rare animals are brown instead of jet black. The isolated southern population centred in Montenegro has been named as *S. a. prenjensis*, although it is not obviously different in appearance. *S. a. aurorae* is distinctive in having whitish, grey, yellowish or intense yellow markings that vary from restricted diffuse spotting to a broad irregular band along the back . It occurs in a small (50 square km) area of mixed deciduous and coniferous forest in north-east Italy (Vicenza province) between Trento and Asiago at altitudes of 1,300–1,800m.

Similar Species Not very likely to be confused with other species but may superficially resemble crested newts (Pl. 4) which have no large paratoid glands, and usually an orange or yellow belly.

Habits A montane species occurring between 400m and 2,800m, but usually encountered from 800m to 2,000m where it can be very abundant. Found in broadleaved and coniferous woodland and above the tree-line in alpine meadows and dwarf heath etc. Mainly nocturnal but sometimes seen in shady places during the day and occasionally in more open situations, especially after rain or in overcast weather. Hides beneath stones and logs, and in holes. At high altitudes may hibernate for 8 months of year. Range does not usually overlap much with Fire Salamander, which is typically found at lower altitudes.

Fully metamorphosed young 4–5cm long are produced, usually in broods of two but sometimes of up to four or even five. Development takes two years at lower altitudes and three at higher ones. The embryos first consume their own yolk, then that from unfertilised eggs that accompany them, and finally cellular tissue from the mother's 'womb'. As might be expected given their slow reproduction, Alpine Salamanders can live more than ten years.

LANZA'S ALPINE SALAMANDER *Salamandra lanzai* Pl. 1

Range Italy (Cottian Alps in western Piedmont), and neighbouring France (Massif du Queyras). Map 4.

Identification Up to about 17cm. Very similar to the Alpine Salamander but adults rather bigger, with a more flattened head, slight webbing between the fingers and toes, and a somewhat longer tail with a rounded tip. No enlarged pores present along centre of back.

Variation None reported.

Similar Species Alpine Salamander (p.33).

Habits Similar to Alpine Salamander. Found from 1,500–2,200m in rocky alpine meadows and on scree slopes, often close to streams and usually above the tree line although it may occur in mixed forest just below this. Usually active in rainy or very humid conditions, when it may be seen by day in quite large numbers. Animals are often conspicuous, sometimes standing on rocks with forelimbs extended. Hides beneath rocks and stones when inactive.

Like the Alpine Salamander, produces fully formed young, the number varying from 1–6.

LUSCHAN'S SALAMANDER *Mertensiella luschani* Pl. 3.

Range Karpathos, Kasos and Saria islands in the south-east Aegean sea between Crete and Rhodes; Megisti Kastellorizon (island off the south-west coast of Asiatic Turkey). Also south-west Asiatic Turkey itself. Map 5.

Identification Up to 14 cm including tail. The only tailed amphibian in the south-east Aegean islands. Rather slender with a thin tail, very prominent eyes, smooth skin, and narrow but prominent paratoid glands. Back and throat with scattered minute spines, especially in males which also have a prominent soft 'spike' on the upper surface of the tail base and a strong cloacal swelling. Dark brownish above, often with small light yellowish spots, flanks pale or whitish-yellow, belly flesh-coloured with lighter markings; throat yellowish or pink and underside of tail orange-yellow.

Variation The populations from Karpathos, Kasos and Saria are named as *M. l. helverseni.*, and those from Megisti as *M. l. basoglui.* Megisti animals are often reddish.

Similar Species None within range.

Habits Quick moving, for a salamander. Often found in quite dry habitats on Karpathos, such as pine woods and open scrub and cultivated areas, sometimes in the vicinity of seasonal springs and streams. Occurs up to 1,000m. Active in winter and early spring, spending the hot dry summer deep in cracks and crevices. Nocturnal but may also be active in the day in wet weather. Hides by day in moist cool places in stone piles, dry-stone walls, under flat stones and in rock cracks which often form part of extensive underground systems of fissures. If threatened may stand high on its legs with its back arched and may also squeak. Capable of shedding its tail if this is grasped.

Mating takes place in spring, the male crawling under the female and using his forelimbs to grasp hers; he then waves his tail from side to side, perhaps stimulating the cloaca of his mate with the fleshy spur on his tail base. A large 1cm high spermatophore is then deposited and the female manoeuvred so that her cloaca can pick it up. Two fully-formed young up to 7cm long are usually produced after an interval of several months, at the beginning of the winter activity period in November.

Other names. *Salamandra luschani.*

SPECTACLED SALAMANDER *Salamandrina terdigitata*   **Pl. 2**

Range Peninsular Italy but absent from much of the east. Map 6.

Identification Females up to 11.5cm including tail, males to 9cm. A small, rather slender salamander; the only European species with four toes on the hind feet. Head rather long with prominent eyes. Tail longer than head and body, cylindrical at base with a distinct dorsal ridge. Skin rather rough; prominent ribs give body 'segmented' appearance. Dull blackish or brownish above with a prominent, roughly triangular yellow or even vermilion marking on head. Underside of body pale, often pinkish, with largely black throat and dark blotches on belly, particularly along sides. Lower surfaces of legs and tail bright red in adults.

Variation Quite uniform over most of its range. A population in Basilicata, southern Italy is small-bodied and light-coloured with a large, whitish head marking.

Similar Species Not likely to be confused with any other species.

Habits Has a very local distribution within its range, and is usually confined to frequently north-facing mountain valleys often well away from human influences. Here it is found in broad-leaved woods with luxuriant undergrowth of herbaceous plants. Often lives in shadowy places close to clear, well-oxygenated, acid streams usually with rocky beds, and occasionally near ponds. It hides under rocks, dead wood, the loose bark on stumps and logs, and in leaf litter. Mainly active in spring and autumn. Nocturnal, but also seen the early morning and evening, and sometimes even at other times of day after rain. May descend deep into the ground in winter. When molested sometimes feigns death, lying on its back and exposing its brightly coloured lower surface; alternatively the tail is curled up exposing its red underside. Usually found between 200m and 700m but may occur from sea level to 1,100m and even 1,500m.

Courtship takes place on land, the male following the female as she walks in a circle, while both wave their tails. He then deposits a spermatophore which she picks up. Females produce 35–60 eggs in a season, laying them in still or slow-moving shadowed water in clumps of up to 20 which are attached to twigs, dead-leaves, roots, stones or water plants. Sometimes several females lay in the same place. The eggs develop into tadpoles in about three weeks and females are sexually mature when they are about 7cm long. Spectacled Salamanders can live 10–12 years.

Note Salamanders of this type were widespread in the distant past and fossils are known from Greece and Sardinia.

GOLDEN-STRIPED SALAMANDER *Chioglossa lusitanica*   **Pl. 2.**

Range Confined to north-west Spain (Galicia and Asturias) and northern and central Portugal as far south as the river Tajo. Map 7.

Identification Adults up to about 16cm, usually 12–13cm, including tail; males rather smaller. Body very slender with a groove along the mid-back and 10–12 grooves on each flank. The extremely long cylindrical tail, may be up to 2.5 times the length of the head and body in adults if undamaged, although it is shorter in young animals. Eyes large and prominent. Upper parts usually dark brownish, dark grey or blackish, typically with two golden or copper-coloured stripes on the back, which join together to continue as single stripe on the tail; the stripes may occasionally be broken into a series of spots. Underside brownish or greyish. Males have a noticeably swollen cloacal region, especially during the breeding season when they develop rough nuptial swellings on the inside of the fore-limbs.

Variation Southern animals are rather smaller with narrower heads and shorter tails and may have the stripes on the back reduced.

Similar Species Not likely to be confused with any other species.

Habits A local species found in wet mountain regions with rainfall above 1,000mm a year. Occurs up to 1,300m, although usually lower, especially in the north. Typically found near clear, well oxygenated rocky streams and springs with dense surrounding growth of moss, shrubs, bushes, broad-leaved trees and sometimes pine; but also occurs in wet caves and abandoned mines. May be very abundant, up to 4–5 adults occurring per metre of stream. Largely nocturnal, hiding under stones and vegetation during most of the day, but sometimes active in light during rain or very humid conditions. Rather lizard-like and can move rapidly for a salamander, sometimes climbing on rock faces. Some animals are quite sedentary but others travel long distances, occasionally over 350m in a single night, and may sometimes be found 700m from water. During summer months often aestivates under stones etc. When disturbed scuttles away into crevices or into water, where it is a strong swimmer. If molested may bring tail forward and hide its head under it; also squeaks and may discharge white skin secretion especially from the tail. All or part of the tail can be shed if this salamander is attacked by predators and will move for some time afterwards; up to a quarter of animals may have broken the tail in the past. It grows again rapidly but tends to be greyer and more uniform than original; the regenerate can also be broken. Prey is caught with the tongue, which is sticky and relatively long.

Males arrive first at the mating places which are often rocky surfaces with a flow of water over them. A male crawls under his mate and holds her arms with his and, after some cloacal rubbing, a spermatophore is transferred. Females lay about 12–20 eggs under often shallow water attaching them to stones, roots or plants, sometimes in the same places as other females. Eggs develop in 6–10 weeks and tadpoles often pass their first winter in water, sometimes even remaining in it for 2–3 years and becoming very big and dark. Newly metamorphosed young are about 7cm long and mature when they reach an age of four years or more (later in females) and a length of around 4cm from snout to vent. Animals can live at least 8 years in wild and over 10 in captivity.

SHARP-RIBBED NEWT *Pleurodeles waltl*   **Pl. 3**

Range Portugal and Spain but not much of north. Also northern Morocco. Map 8.

Identification Adults normally between 15cm and 30cm, including tail; one of the largest tailed amphibians in Europe. A row of 7–10 prominent warts (poison glands) along each flank, that coincide with rib tips which are sharp and often project right through them Head broad and very flat, with small eyes. Body heavily built with a coarsely granular skin, the granules each having a black horny tip. Tail at least as long as the body and flattened from side to side; the amount of cresting varies. Upper side yellowish-grey to olive or brownish, the colour darkening with age, with dark brown often fairly regular spotting when in water; warts on sides are orange, yellow or whitish. Belly yellow, orange, whitish or grey, usually with irregular dark markings. Males tend to be more rufous with longer deeper tail and when breeding have dark, rough nuptial pads on the insides of their fore-legs.

Variation Head and tail shape rather variable but no definite trends.

Similar Species Young Sharp-ribbed Newts can look like the Pyrenean Brook Newt but this species is confined to the Pyrenees. It also has a longer head, lacks rows of warts on the flanks and has a conical cloacal swelling in females.

Habits Very aquatic, occurring in a range of mainly Mediterranean habitats including scrub, open woods and cultivated land, occasionally up to 1,500m. Found in ditches, ponds, reservoirs, wells, cisterns, slow moving rivers and irrigation channels, often with rich vegetation. If these dry up, this salamander aestivates beneath stones, in crevices or buried in mud, or it may migrate across land to other waters. Occasionally found on land, especially small individuals, which can be encountered under rocks and logs etc up to 200m from water. Active mainly in the evening and at night but also encountered in the day, especially in rainy weather. Often lives largely or wholly on the bottom of ponds in deep water rising only to breath. At night hunts largely by scent, in the water but sometimes also on its banks. Food consists of invertebrates but also small fish and amphibians, including its own species. There are a range of defensive responses. Salamanders may bite and make deep noises if attacked. They also expose their sharp ribs, arch their backs and wave their bodies from side to side and produce a characteristic smell. Predators trying to eat this salamander are likely to suffer punctures from the sharp ribs through which venom from the associated poison glands can enter. Young animals especially may raise the tail vertically and wave it.

Courtship occurs in water, the male gripping the arms of the female from beneath with his own and then swimming with her on his back for hours or even days. Eventually he partially releases her and deposits a spermatophore 12mm across in front of her snout and then turns her through 180° so that she can take it into her cloaca. A total of 6–7 spermatophores may be produced. Egg-laying begins 2–3 days later, 150–1,300 being deposited over a few days, the number depending on the size of the female. They are scattered in small groups of about 10–20 on aquatic plants and sometimes on stones and pond bottoms. Hatching occurs in 1–2 weeks. Metamorphosed animals can mature in 18 months and may not have a terrestrial phase. Neoteny may occur in this species which has lived 25 years in captivity.

Note Sharp-ribbed Salamanders have a long history in Europe going back at least 13–14 million years.

PYRENEAN BROOK NEWT *Euproctus asper*   **Pl. 3.**

Range Confined to Pyrenees and nearby mountains. Map 9.

Identification Adults normally up to about 16cm including tail, occasionally longer; females bigger than males. Fairly robust with a flattened head, small eyes and only slight neck. Paratoid glands absent. Tail about as long as head and body and compressed from side, deeper in males. Skin is covered with small tubercles bearing horny tips and is much rougher than in other salamanders and newts within the range of this species. Usually muddy brownish, greyish, olive or blackish above, often with some lighter, usually yellow markings that may include a vertebral stripe, especially in younger animals. Belly with dark markings at sides and a yellow, orange or red central band that is sometimes unspotted. Cloacal swelling rounded in males and conical in females; males tend to have a more brightly coloured underside than females.

Variation Considerable variation in colour and pattern.

Similar Species Closely related to Corsican and Sardinian Brook newts, but not likely to be confused with any species within its range. Young of Sharp-ribbed Newt look rather similar but have a row of large tubercles on each flank and are absent from the Pyrenees.

Habits A largely aquatic mountain species. Normally found between 700m and 2,500m, but sometimes down to 140m and up to 2,600m. Particularly abundant around 2,000m in or near well-oxygenated, cold streams that are often acid and slow flowing, and in mountain lakes. Water inhabited usually has temperatures less than 15°C, little vegetation, and stony or rocky bottoms. Also sometimes found in caves. Often encountered in damp gutters, runnels etc., or when turning stones at edge of water. Active by day and night, but usually seen moving on land only after dark. At lower altitudes almost entirely aquatic although it may aestivate on land. Towards its upper limits it may hibernate for several months on land. Tends to be rare where trout, one of its main predators, are abundant.

Mating, which may take many hours, occurs in water. The male raises his tail to display the bright colour along its lower edge to the female. He later uses it to hold her around the loins, the two animals facing in the same direction, and stimulates her cloaca with his hind feet before transferring 1–4 spermatophores directly to it. Sometimes other males attempt to interrupt courting pairs, and females may even go through an imitation of the mating process together. Fertilised females produce 20–40 eggs in a year which are inserted into crevices and attached to rock surfaces using the extensible cloaca. They hatch in about 6 weeks, often spending one winter as larvae before metamorphosing and sometimes two at high altitudes. Males mature in 2–3 years depending on altitude and females take rather longer. This newt has been known to live 26 years.

CORSICAN BROOK NEWT *Euproctus montanus*   **Pl. 3.**

Range Corsica. Map 10.

Identification Adults up to 13cm including tail, but usually less. Very like Pyrenean Brook Newt but smaller and smoother skinned with more or less distinct paratoid glands, no dark spotting on throat and often a relatively uniform underside. Brown or olive above, sometimes with lighter greenish, yellowish or reddish markings that often form a vertebral line. Throat and belly whitish, grey or brown, sometimes with white flecks. Male has spurs on hind legs and a conical cloacal swelling.

Variation Some variation in colouring.

Similar Species Closely related to other Brook newts, but no really similar species within its range. The Corsican Fire Salamander (p. 33), the only other tailed amphibian on Corsica, is black and yellow.

Habits Found from sea level to 2,260m in clear, often acid streams, ponds and lakes, frequently in rocky terrain in open maccia and woods, especially of sweet chestnut, between 600m and 1,500m. Mainly crepuscular and nocturnal and often encountered when turning stones in the water or at its edges. Prefers situations without much water movement and avoids current by sheltering among stones. This and the next species lack lungs or have them greatly reduced. Lung reduction is common in salamanders that live in cool oxygen-rich situations where they can breath entirely through the skin (see also Cave salamanders, p. 50). When on land, animals hide among roots, under stones and in or beneath dead tree stumps and fallen logs, usually close to water. Hibernates on land and may also retreat there in summer.

Mating, which can take some hours, occurs in water, the male grasping the female with his jaws and by coiling his tail round her. His pointed cloaca is pushed against hers and one or two spermatophores are transferred directly, sometimes with the help of the hind legs. Females usually lay 20 or so eggs, and up to a maximum of 60 per season. These are deposited under stones and guarded by their mother until they hatch. Males may develop their reproductive spurs only 6 months after metamorphosis. Has lived 7 years in captivity.

SARDINIAN BROOK NEWT *Euproctus platycephalus* **Pl. 3**

Range Sardinia; most localities where it survives are in mountain areas in the east of the island. Map 11.

Identification Adults up to about 15cm including tail, but usually smaller. Very like Corsican Brook Newt but head often longer and flatter and the upper jaw overhangs more; the paratoid glands are generally not so distinct and the throat is usually spotted. Typically brown or olive above, often with lighter green or yellow blotches and spots and a reddish vertebral stripe. Underside frequently yellowish or reddish, particularly along the centre of belly, and usually dark spotted, especially in males which have spurs on hind legs. Both sexes have a more or less conical cloacal swelling.

Variation Considerable variation in head shape and colouring.

Similar Species None. The only other salamanders on Sardinia are Cave salamanders (p. 50).

Habits Very similar to other Brook newts. Found from 50–1,800m in still or more usually running water in mountainous areas, including small and large rivers and lakes. Most abundant above 600m and shows a preference for relatively calm sections of rivers and streams. On land found among roots and under stones and logs, usually close to its aquatic haunts.

Courtship, which may take some hours, occurs in water: the male holds the body of the female in his mouth and may carry her off to a suitable place for mating. His tail is twisted about her pelvic region and he places his cloaca against hers before transferring a single spermatophore, sometimes with the help of the hind legs. Females lay 50–220 eggs, occasionally over a period of several months. These are attached to the underside of stones, the female turning on her back and extending her cloaca into a tube that places the eggs in position.

Note The Sardinian Brook Newt is endangered and has disappeared from many of the localities where it was formerly known. This appears to be a result of damage to its habitat by pollution, especially with pesticides, and perhaps predation by introduced trout.

Pond Newts
(Triturus)

A clearly defined group of which 11 of the 12 recognized species are found in the area covered by this book. Pond newts are usually terrestrial for part of the year, but they enter still or, more rarely, slow-flowing water in spring to breed. At this time the males develop a characteristic breeding dress usually consisting of bright colouring and a flexible cutaneous crest on the tail and often on the back as well. Also, the toes on the hind feet may become elongated, fringed or webbed. Both sexes may develop a glandular swelling on each side of the back and these often give the body a squarish cross-section, especially in males.

The courtship is elaborate, the male displaying before his largely passive mate. Display usually consists of the male standing in front of the female, vibrating his reflexed tail-tip to direct pheromones from his cloaca towards her and intermittently lashing her with it. After some minutes, he emits at least one spermatophore, over which the female walks, or is led, and picks it up in her cloaca. Details of the display vary considerably but are characteristic for each species. Females usually lay their eggs singly, attaching them to aquatic vegetation and often using their hind feet to eggs singly, attaching them to aquatic vegetation and often using their hind feet to wrap leaves round them; more rarely eggs are attached to stones etc. The laying period is often fairly extended being spread over several days or weeks.

In areas where more than one species of Pond newt occur together, females seem to recognise their own males partly by their characteristic breeding dress. The crests, where present, also increase the surface area of the male, enabling him to take up extra oxygen from the water. This means he does not have to interrupt his vigorous courtship by coming to the surface to breath.

On land, Pond newts are secretive and nocturnal and usually only found when turning over stones, logs, dead bark and vegetation in damp places; they may also use the burrows of other animals and hibernation often takes place away from water. Most immature newts spend their time on land although some, like the Northern Crested Newt, often return to the water each spring even though they do not breed, and other species may go to water the autumn before their first reproductive season. Pond newts are more diurnal when in their breeding ponds and can often be seen as they come up to take in air. However a lot of their activity still occurs after dark and they may be more active then than in the day and use more open areas of water. Occasional adult animals may spend the breeding season on land but some newt populations seem to be aquatic for most of the year. In a few cases neoteny occurs: tadpoles do not metamorphose into adults, but grow to a large size and are often capable of breeding in this condition. This occurs most commonly in deep water or at high altitudes.

Pond newts feed on a wide variety of small invertebrates, both in and out of the water, and some, particularly the larger species, are known to prey on small fish and other amphibians and their eggs. Like many other amphibians, Pond newts often eat their shed skins and, as with other newts and salamanders, regenerate well after injury, tails and limbs often re-growing quite quickly.

Identification Because their courtship dress probably partly functions as a means of clear and unequivocal recognition for females, breeding male Pond newts are generally easy to identify. Non-breeding males and females, especially of the smaller species, are more difficult to run down, although with practice it is often possible to distinguish even nondescript individuals by 'jizz', as with so many problem groups. Because of the wide range of variation found in some species (seasonal and sexual differences, as well as regional ones), it is not always feasible to identify newts with certainty using only a key. Identification should therefore always be carefully checked against the relevant description of the species.

The descriptions of breeding male Pond newts refer to animals in full breeding dress. When developing or losing this, they tend to have lower, less elaborate crests and less obvious tail filaments, toe fringes etc., and in some cases might be mistaken for other species unless examined carefully.

As in most other newts and salamanders, male Pond newts have a bigger cloacal swelling than females.

Some Pond newts have shallow grooves on top of the head which are occasionally helpful in identification; their usual positions are shown in the Fig., adjacent.

Top of newt's head

Key to Pond newts
(*Triturus*)

1. South and west France and Iberian peninsula. Upper parts with extensive, bright green markings. Underside relatively dull, never orange to red

Marbled Newt, *T. marmoratus* (p. 42, Pl. 4)

Southern Marbled Newt, *T. pygmaeus* (p. 42)

Upper parts without extensive, bright green markings. Ground colour of underside pale silvery yellow to red (exceptionally black), at least in centre   **2**

2. Up to 12cm. Belly uniform deep yellow to red; unspotted (with rare exceptions). Frequently, numerous spots on flanks, often forming a complex lattice-like pattern in males. No grooves or stripes on head. Upper parts usually grey to blackish or brownish, often marbled with black in males. Breeding males have smooth, low, pale crest on body, spotted or barred with black

Alpine Newt, *T. alpestris* (p. 44, Pl. 6)

Not like Alpine Newt. Belly often with at least some dark spots. If not, head may have grooves, bright colour on belly may be confined to centre   **3**

3. Absent from Iberian peninsula and south-west France. Usually large: adults over 11 cm. Skin, often rough; moist in terrestrial animals. Underside uniform bright yellow to reddish-orange with a strongly contrasting pattern of bold dark spots or blotches (exceptionally, underside is entirely black). Dark or reddish above with darker blotches. Breeding males have dark, usually spiky crests often indented at base of tail

Crested newts (p. 43–44, Pl. 4)

Not particularly large: usually under 11 cm. Skin often smooth; tends to be dry in terrestrial animals. Belly pattern less bold than in Crested newts, bright colour often limited to centre of belly. Head may be grooved.   **4**

4. Carpathian and Tatras mountains only. Underside uniform yellow to reddish-orange, often unspotted in centre and on throat, spots at sides often sparse or even absent. Grooves on head, but no obvious stripes. Breeding males have almost square cross section and only poorly developed crests

Montandon's Newt , *Triturus montandoni* (p. 45, Pl. 6)

South Italy only, Very small: most under 8cm. Dark colour on side of head and neck usually has characteristic lower border (see Fig.,p. 50). Underside with dark spots; throat yellow to orange, belly paler. Breeding males have no cresting on body and lack fringes on toes

Italian Newt, *Triturus italicus* (p. 49, Pl. 6)

West and central Iberia only. Belly bright orange, often flanked by pale streaks, with dark spots present at least at sides. Sometimes a single groove on mid-line of snout; no stripes on head. Breeding males have no crest on body and tail crest poorly developed; toes not webbed or fringed. Female has a conical cloacal swelling

Bosca's Newt, *Triturus boscai* (p. 48, Pl. 5)

Absent from Iberia and south France. Underside usually spotted, including throat. Yellow to red colour on belly confined to central area; throat pale.

Head typically has grooves. Breeding males usually with fairly large crests and fringed hind feet; tail ends in filament only in eastern parts of range

Common Newt, *Triturus vulgaris* (p. 46, Pl. 5)

West Europe only. Spots generally small and weak on belly and absent from throat. Colour on belly confined to centre and usually pale yellow or silvery-orange. Grooves present on head. Breeding males have tail ending in filament and hind feet usually dark and webbed

Palmate Newt, *Triturus helveticus* (p. 48, Pl. 5)

MARBLED NEWT *Triturus marmoratus*   Pl. 4

Range N Portugal, N Spain (but not much of E), S, W and central France. Map 12.
Identification Adults up to 16cm including tail. Easily distinguished from nearly all other European newts by combination of extensive, bright green, dorsal colouring and fairly sombre belly, which is black, grey or brown, often with tiny whitish spots. A large green newt, marbled with black or grey above. Adult females and young have an orange vertebral stripe. Terrestrial animals are brighter green than aquatic ones and have a dry, velvety skin.

Breeding males develop a dorsal crest with an often rather wavy upper border, that is regularly barred with black and is usually indented at the level of the vent; a light streak is usually present on side of the tail.
Variation Marbled Newts occasionally hybridise with the Northern Crested Newt where the two species co-exist in north-west and central France. Hybrids, originally described as a separate species, *Triturus blasii*, are intermediate between the parent forms and often large.
Similar Species Unlikely to be confused with any other European newt or sala-mander except the Southern Marbled Newt.
Habits Similar to the Northern Crested Newt, but in cooler areas is rather less aquatic and, outside the breeding season, may sometimes be found in fairly dry woods and heathland where it hides under stones and logs and in dry-stone walls. In hot dry areas it is more closely associated with water. Occurs up to 2,100m in south of range. In area of overlap with Northern Crested Newt tends to be in hillier areas and reproduces in small forest pools. Breeds in a variety of permanent and tempo-rary water, the males going to the ponds first and defending small vegetation-free territories on the bottom of the pools where mating takes place. Females lay 200–400 eggs. Young animals may not breed until they are 5 years old but can then do so for a decade. This species has lived 25 years in captivity.

SOUTHERN MARBLED NEWT *Triturus pygmaeus*

Range Southern Portugal and south-west Spain as far north as the central moun-tains (Sierra de Guadarrama to Sierra de Gata). Map 13.
Identification Adults less than 12cm including tail. Similar to the Marbled Newt but smaller and underside often creamy yellow with dark spots. In breeding males, the crest is smooth-edged and not indented at the base of the tail.
Variation Southern animals tend to be marbled above while northern ones have dis-crete dark spots, which helps distinguish them from Marbled Newts in the area of contact with this form.
Habits Generally similar to the Marbled Newt.
Similar species Marbled Newt.
Other Names Until recently was regarded as a subspecies of the Marbled newt, as *Triturus marmoratus pygmaeus*.

NORTHERN CRESTED NEWT *Triturus cristatus*   Pl. 4.

Range Northern and central Europe north of Alps, from Britain and France eastwards to Russia and Ukraine, but not Ireland or much of Scandinavia. Also eastwards to central Asia. Map 14.

Identification Usually up to 15cm including tail; females may occasionally reach 18cm. A large, dark, very coarse skinned newt, always lacking green colouring of Marbled Newt. Typically dark brownish or blackish above with darker spots and conspicuous white stippling along flanks. Belly usually yellow or orange, with a variable pattern of dark spots or blotches; throat dark. Terrestrial animals have a moist skin and may appear almost jet-black above, contrasting strongly with the usually bright underside. Babies may have a yellow stripe along the back.

Breeding males develop a high, usually spiky crest, indented at base of tail, which has a whitish or bluish streak along its side; in females the bottom edge of the tail is orange.

Variation Belly sometimes completely black in Scandinavia. The range of the Northern Crested Newt contacts all three other Crested newt species, which were once considered subspecies of it and with which it sometimes interbreeds. In France it also interbreeds with the Marbled Newt (see opposite).

Similar Species Apart from other Crested newts, this species is unlikely to be confused with any other species, except possibly the Alpine Newt (p. 44) if dark markings on belly are very reduced, and the Alpine Salamander (p. 33) if belly is all black.

Habits Northern Crested Newts are fairly aquatic and may occasionally be encountered in water throughout the year, although they are usually terrestrial outside the breeding season, sometimes occurring several hundred metres from water, especially in broad-leafed woodland where they can reach densities of 600 per hectare; also found in agricultural areas. Usually breeds in relatively deep (often more than 0.5m) still, or slow-flowing, water with good weed growth and some exposure to the sun. Tends to use larger water bodies than smaller newts, including farm ponds and mature gravel pits but can sometimes even occur in ditches and flooded wheel tracks and brackish water. When disturbed may produce noxious whitish skin secretion with characteristic smell; the body may also be twisted round objects and the tail rolled sideways, this position being held with eyes closed and without breathing for some seconds. Mainly a lowland form found below 1,000m, but more montane in the south of its range and occurring up to 1,750m in parts of Austria.

Northern Crested Newts are more nocturnal when breeding than the smaller newt species with which they sometimes occur. Males may gather to display to each other using similar movements to those employed in courtship. Females come to such leks and are courted there. Sometimes males 'steal' females from others that are already displaying to a potential mate. Females lay 200–400, exceptionally 700, eggs in a season. Interestingly, a chromosome anomaly prevents half the eggs developing, something that occurs in other species of Crested newt and in the closely related Marbled Newt. Viable eggs develop into larvae in 1.5–3 weeks and these tend to swim in open water, in contrast to the adults which are more bottom dwelling. They often metamorphose after 3–4 months although overwintering may take place in cold areas; neoteny occasionally occurs, at least in Sweden and eastern Germany. Newly metamorphosed newts are 5–8cm long and sexual maturity is reached at a length of about 12–13cm, after 2–3 years in males with females often maturing rather later. Average life of wild adults is around 7–8 years in Britain, although they can live twice as long, at least in captivity.

Other names. Warty Newt.

ITALIAN CRESTED NEWT *Triturus carnifex*   **Pl. 4**

Range Italy, extreme southern Switzerland, Slovenia parts of Austria, extreme southern Germany, and west Balkan area; also reported from the Czech Republic; an isolated population is found around Lake Geneva. Map 15.

Identification Usually up to 15cm, females occasionally to 18cm. Rather smoother-skinned than Northern Crested Newt. Dark brown, grey or yellowish above often with well marked darker spots and with little or no white stippling on flanks. Belly often more orange than in Northern Crested Newt, with large, well-defined, dark greyish spots; sometimes completely black beneath; throat dark. Female often has a bright yellow stripe along the back. Upper edge of the crest of breeding males often relatively smooth.

Variation Interbreeds with other species of Crested newts.

Habits Generally similar to Northern Crested Newt. Found in a wide range of habitats varying from beech woods to quite arid Mediterranean areas and extending up to 1,800m. Neoteny occurs.

BALKAN CRESTED NEWT *Triturus karelinii*   **Pl. 4**

Range Balkan peninsula, except western parts; also found in the Crimea, northern Asiatic Turkey, parts of the Caucasus and north-west Iran. Map 16.

Identification Up to 16cm in females. Relatively smooth-skinned, with limited white stippling along flanks and often a bluish sheen. Spots on belly tend to be small and the throat is pale with darker spots. Females may have a yellow or orange stripe along the back. Upper edge of the crest of breeding males often relatively smooth, especially on tail, and not fully indented at the base of this.

Habits Similar to Northern Crested Newt; often found at some altitude (up to 1,300m).

DANUBE CRESTED NEWT *Triturus dobrogicus*   **Pl. 4**

Range Plain of Tisza and lower Danube rivers and Danube Delta. Occurs in eastern Austria, Hungary, north Serbia, Moldavia, south Romania and adjoining Bulgaria. Map 17.

Identification Usually up to about 13cm, females occasionally to 16cm. An often modest-sized, slender, small-headed Crested newt with short limbs and very coarse skin. Brown or even reddish above, often with very distinct darker spots and little or no white stippling on flanks. Belly often reddish-orange with blackish-brown spots which may join together to form one or two bands; throat black with fine white spots. Female frequently has yellow vertebral stripe.

   Breeding males may have a very high and jagged crest on the body which is separated by an indentation from the smoother crest on the tail.

Variation Often intergrades with other species of Crested newts.

Habits Found mainly in lowland river systems where it lives in slow-flowing and still water in the rivers themselves and in lakes, ponds and ditches, often with rich aquatic vegetation. Usually found below 600m. Frequently coexists with fish and spends long periods in water. May live 8 years or more.

ALPINE NEWT *Triturus alpestris*   **Pl. 6.**

Range More widely distributed than its name suggests. From north, central and east France eastwards to extreme western Ukraine and Romania and from southern Denmark south to Greece and northern Italy. Isolated populations exist in southern Italy (Calabria), north-west Spain (Cantabrian mountains), and central Spain

(Guadarrama mountains, where possibly introduced). Map 18.

Identification Females up to about 12cm including tail, males smaller. A medium-sized newt, usually identifiable by distinctive colouring: dark above, with uniform deep yellow to red belly. Head rather flat, without grooves. Males grey to blackish above, usually with darker markings, females often browner and more uniform with grey or blue-green marbling. Frequently numerous, small spots along the flanks, often set on a light ground, especially in males, giving a lattice-like effect. Flanks may occasionally have a white stipple. Throat often plain, but can be spotted; belly almost always unmarked (rarely with a few small spots, for instance some animals from the southern Balkans and Italy, and exceptionally elsewhere). Skin of terrestrial animals velvety; smooth or granular in water.

Breeding males have a low, smooth-edged, yellowish crest, barred or spotted with black; cresting in females is restricted to the tail.

Variation There are some differences in shape of the head and in colouring. Most populations are assigned to *T. a. alpestris* but southern ones show some differentiation. The isolated Spanish populations are named *T. a. cyreni*. Ones from extreme south-eastern France and northern Italy west and south of the Po valley, *T. a. apunaus*, have heavily spotted throats and bright red bellies and the isolated population in Calabria is named *T. a. inexpectatus. T. a. veluchiensis* occurs in the Veluchi mountains of central Greece and is also found in the northern Peloponnese. Neotenous, or partly neotenous, populations occur particularly in mountain lakes in Slovenia, Bosnia and Montenegro. Often, these have more or less dark colouring, like adults, but may also have well developed feathery gills. Most neotenous populations appear to be a result of local environmental conditions and are closely related to the more normal populations surrounding them, but that in Zminicko lake at 1,285m in the Durmitor mountains of Montenegro is genetically distinct and named as *T. a. serdarus*.

Similar Species Quite distinctive, but might be confused with Marbled Newt (p. 42) or Crested newts (p. 43–44).

Habits Very aquatic and nearly always found in or quite near water. In north of range is frequently encountered in lowlands in cool shady ponds in woods but occurs in wide variety of other habitats, including shallow, open waters, wells and even flooded tyre tracks. Common in cold, almost plantless pools, lakes and slow-flowing streams in mountain regions, often above the tree-line. Strictly montane in southern parts of range, where it may occur at 2,500m in the Alps and Albania. When on land, usually found in very cool, moist places.

Breeds early in spring, mature females laying around 250 eggs in a season but sometimes as many as 530. These hatch in 2–4 weeks. If ponds dry up, the newts may return to breed again once they fill again. At high altitudes in Alps, animals may breed only every other year. Newly metamorphosed newts are 3–5cm long and often have an orange stripe on the back. They mature in 2–4 years, neotenous populations tending to take longer. Known to have lived 11 years in the wild and 20 years in captivity.

MONTANDON'S NEWT *Triturus montandoni*   Pl. 6.

Range Carpathian and Tatras Mountain, from Romania through the western Ukraine to the borders of Poland with Slovakia and the Czech Republic. Map 19.

Identification Females up to 10cm including tail, males less. A relatively small newt, with three grooves on the head. Upper parts yellowish-brown, greenish-brown or olive, usually with darker spots or mottling. Underside almost entirely yellow to reddish-orange, often with some small, black spots at sides; occasionally, these spots are more widespread, but they are often absent from the throat. Skin of terrestrial animals usually noticeably granular; rather smoother in water.

Breeding males have crest on tail, which ends in a filament, but only a ridge along the centre of the back. Body is very square in cross section and hind toes are not obviously fringed. The lower edge of the tail is whitish to orange with black spots.

Variation No marked geographical variation recorded. Montandon's Newt hybridises easily with the Common Newt, its nearest relative, and the offspring tend to be intermediate in appearance between the two parent species.

Similar Species Apart from the distinctive breeding dress of the male, the Alpine Newt (p. 44) is larger than Montandon's Newt, usually darker and lacks grooves on head, while the Common Newt (below), has a distinctive belly pattern and often a striped head or at least a vague dark streak on each side.

Habits A montane species usually found from 500–1,500m but ranging from 200–2,000m. Often in coniferous woodland and breeds in a wide variety of waters, including lakes, ponds, slow-moving streams, ditches and even puddles and flooded wheel-ruts. It prefers clear, cold, acid ponds, but is also found in many other situations, even polluted water. On land, often lives near breeding ponds and in wooded country occurs under rocks, fallen trees, and pieces of loose bark, even in sunny situations. Frequently encountered with the Alpine Newt at high altitudes and with the Common Newt lower down.

Females lay 35–250 eggs in a season which often hatch in 1.5–2 weeks but may take 4 weeks at high altitudes, where larvae regularly overwinter. Elsewhere metamorphosis often occurs after about 12 weeks and newly emerged newts are around 3cm long. However, at high water temperatures, development can be completed in 4 weeks and baby newts are then much smaller sometimes being just 1–1.5cm long. Sexual maturity occurs in about 3 years.

COMMON NEWT *Triturus vulgaris*   **Pl. 5.**

Range Widely distributed; occurring over most of Europe, but not southern France, the Iberian peninsula, southern Italy and most Mediterranean islands. Also occurs in Western Asia. Map 20.

Identification Adults up to 11cm including tail, but in some areas (especially southern Balkans) considerably smaller; males tend to be slightly larger than females. Very widespread and the commonest newt over much of its range. Small and smooth-skinned, often with a characteristic pattern on underside; three grooves usually visible on head and a quite deep hollow present between the nostril and the eye. Terrestrial animals and breeding females yellow-brown, olive or brown above, often with small, dark spots (which may fuse into two lines on back, especially in females) and a vague stripe on side of head. Belly typically with well developed dark spots or blotches which usually extend onto the paler throat; bright orange, yellow or even red pigment present, but confined to a central stripe; mature males tend to have larger belly spots than females. Terrestrial animals have dry, velvety skin.

Breeding males develop large dark spots above, clear head stripes, a continuous crest on tail and body and fringes on toes of hind feet; the lower edge of the tail is usually orange with a light bluish streak above it. For other details of breeding dress see below.

Variation Considerable variation in size, amount of spotting on belly (see Fig., opposite) and male breeding dress. Several subspecies have been described; intermediates between them are known, especially in the eastern coastal areas of the Adriatic sea. North and west Europe is occupied by a large form with a relatively high, notched or undulating crest (*T. v. vulgaris*). However the crest is almost smooth-edged when developing or when being resorbed after the breeding season. In Italy, Slovenia and north-west Croatia (*T. v. meridionalis*) the crest is fairly low with an almost smooth upper edge on the body, which is rather square in cross section, and the tail ends in filament. Some north-west Romanian animals are also rather like this and are known as *T. v. ampelensis*. In Greece, Macedonia and the Ionian islands north through Albania to southern Bosnia and extreme southern Croatia (*T. v. graecus*), Common Newts are rather small, frequently under 7.5cm in length, with a low central body crest that is often flanked on each side by an additional crest or ridge, the tail again ends in filament, the centre of the belly is orange or reddish and females are quite heavily spotted.

Similar Species The Common Newt occasionally lacks spots on the throat and such animals can be confused with Palmate Newts (p. 48), especially when subadult. It is also confusable with Montandon's Newt (p. 45) in the Carpathian and Tatras mountains, and with the Italian Newt (p. 49).

Habits More terrestrial than many other species of Pond newts. On land occurs in a wide variety of damp habitats, including cultivated land, gardens, woods, field edges, stone piles etc. Breeds in still, frequently shallow water, preferring often small weedy ponds, ditches etc. and avoiding heavily shadowed and very exposed sites or very weed-clogged ones. Also breeds near the edges of bigger water bodies, and even in brackish conditions. Largely a lowland species, but reaches 1,000m in southern parts of the range and 2,150m in Austria.

Females lay 200–300 eggs in a season which hatch in 1–3 weeks. The larvae live near the bottom of ponds, in contrast to the adults which often approach the surface. They sometimes overwinter and neoteny occasionally occurs. The newly metamorphosed newts are about 4cm long, sometimes with an orange stripe on the back that tends to fade on the hind-body. Males reach maturity in 2–3 years, females taking rather longer. In Britain wild adult Common Newts suffer a 50 per cent mortality each year and rarely survive more than 7 years, but the species has been known to live 28 years in captivity.

Other names. Smooth Newt.

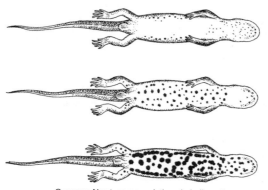

Common Newt, some variations in belly pattern

PALMATE NEWT *Triturus helveticus*   **Pl. 5.**
Range West Europe, south to northern Spain and central Portugal, north to
Scotland and east to Germany, Switzerland and the extreme west Czech Republic,
absent from Ireland. Map 21.
Identification Adults up to 9.5 cm including tail, usually less; males smaller than
females. A small smooth-skinned newt, with three grooves often visible on head,
which may be rather short and rounded. Yellowish, olive or pale brownish above, frequently with small spots which may form two lines on the back, especially in females;
often a dark stripe on side of head. The belly may be almost immaculate but some
spots usually present though often small. Bright colour on underside always restricted to a central stripe on belly which is usually yellow or silvery orange. Throat
unspotted, often translucent pinkish. Skin dry and velvety in terrestrial animals.

Breeding males have a low smooth-edged crest on the body and well-developed
smooth-edged crest on tail, which is squared towards its tip but continues as a dark
filament; hind feet dark and strongly webbed. Head and body have heavy dark markings. Central band on side of tail orange, bordered by two rows of large spots.
Variation North-west Spanish and Portuguese animals have been described as a
separate subspecies, *T. h. alonsoi*. They tend to be small and the hind feet of breeding males may not be fully webbed. Other regional differences exist in the
Cantabrian region and animals from Pozo Negro in the Sierra de la Demanda
(Burgos province) are large with small dark spots and have been named as *T. h. punc-
tillatus*. Hybridisation with the Common Newt is known but extremely rare.
Similar Species Bosca's Newt (below); Common Newt (p. 46). In areas where the
Palmate Newt occurs with these species, aquatic males are easily recognised by their
characteristic colour and pattern, and traces of these features may remain on land.
Terrestrial males and all females are more difficult to distinguish but have more delicate belly colouring and a usually pinker, unspotted throat. Other features that help
distinguish the Palmate Newt from the Common Newt are its lack of a marked hollow between the nostril and the eye and the frequent presence of a light spot near the
origin of the hind leg and two light spots under the hind foot.
Habits Fairly terrestrial, but usually rather more aquatic than the Common Newt.
Breeds in a wide variety of still or occasionally slow running, often shallow waters,
including puddles, flooded tyre tracks, ditches, ponds, heath and woodland pools
(especially in broad-leafed woods), edges of mountain lakes and even brackish ponds
near the sea. In areas of overlap tends to be found in more shaded, clearer, less rich,
and more acid water than the Common Newt. Palmate Newts may be encountered
in a wide variety of habitats near their breeding ponds. They typically occur up to
500m, often in hilly country, but range up to 1,500m in Alps and 2,200m in the
Pyrenees.

Females lay 290–460 eggs that develop into larvae in 8–14 days; these sometimes
overwinter before metamorphosing. Newly emerged newts are 3–4cm long and often
have an orange stripe that extends right along the back. Males reach sexual maturity in 2 years in north of range and females sometimes take a year longer. Neoteny
occasionally occurs; this species has lived 12 years in captivity.

BOSCA'S NEWT *Triturus boscai*   **Pl. 5**
Range Portugal and west and central Spain. Map 22.
Identification Adults up to 10cm including tail, but often less; males are smaller
than females. Sometimes a single longitudinal groove present on snout. In females

the cloacal swelling is conical. Colour rather variable: fairly pale yellow-brown, olive to grey-brown above, usually with more or less distinct dark spots or marbling. Sometimes a weakly defined paler stripe along the mid-back. No stripe on side of head. Underside with bright orange-yellow to orange-red colouring, this pigment often confined to a broad central stripe down belly, in which case the sides are conspicuously paler and whitish forming a light band. Dark, frequently large spots nearly always present on belly, often arranged in an irregular row on each side and sometimes extending onto throat. Lower edge of tail often orange, frequently with dark spots. On land, some animals may be particularly dark. and sometimes the paratoid glands are yellow or reddish and there is a reddish stripe along the back. The skin is smooth in water but granular in terrestrial animals.

Breeding males are not very different from females and have no crest on the body and only a fairly low one on the tail which may end in a short filament; the toes are not webbed or fringed. When present, the light band on each side of the body is conspicuous; often a light band on the tail which becomes lighter towards its tip.

Variation Shows considerable local variation in adult size, colouring and habits.

Similar Species The only small newt species over most of its range, but may be confused with Palmate and Alpine Newts in the north. At all seasons the Palmate Newt is more delicately coloured and the belly is yellow to pale silvery orange, not bright orange. The spots on its underside tend to be finer and never extend onto the throat, three grooves are also often visible on head, and there may be dark stripe through the eye. The Alpine Newt is larger and its belly is uniform orange, nearly always lacking dark markings, the spots along the flanks are usually small and very numerous. In both these species, the shape of the female cloacal swelling differs from that in Bosca's Newt.

Habits Breeds in small ponds, streams, pools in rivers, cattle troughs, lagoons and even waters in open caves; most frequently found in clean water, often but by no means always without much vegetation. Usually spends some time on land but may sometimes be completely aquatic. Active at night and also by day, especially when breeding. Found up to 1,800m in Galicia.

Females lay 100–250 eggs over the breeding season that often hatch in 1–3 weeks; at high altitudes larvae may overwinter. Newly metamorphosed animals are around 3.0–3.5cm long and mature in 2–4 years. In the wild males may live 6 years and females 9 years.

ITALIAN NEWT *Triturus italicus* Pl. 6.

Range Restricted to southern Italy, east of a line from Ancona to Southern Lazio. Map 23.

Identification Adults up to 8 cm including tail but usually about 6cm; males smaller than females. A diminutive newt easily distinguished from the Common Newt, the only other small species extending into its range, by size and colouring, especially that of the throat, pattern on side of head, and breeding dress. Head with no more than a single slight central groove and without striping. Brownish to olive above and flanks often lighter, the ground colour often heavily overlaid with dark spots, some of which may join to form a line between the fore and hind legs. Underside orange or yellow, the throat darker than the belly (throat of Common Newt is paler than belly). Bright belly pigment is bordered by pale bands and the whole area has scattered dark spots. On side of head, light colour usually extends upwards as a prominent narrow band towards eye, while dark dorsal colouring extends well down (see

Fig. on p. 50 for typical pattern).

Breeding animals have no crest on body, but tail is crested and may end in short filament in both sexes; a ridge on each side of the back gives the body a squarish cross section, especially in males; toes not fringed. Both sexes develop gold spots on flanks and a gold blotch behind the eye during the breeding season.

Variation Animals from higher altitudes tend to be larger.

Similar Species Common Newt (see p. 46).

Habits In breeding season found in still, often small and temporary ponds, and artificial irrigation features such as cisterns and irrigation ditches. On land encountered under stones, logs etc. Commonest below 800m but occurs up to 1,525m.

Eggs hatch very rapidly and tadpoles can mature in 4–6 weeks, although neoteny may occur quite commonly in permanent water. Newly metamorphosed animals are about 2.5cm long

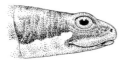

Italian

Common Newt

Small newts in Italy, head patterns

Cave Salamanders
(Family *Plethodontidae*)

European cave salamanders belong to an otherwise New World group, the Lungless salamanders (*Plethodontidae*) which range across north and tropical America and have approximately 266 species, about half of all the salamander species living at the present time. The European species (*Speleomantes*) are closely related to the Web-toed Salamanders of the Sierra Nevada mountains of California and, until quite recently, were placed in the same genus, *Hydromantes*. Although there is now a huge geographical gap between the two groups, lungless salamanders may once have occurred in the intervening area in northern Asia. As their common name suggests, these salamanders lack lungs, more or less, and they breathe through the skin and the lining of the mouth. The European species have a tail that is shorter than the body and well-developed limbs with short, often blunt toes that are partly webbed, especially in adults. The head is broad and well defined with prominent eyes and there is a characteristic groove from the nostril to the edge of the lip (the nasolabial groove). The sticky tongue is mushroom-shaped with a long stem and is shot forward to catch prey. European cave salamanders are often abundant, being found especially in limestone country and, in spite of their name, also occur outside caves, being found on the surface especially in leaf litter, and detritus close to streams in cool moist often north-facing wooded valleys. Although often active by day in dark caves, these salamanders are nocturnal outside them, appearing mainly in wet weather and retreating into caves and crevices and beneath stones and logs in dry periods. Provided conditions are right, they may be seen throughout the year. Cave salamanders are agile and climb readily on even vertical cave walls and outcrops, using the tail as well as the feet. As with many other amphibians, their poisonous skin secretion gives them some protection from predators and this distastefulness may possibly be advertised by their often striking colouring, and by rolling the tail upwards in the presence of enemies.

Male cave salamanders tend to be smaller than females and have a tiny tentacle associated with each nasolabial groove, long teeth at the front of the upper jaw and a gland under the chin (the mental gland); as in other salamanders, the cloacal swelling is larger in males. Cave salamanders mate on land, the male first often waving his tail and later stroking the female with his chin, smearing her with the secretion of the mental gland which apparently contains pheromones. The long teeth at the front of his upper jaw may perhaps be used to scratch the skin of the female, ensuring that the secretion gets into her circulation. Later, the male mounts his mate, holding on with his limbs and stroking her head with his chin. The pair may then advance, the male waving his tail from side to side and eventually bringing his hindparts forwards and depositing a spermatophore which is eventually taken up by the female.

Females lay 5–10, occasionally up to 14, large eggs that are about 5–6.5mm in diameter and are deposited in damp rock crevices. The mother guards them from other Cave salamanders, which often eat eggs and young, although she is believed to sometimes eat a proportion of them herself. This is perhaps not entirely surprising as she does not otherwise feed during the long incubation period which may last six months or a year. Snacking off one's offspring seems more reasonable in these circumstances. The eggs hatch into fully formed baby salamanders around 2cm long that may stop close to their mother for a time.

The species of cave salamanders are quite similar yet each highly variable. They are best identified largely by their geographic range.

AMBROSI'S CAVE SALAMANDER *Speleomantes ambrosii* Pl. 2.
Range South-east France and north-west Italy (Liguria). Map 24.
Identification Up to 12.5cm including tail. Snout with distinct canthi between the sides and top; toes sometimes relatively pointed; tongue shorter than in Sardinian species of cave salamanders. Brown to black, and flecked, marbled or striped with grey, green, yellow, pink or darker red or brown; uniformly coloured animals also exist. Belly often dark with lighter markings.
Variation Sometimes treated as two species, populations in south-east France and Italy west of La Spezia being called *Speleomantes strinatii*. Various subspecies have also been described.
Similar Species Italian Cave Salamander.
Habits Found in scattered localities and sometimes quite common. May occasionally climb in vegetation. Occurs up to 1,250m Males mature at about 4 years and females at about 5. May live up to 17 years.

ITALIAN CAVE SALAMANDER *Speleomantes italicus* Pl. 2.
Range Montane areas of northern peninsular Italy. Map 25.
Identification Up to 12.5cm including tail. Similar to Ambrosi's Salamander and,

like that species, has canthi, sometimes relatively pointed toes and a generally dark belly. Over most of range is usually dark with red or yellowish spots or marblings, but in north-west is very variable in colouring, like Ambrosi's Salamander with which it hybridises over a narrow area.
Habits Often very common. Usually on limestone but occasionally on sandstone or ophiolytic rock. Found from 80–1,600m.

MONTE ALBO CAVE SALAMANDER *Speleomantes flavus*   Pl. 2
Range North-east Sardinia, between Siniscola and Lula. Map 26.
Identification Up to 15cm including tail; the largest European cave salamander. Snout rounded without distinct canthi between the sides and top; fingers stubby. Yellow, grey-green, brown or black above, often with contrasting spots. Sometimes largely yellow and some animals may be dark with pink legs. Underside typically whitish and unspotted. Tongue is longer than in continental species.
Similar Species Only other tailed amphibian within range is the Sardinian Brook Newt (p. 39).
Habits Occurs up to 1,050m. Reaches sexual maturity at a length of 7–9cm.

SUPRAMONTANE CAVE SALAMANDER   Pl. 2
Speleomantes supramontis
Range East Sardinia, inland from the Gulf of Orosei. Map 26.
Identification Up to 13.5cm including tail. Generally like Monte Albo Salamander. Brown to black with yellow, grey-green or olive-green spots or marbling; some animals generally yellowish or greenish. Underside often whitish with dark spots.
Habits When encountered on surface is often in places with a good growth of moss, where it may reach densities of 300 animals per hectare. Found from 100–1,360m. Matures in 2–3 years.

SCENTED CAVE SALAMANDER *Speleomantes imperialis*   Pl. 2
Range South-east Sardinia, south of range of Supramontane Cave Salamander. Map 26.
Identification Up to 12.5 cm including tail. Generally like Monte Albo Cave Salamander. Purple brown above, usually with scattered, yellow, greenish or gold spots or marbling. Belly pale and without spots in adults.
Habits Distinctive in producing an aromatic smell when handled. Occurs up to 1,170m.

GENE'S CAVE SALAMANDER *Speleomantes genei*   Pl. 2
Range South-west Sardinia only. Map 26.
Identification Up to 11.5cm including tail. The smallest and generally darkest European cave salamander. Dark brown to blackish above, often with small scattered greenish or whitish, occasionally yellowish, spotting. Belly whitish, often with darkish speckling. Also differs from other Sardinian cave salamanders in occasionally having slight canthi on snout and a rather shorter tongue, although this is still longer than in continental species.
Habits Often very common; occurs up to 650m.

Olm
(Family *Proteidae*)

The aquatic family Proteidae consists of only one cave-dwelling species in Europe and five open-water species in the eastern USA. All of them are permanent larvae with characteristic feathery gills. However, unlike normal salamander larvae, they can breed in this state.

OLM *Proteus anguinus*   **Pl. 3.**

Range Southern Slovenia and adjoining extreme north-east Italy, through coastal areas of Croatia and a narrow adjoining area of western Bosnia; an introduced population occurs in north-east Italy, near Valstagna, and there is also a small introduction in the French Pyrenees. Map 27.

Identification Adults usually 20–25cm long including the tail, rarely up to 35cm. Unmistakable; a large aquatic salamander that is nearly always pale with large bushy salmon-pink gills that are retained throughout life. Body long, slender and cylindrical, and tail flattened from side to side. Limbs small with three toes on fore-limbs and two on hind-limbs. Eyes very small and placed beneath the skin, although they may still be light sensitive. Usually whitish, though occasionally greyish, pinkish or yellowish, with faint blotches particularly in young animals.

Variation Most populations are assigned to the subspecies *P. a. anguinus*. Animals from Bela Krajina in south-east Slovenia have been named as *P. a. parkelj* and are distinctive in being black or dark brown with more visible eyes and relatively short head, limbs and tail. However they are genetically like nearby pale populations. Typically pale Olms will become dark if kept in light.

Similar Species Not likely to be confused with any other salamander.

Habits Lives exclusively in underground streams and lakes in caves in karst limestone country. It may occur very deep underground but also in cave entrances. Normally found in waters with a temperature ranging between 5°C and 15°C. Most readily seen in 'tourist' caves such as those open to the public at Postojna, Slovenia. May occasionally be swept into open streams by floods. The dark population *P. a. parkelj* is found more commonly in surface waters. Olms hide in crevices or in bottom sediment when disturbed. Food consists of aquatic invertebrates, especially insect larvae and crustaceans such as water hog-lice (*Asellus*) and shrimps, which it detects by the water movements they make and by chemical cues. Olms are capable of surviving for very long periods without food.

Males defend a territory, pursuing and biting others. During courtship they fan females with their tails, eventually depositing a spermatophore that is taken up by their mate. Up to 70 eggs may be laid over a period of some weeks and are attached to rocks, where they are guarded by the mother during the several months they may take to hatch. Young from such eggs are about 2cm long but females sometimes retain the eggs within the body, eventually producing just two babies up to12cm in length. Juvenile Olms have prominent eyes which soon become reduced. Animals take 7 or more years to reach sexual maturity and live perhaps 30–40 years.

Note Olms are seriously threatened by the pollution of many of their habitats.

FROGS AND TOADS
(Tailless Amphibians, Anura)

There are about 5,000 species of frogs and toads of which only 44 are found in the area covered by this book. They are all tailless, with a short body, quite long hind legs and usually moist skin without scales.

The English terms *frog* and *toad* originally referred to the two basic types of anurans found in Britain, Typical frogs (*Rana*) and Typical toads (*Bufo*), but they are now used for members of other groups in a rather arbitrary way. Frog is usually employed for the more graceful animals with wetter, smoother skins, while toad is used for the drier, warty, stouter forms. As animals of these two general types are found together in a number of separate families, neither frog nor toad indicates a closely related group of animals.

Identification Some of the most important features for identifying frogs and toads are shown in the figure below and in the key and texts. Other useful points to look for are general proportions, especially leg-length, skin texture and colour. When checking the leg-length of frogs, the leg should be gently turned forwards along the body, and the 'hump' in the back carefully pressed down (see Fig., opposite). For identification of eggs and larvae, see p. 238 and p. 248.

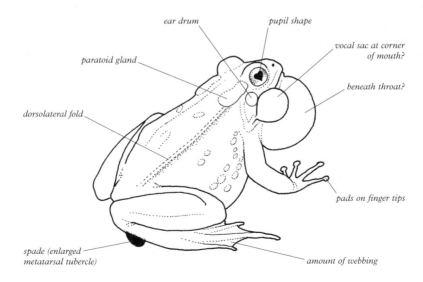

Some features to check when identifying frogs and toads

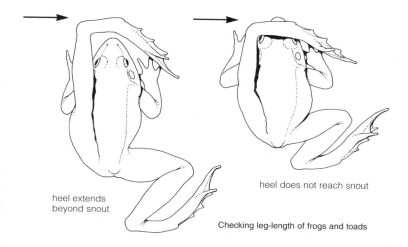

heel extends
beyond snout

heel does not reach snout

Checking leg-length of frogs and toads

Key to Frogs and Toads

1. Fingers with round pads at tips; climbs well. Often bright green   **2**
   Fingers without round pads at tips   **3**

2. **TREE FROGS** *Hyla*
   Dark stripe on flank with upward branch on hip
   > Common Tree Frog, *Hyla arborea* (p. 77, Pl. 11)
   > Italian Tree Frog, *Hyla intermedia* (p. 78, Pl. 11)

   Dark stripe on flank weak without upward branch; often spotted above;
   Corsica and Sardinia etc. only.
   > Tyrrhenian Tree Frog, *Hyla sarda* (p. 78, Pl. 11)

   No dark stripe on flank in adults; south-west Europe
   > Stripeless Tree Frog, *Hyla meridionalis* (p. 79, Pl. 11)

3. Underside vividly patterned: dark with yellow, orange or red areas;
   upperside sombre, warty; pupil heart-shaped, round, or triangular   **4**
   Underside lacks bright contrasting pattern   **5**

4. **FIRE-BELLIED TOADS** *Bombina*
   Underside yellow or orange with dark markings. Bright patch of colour on
   palm of hand extends on to thumb (Fig. p. 61)
   > Yellow-bellied Toad, *Bombina variegata* (p. 60, Pl. 11)

   Underside red or red-orange with dark markings enclosing numerous white
   dots. Bright patch of colour on palm of hand does not extend on to thumb
   > Fire-bellied Toad, *Bombina bombina* (p. 61, Pl. 11)

5. Pupil vertical, slit-shaped in good light   **6**
   Pupil not vertical, may be horizontal, round or triangular   **12**

6. Prominent spade on hind foot (Fig., p. 71). Ear-drum not visible   **7**
   No spade on hind foot. Ear-drum visible in most cases, though sometimes
   not very obvious   **8**

7. **SPADEFOOTS** *Pelobates*
   Spade on hind foot black. South-west Europe only
                     Western Spadefoot, *Pelobates cultripes* (p. 68, Pl. 9)
   Spade on hind foot pale. Webbing between hind toes indented (Fig., p. 71).
   Back of head relatively flat. East Europe only
                     Eastern Spadefoot, *Pelobates syriacus* (p. 70, Pl. 9)
   Spade on hind foot pale. Webbing between hind toes not indented (Fig., p.
   71) or only slightly. Head behind eyes domed
                     Common Spadefoot, *Pelobates fuscus* (p. 69, Pl. 9)

8. Rather slender often with small green or olive spots. Heel usually reaches to
   eye or beyond when hind limb is laid forward along body. Body length may be
   nearly three times width of head.   **9**
   Plump. Heel often does not reach beyond eye when hind limb is laid forward
   along body. Body length may be only just over twice width of head   **10**

9. **PARSLEY FROGS** *Pelodytes*
   Tubercles under hand at base of fingers not conical (Fig. p. 72)
                     Parsley Frog, *Pelodytes punctatus*, (p. 71, Pl. 8)
   Tubercles under hand at base of fingers conical
                     Iberian Parsley Frog, *Pelodytes ibericus* (p. 72)

10. **MIDWIFE TOADS** *Alytes*
    Three tubercles on palm of hand   **11**
    Two tubercles on palm of hand (Fig., p. 66); Iberian peninsula
                     Iberian Midwife Toad, *Alytes cisternasii* (p. 66, Pl. 8)

11. Slender, Majorca only.
                     Majorcan Midwife Toad, *Alytes muletenesis* (p. 67, Pl. 8)
    Robust, West European mainland
                     Common Midwife Toad, *Alytes obstetricans* (p. 64, Pl. 8)
                     Southern Midwife Toad, *Alytes dickhelleni* (p. 67, Pl. 8)

12. Warty with large, obvious paratoid glands (Fig., p. 74)   **13**
    Smooth-skinned, or not particularly warty; no paratoid glands   **14**

13. **TYPICAL TOADS** *Bufo*
    Paratoid glands oblique (Fig., p. 74). Colour not green (occasionally olive);
    often lacks clear patterning. Tubercles under longest hind toe usually paired.
    Eye usually deep gold or copper coloured
                     Common Toad, *Bufo bufo*, (p. 73, Pl. 10)

Paratoids roughly parallel. Often olive, grey or greenish; usually without very distinct pattern except for a yellow vertebral stripe (only rarely absent). Tubercles under longest hind toe usually paired (Fig., p. 75). Eye pale gold. Not much of eastern Europe, Italy etc.

Natterjack, *Bufo calamita* (p. 74, Pl. 10)

Paratoids roughly parallel. Upper side usually boldly marked with greenish, often dark-edged 'islands' on pale background; rarely has yellow vertebral stripe. Tubercles under longest hind toe normally single (Fig., p. 75). Eye pale gold. Not most of west Europe

Green Toad, *Bufo viridis* (p. 76, Pl.10)

14. Pupil not horizontal, often more or less round, or triangular. Ear drum often not obvious **15**

Pupil horizontal. Ear-drum usually fairly obvious **18**

15. **PAINTED FROGS** *Discoglossus*

Corsica only; tip of fourth finger wider than base. Widened rather than narrowed

Corsican Painted Frog, *Discoglossus montalentii* (p. 64, Pl. 7)

Tip of fourth finger narrowed **16**

16. Ear drum usually visible **17**

Ear drum hidden; Iberian peninsula only

West Iberian Painted Frog, *Discoglossus galganoi* (p. 62)

East Iberian Painted Frog, *Discoglossus jeanneae* (p. 63)

17. Malta, Gozo, Sicily, north-east Spain and adjoining France

Painted Frog, *Discoglossus pictus* (p. 62, Pl. 7)

Corsica and Sardinia etc.

Tyrrhenian Painted Frog, *Discoglossus sardus* (p. 63, Pl. 7)

18. **TYPICAL FROGS** *Rana*

No continuous dorsolateral folds (Fig., p. 58) but short fold just behind ear. Ear-drum as large as eye (females) or much larger (males). Introduced into north Italy, France, Spain and elsewhere.

American Bullfrog, *Rana catesbeiana* (p. 97, Pl. 14)

Continuous dorsolateral folds always present. Ear-drum not bigger than eye **19**

19. Eyes close together (Fig., below). Often green or greenish. Males have vocal sacs at corners of mouth

Water Frogs (p. 95, Pl. 14)

Eyes well separated (Fig., below). No bright green in dorsal colouring. Dark facial mask nearly always present. Males lack external vocal sacs **20**

eyes close
together

Water Frog

eyes well
separated

Brown Frog

20. **BROWN FROGS**

The species of Brown Frogs are all very similar and rather variable which makes it difficult to produce a key that will identify them all with certainty. Couplet 21 will not always place frogs correctly, so if a frog is taken through the key but does not fit the relevant description, return to that point and follow the other branch of the couplet.

Pyrenees only; small, 5cm or less. Nostrils well separated (see Fig, p. 85), heel reaches snout tip, dorsolateral folds closely spaced, ear drum small or invisible.

Pyrenean Frog, Rana pyrenaica (p. 81, Pl. 12)

Pyrenees and most other areas; adults larger. Not this combination of features **21**

21. Legs short: heel usually does not extend beyond snout (see Fig. p. 55). Dorsolateral folds close together (distance between them, measured just behind forelimbs, goes 5.5–7 times into length of animal). **22**
Legs long: heel usually extends to, or beyond, snout. Dorsolateral folds well separated (distance between them, measured just behind forelimbs, goes not more than 5, rarely 5.5, times into length of animal) **23**

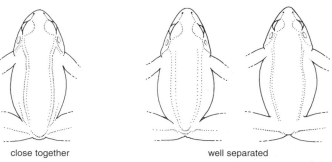

close together   well separated

Dorsolateral folds

22. Metatarsal tubercle small and soft (Fig., p. 82); snout often blunt. Almost never striped above.

Common Frog Rana temporaria (p. 80, Pl. 12)

Metatarsal tubercle large and hard, sometimes sharp-edged, half to two-thirds length of first hind toe; snout usually pointed. Often striped above.

Moor Frog, Rana arvalis (p. 82, Pl.12)

23. Underside usually unmarked or with stippling only at sides of throat in dark individuals. Ear-drum large and usually very close to eye. Metatarsal tubercle rather prominent.

Agile Frog, Rana dalmatina (p. 83, Pl. 13)

Throat usually spotted or marbled, and often belly as well. Ear-drum not very close to eye. **24**

24. Iberian peninsula only. A small frog, frequently with dark throat, weak metatarsal tubercle and often a pinkish flush on belly of breeding males.

Iberian Frog, *Rana iberica* (p. 87, Pl. 13)

North Italy only. Distance between nostrils less than distance between nostril and eye (Fig., p. 85). Ear-drum prominent. Often some reddish or orange colouring beneath in breeding males.

Italian Agile Frog, *R. latastei* (p. 84, Pl.13)

Distance between nostrils more than distance between nostril and eye. Ear-drum not prominent. No pink on underside.

Balkan Stream Frog, *Rana graeca* (p. 85, Pl. 13)
Italian Stream Frog, *Rana italica* (p. 86)

round-triangular   heart-shaped   horizontal

vertical, cat-like

Pupil shapes in frogs and toads

Fire-bellied Toads, Painted Frogs and Midwife Toads
(Family *Discoglossidae*)

A small, rather primitive family of about 18 species, 11of which are found in Europe and neighbouring areas and seven in East and South-east Asia. Discoglossids are the only predominantly European family of amphibians or reptiles, apart from Parsley Frogs. All species have a disc-shaped tongue that cannot be protruded to catch prey as it can be in many other frogs and toads. The three main types of discoglossid found in Europe differ considerably in appearance and habits, but are alike in having relatively subdued voices and being able to breed two or more times a year.

Fire-bellied toads (*Bombina*) and Painted frogs (*Discoglossus*), are largely aquatic and breed in water, the males grasping the females around the loins during amplexus. Unlike the Midwife toads (*Alytes*), the males of these species have nuptial pads. Midwife toads also usually grasp the females around the loins but they mate on land and the large eggs are produced in strings that the male carries wrapped around his hind legs. He keeps the eggs moist, sometimes by visiting pools and streams, until they are ready to hatch, when he deposits them in shallow water. The newly metamorphosed young of Painted frogs are small, but those of other European discoglossids are quite large. They all often reach maturity rapidly.

YELLOW-BELLIED TOAD *Bombina variegata*   Pl. 11

Range France to Romania and Moldova, north to Belgium and central Germany, and south to Italy and much of Greece. Populations in Carpathian mountains are largely separated from the main range of the species. Map 28.

Identification Adults usually less than 5cm. A small warty, aquatic toad with a flattened body and round, heart-shaped or triangular pupils. Underside brightly coloured: typically yellow or orange with blue-grey or blackish markings that only enclose few small white spots. Yellow or orange colour on palm of hand extends on to first finger and its tip is light above as are often those of the other fingers (Fig., opposite). Bright patch on groin joins that on underside of thigh and this may extend backwards to be visible from behind. Back grey, brown, yellowish or even olive, with prominent warts, often ending in black spiny points. No vocal sac. Breeding males have dark nuptial pads on the inside of the forearms and on the three inner fingers.

Variation Occasional animals have the belly almost entirely dark or entirely yellow or orange. Nearly all the characteristic features of this species are subject to occasional exceptions. Italian individuals (*B. v. pachypus*) often have dark throats and breasts and predominantly yellow or orange bellies; Southern Balkan animals (*B. v. scabra*) frequently have many dark points which are often spiny and may extend onto the belly; Dalmatian animals are similar but are often known as *B. v. kolambatovici*. For hybridisation with Fire-bellied Toad see that species.

Similar Species Confusable with the closely related Fire-bellied Toad. Range is often useful in identifying untypical animals.

Voice Calls in chorus, both by day and night, although often heard most obviously in the evening. Song a rather musical 'poop ... poop ... poop', brighter and faster than the call of Fire-bellied Toad; usually more than 40 calls a minute and sometimes one or two calls a second. Can be mistaken for the Scops Owl and, at a distance, may sound like even more distant chickens.

Habits Largely diurnal, although also active at night. Found in open often sunny, shallow and often temporary water that may have little vegetation: small ponds, drainage ditches, pools in marshes and woods, clay pits, shallows at the edges of rivers, drinking troughs, flooded tyre tracks and pools in disturbed areas; also streams, particularly in south. Especially where it occurs with the Fire-bellied Toad, the Yellow-bellied Toad tends to be found in more mountainous or hilly areas and can reach 2,100m in the south of its range. Very aquatic and sociable, many animals being found together in small areas of water. A lively, active toad, often seen floating with legs spread on surface of water. Also hunts on land, especially at night and after rain, and babies can occur over a kilometre from water. Sometimes encountered here when turning stones and logs etc and also hibernates away from water. Will dive into mud at bottom of ponds when disturbed. If molested, will arch back strongly and throw up limbs to show bright belly and may even turn itself upside down. This 'unken reflex' warns of the poisonous nature of the whitish skin secretions of this toad, which may be produced at the same time and have a strong smell.

May breed suddenly, often after rain and this can happen about three times a year in the North. Breeding males defend territories about a metre or more across. Females lay 120–170 eggs in clumps of up to 30 among grassy or other aquatic vegetation. They hatch in 2–3 days and the tadpoles metamorphose in 6–10 weeks to produce toads that are 1–1.5cm long and mature in the first or sometimes the second year after hatching. This species has lived 27 years in captivity.

Yellow-bellied toad   Fire-bellied toad

Palms of left hands of Yellow-bellied and Fire-bellied Toads, showing characteristic patterns

FIRE-BELLIED TOAD *Bombina bombina* Pl. 11

Range East Europe as far north as Denmark, south Sweden (where reintroduced after extinction) and Russia, and south to north Serbia and Turkey; also in small area of north-west Anatolia. Map 29.

Identification Adults usually less than 5cm. Very like Yellow-bellied Toad (opposite): a small, warty, aquatic toad with characteristically flattened body and brilliantly coloured underside; pupils usually round or triangular. Differs from Yellow-bellied Toad in often having bright red-orange or red markings on belly combined with dark grey or blackish patches that enclose often numerous white dots. Bright patch on palm of hand does not extend on to first finger (Fig., above) and its tip is dark above or at least not brightly coloured, something that may be true of other fingers. Bright patch on groin does not join that under thigh and there is usually no bright colour on the back of the upper hind leg. The Fire-bellied Toad tends to be less robust than the Yellow-bellied Toad with a narrower head, its back is often darker or sometimes green or with green spots, and the skin is more delicate with smaller and smoother warts that are sometimes rough tipped but usually not really spiny. An internal vocal sac is present. Breeding males have dark nuptial pads on the inside of the forearm.

Variation As with Yellow-bellied Toad, nearly all the characteristic features of this species are subject to occasional exceptions: in some animals, the amount of black on the belly is reduced. In areas where the ranges of Fire and Yellow bellied Toads meet, particularly along the base of mountain ranges around the Danube basin, animals with intermediate characters are often found. In some places such individuals result from hybridisation where the two species meet and typical Fire – and Yellow-bellied Toads may occur nearby. At other localities where hybridisation occurred in the past, whole populations may be intermediate in appearance and 'pure' animals of one or both species are absent.

Similar Species Only likely to be confused with Yellow-bellied Toad (opposite). Range is often useful in identifying untypical animals.

Voice Calls usually in chorus, especially by day and in evening. A rather musical, although mournful, 'oop ... oop ... oop' that is louder than that of the Yellow-bellied Toad. Call speed varies, but usually slower than this species, with less than 40 calls a minute and often about one call every 1.5–4 seconds. The call has been confused with that of Scops owl and Pygmy owl and choruses can sound like distant bells.

Habits Very similar to Yellow-bellied Toad but tends to be exclusively a lowland animal usually found well below 250m, but up to 730m in the Czech Republic. Occurs in shallow exposed water in meadows, farmland, wood edges, marshes and flood plains, and often encountered at the edges of lakes and ponds. May occasionally be found in temporary waters but prefers places with a good growth of subaquatic vegetation. Also encountered on land where it hibernates and may be found under stones and logs and among roots; babies may occur 500m from water. Defensive behaviour is like Yellow-bellied Toad.

Usually mates in late spring, especially after breeding pools have been cooled by rain. Females may produce 300 eggs in a season in clumps of up to 30 which are laid in aquatic vegetation. They develop in 2–5 weeks to produce baby toads 1–1.5 cm long. Males usually mature the next year and females the year after that. Has lived 20 years in captivity.

PAINTED FROG *Discoglossus pictus*   Pl. 7

Range Sicily, Malta and Gozo; southern France (Pyrenees Orientales, Aude) and adjoining north-east Spain (Gerona); also north-west Africa. The French and Spanish populations spread from an introduction of African animals around Banyuls-sur-Mer, probably made in the late 19th Century; a small introduction is also present near Amboise in the Loire valley of France. Map 30.

Identification Adults up to 8cm but most animals considerably smaller; males may be bigger than females. A medium-sized, shiny, plump frog with a flat head and pointed snout. Pupils are usually roundish or triangular. Ear-drum, while usually visible, is inconspicuous in most cases. Skin smooth often with small warts that may form lines. Colour extremely variable and ground colour may be grey, olive, yellowish, greenish, brown or even red above. In many animals there are darker, often light edged spots, while a minority have one or three light stripes on the back; these two colour forms often occur in the same populations. Belly whitish, sometimes with brown speckling. Males tend to have more extensive webbing on the hind feet than females and, when breeding, have black nuptial pads on the two inner fingers and often similar dark rough structures on the throat, belly and toe webbing

Variation Animals from Sicily, Malta and Gozo are assigned to the subspecies *D. p. pictus*, while those introduced in France and Spain are probably the largely Algerian form, *D. p. auritus*.

Similar Species Other Painted frogs; also some similarity to Typical frogs (*Rana*) but can be distinguished from these by pupil shape (not horizontal), inconspicuous ear-drums and disc-shaped tongue.

Voice A rather quiet, rolling 'laugh'; a feeble, rapid, growling 'rar-rar-rar' produced on surface or when submerged.

Habits Occurs in a variety of Mediterranean situations such as open sandy coastal areas, pastures, wetlands, vineyards, woods and forests. Often found in and around dense vegetation and nearly always close to often shallow water which may be still or running (such as pools, streams and river edges) and may be seasonal or even brackish. In Sicily, frequently occurs in man-made cisterns, ditches and other irrigation features. Active by day and night. Found up to 1,500m in Sicily but usually close to sea level in France and Spain (rarely to 500m). Often seen sitting with head just above water. May occur with Water frogs (p. 88). Reproduction is similar to the West Iberian Painted Frog.

WEST IBERIAN PAINTED FROG *Discoglossus galganoi*

Range Portugal and much of western Spain. Map 31.

Identification Up to 8cm but usually to about 6cm. Very similar to the Painted Frog in form and colouring. Unlike in this species, the ear of the West Iberian Painted Frog is not visible or scarcely so. Most animals are spotted although striped and, rarely, plain individuals also occur. Nuptial pads etc. are similar to those of Painted Frog.

Similar Species Best distinguished from the Painted Frog by lack of visible ear and by range.

Voice A weak, fast, often repeated 'rah ... rah ... rah', mainly at night. The individual croaks are distinctly longer than those of the Painted Frog.

Habits Occurs on soils derived from granite and other metamorphic rocks rather than limestone. Can occur up to 1,600m but usually under 500m. Behaviour generally similar to Painted Frog. Typically encountered in or near still or slowly moving, mainly shallow water where it is often found in dense herbaceous cover. Frequently occurs in small water bodies, even drinking troughs. Found in sandy areas near the sea, meadows, thickets, gulleys, woods including those along rivers. Often crepuscular but young animals especially may be out in the day particularly in rainy or humid weather. Breeding takes place in winter and spring. Individual females go repeatedly to the water, often with different males, depositing 20–50 eggs each time and laying up to 1,500 in the course of a single breeding period and more than 5,000 in a year. The eggs are scattered in small groups on the bottom of water bodies or in aquatic vegetation. They hatch in 2–9 days and development usually takes about 4–9 weeks. The newly metamorphosed baby frogs are about 1cm long and mature in 3–5 years. Some individuals can survive for 10 years.

EAST IBERIAN PAINTED FROG *Discoglossus jeanneae*

Range Southern and eastern Spain but apparently restricted to a number of isolated areas. Map 32.
Identification Up to about 6cm. Extremely similar to West Iberian Painted Frog and usually best identified by range.
Similar Species West Iberian Painted Frog. Painted Frog usually has visible ear drum.
Voice Like West Iberian Painted Frog.
Habits Generally like West Iberian Painted Frog but usually found on limestone and gypsum soils.

TYRRHENIAN PAINTED FROG *Discoglossus sardus*   **Pl.7**

Range Corsica, Sardinia (and offshore islands), the Tuscan islands of Giglio and Monte Cristo, Monte Argentario on the west Italian coast and the Iles d'Hyères (Port-Cros and Ile du Levant) off southern France. Map 33.
Identification Usually up to about 7cm, sometimes larger. Similar to Painted Frog, but tends to be more robust with a broader head. Pattern often consists of dark spots (frequently less well defined than in Painted Frog and lacking light borders) and there is usually a pale blotch on the back; some animals are uniform, but none are striped.
Voice Similar to Painted Frog.
Habits Occurs in a wide range of habitats, from slightly brackish pools on bare stretches of coast to sunny marshes and ponds and clear, often slow-flowing streams. Can occur in maccia and coniferous and broad-leaved woods, especially in the mountains. Found from sea level to 1,300m on Corsica and to 1,770m on Sardinia. On Corsica differs in its broad habitat range from the Italian Pool Frog (p. 92) which is largely confined to low altitudes and still water often with a lot of vegetation.
Variation Some variation in shape and pattern.
Similar Species Italian Pool Frog (p. 94); Corsican Painted Frog (p. 64).
Note The presence of this species on widely separated islands just off the coast of continental Europe, but not on the mainland itself, suggests it may have recently been present there but has become extinct, something supported by finds of fossil Painted frogs in peninsular Italy. The European Leaf-toed Gecko (p. 124) is a similar

lar case. Such loss of range may be true of Painted frogs as a whole. This is a long-standing group that is now restricted to the European and African countries of the western Mediterranean area. In the quite recent past Painted frogs extended far to the east. The Levant Painted Frog (*Discoglossus nigriventer*) only became extinct in Israel in the mid-20th Century and recent fossils show the group was also once present on Crete. Painted frogs may have lost range in response to the spread and diversification in Europe of Typical frogs (*Rana*).

Tyrrhenian
Painted Frog

Corsican
Painted Frog

Painted Frogs, differences in snout profile

CORSICAN PAINTED FROG *Discoglossus montalentii*   Pl. 7
Range Corsica. Map 34.
Identification Up to about 6.5cm. Very similar to Tyrrhenian Painted Frog but the snout is rounded when seen in side view and has a horizontal upper profile (see Fig., above), the toes are also blunter (tip of fourth finger wider than its base rather than narrower) and the hind legs are rather longer. Some animals have dark spots on back; while others are more or less uniform grey, brownish or reddish; belly is yellowish white.
Similar Species Tyrrhenian Painted Frog (p. 63).
Voice More musical than other painted frogs, including the Tyrrhenian Painted Frog which occurs within its range, and rather like the call of the Yellow-bellied Toad: 'poop ... poop ... poop'.
Habits Found in pristine habitats in hilly and mountain regions from about 300m to 1,900m. It often occurs at higher altitudes than the Tyrrhenian Painted Frog and is absent from the coastal lowlands. Usually encountered in still or running fresh water, sometimes in open areas but especially in and around often precipitous streams in woods and forests. An agile frog that often leaps between boulders in water courses. It may occasionally be found living alongside the Tyrrhenian Painted Frog.

COMMON MIDWIFE TOAD *Alytes obstetricans*   Pl. 8
Range Much of Iberian peninsula, through France and Belgium to extreme southern Netherlands and eastwards to north-central and south-west Germany and Switzerland. A few small introductions have been made in England. Map 35.
Identification Adults up to 5.5cm but usually less; males smaller than females. A small, plump toad with prominent eyes and vertical pupils. Skin quite smooth but usually with some small warts and granules, some of which may form a line along each side of the back. Colour variable, but typically grey, olive or brownish with some small darker, often greenish or brown markings. The warts are frequently red or yel-

low. In summer, males may be seen with strings of yellowish eggs wound around the hind limbs.

Variation The most widespread form (*A. o. obstetricans*) is usually greyish or brownish above with diffuse dark spots, and the underside is whitish grey. Animals in north-eastern Spain (*A. o. almogavarii*) are similar but tend to be brownish or yellowish with a marbled pattern of green or brown spots. Populations in north and west Spain and Portugal (*A. o. boscai*) usually lack warts and are whitish with well-defined green or brown spots and a white belly. Toads may be similar in central and eastern Spain *A. o. pertinax* but the throat may lack spots.

Similar Species In the Iberian peninsula can be confused with Southern Midwife Toad (p. 67); Iberian Midwife Toad (p. 66) has two tubercles on palm instead of three. Spadefoots (p. 68) have very large metatarsal tubercles; Parsley frogs (p. 71) can be distinguished by their longer legs and virtual lack of webbing on the hind feet.

Voice Calls mainly at night, but also by day when conditions are cloudy and humid, for instance before a storm, and at high altitudes. Often sings from its refuges away from water. The call is a high-pitched, plosive, musical 'poo … poo … poo …' about one call every 1–3 secs; usually higher and shorter than Yellow-bellied Toad (p. 60). Individuals answer each other and vary considerably in pitch according to their size. Females also sing during courtship and may alternate calls with their potential mate. Away from water, midwife toads frequently call from cover and can be very difficult to locate. At a distance a chorus may sound like far-off church bells. Single calling individuals could be confused with Scops and Pygmy owls.

Habits Found often in sunny situations, especially in the north, and frequently in rather hilly country. Occurs in woodlands, cultivated land, quarries, sand pits, rock slides, stony slopes, embankments and a variety of similar habitats at altitudes of up to 2,400m in the Pyrenees. Usually found fairly near water, although it can occur 500m from it, and is often associated with human habitations, penetrating into villages and urban areas where it lives in parks, churchyards and gardens. Mainly active at dusk and at night. By day hides in crevices, under logs, stones, in dry-stone walls, under paving, and often in burrows which it may excavate using the forelimbs. Hibernation takes place on land in burrows or crevices and sometimes even in damp cellars. When threatened, Common Midwife Toads inflate themselves and pull the limbs close to the body. Less frequently they take up a characteristic 'fright posture', standing high on their legs with the rump raised, the eyes shut and the head turned down between the forelimbs.

In the south, females may lay 2–3 clutches a year. A clutch contains up to 80 eggs but usually far fewer. The eggs are arranged in a rosary-like string and eventually swell to a diameter approaching 5mm. After fertilisation, the male wraps the string around his hind legs and may court more than one female in succession, so he often carries clutches from two of them and sometimes as many as four. If possible, females select large males, recognising them initially by their deeper voices. Males not carrying eggs are also preferred and, if a chosen male is already burdened with a clutch from another female, relatively few eggs will be produced. The male often retreats to a humid burrow to keep the eggs moist. In very dry conditions a male may travel to ponds etc. to maintain his and their water balance. The skin secretion of the male also appears to protect the eggs from fungal and other infections. Egg development lasts 3–8 weeks after which the male takes them to water just before they are ready to hatch into tadpoles. The water chosen tends to be permanent and cool and includes farm and village ponds, springs, drinking troughs, and quiet stretches of

tyre tracks, is occasionally used. Some tadpoles may overwinter and then often grow very large. The newly metamorphosed toads are around 2–2.5cm long and become mature 2–3 years after birth. Common Midwife Toads may live five years or more.

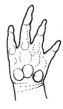

Most species   Iberian Midwife

Midwife Toads, palms of hands

IBERIAN MIDWIFE TOAD *Alytes cisternasii*   Pl. 8

Range South and east Portugal and western and central Spain. Map 36.

Identification Adults up to 5cm, but usually smaller; females bigger than males. Similar to Common Midwife Toad but more robust with a bigger head and shorter legs, a short thick outermost finger, and rather smoother skin, although there is often a row of small warts above the eye. Most distinctively there are two instead of three tubercles on the palm of the hand (Fig., above). The ground colour is whitish, cream or brown, sometimes with a greenish tinge, with spots, blotches or marbling. Scattered small orange granules are usually present, especially on the flanks and there is often a light ∧-shaped patch on mid-back behind the head. The underside is white to yellowish without spots.

Variation No obvious variation recorded.

Similar Species Common and Southern midwife toads (p. 64 and opposite). Western Spadefoot (p. 68) and Parsley frogs (p. 71).

Voice Very similar to Common Midwife Toad and produced mainly at night, especially soon after dusk. This toad may also call from its daytime refuges and is perhaps more likely be heard in winter than the Common Midwife Toad.

Habits Found in quite dry areas with low rainfall, hot summers and mild winters, usually in the lowlands although it may reach 1,200m. Occurs in both wild and cultivated areas with open vegetation of, for instance, holm, cork and other oaks, olives and brushwood. Often found on sunny slopes and relatively close to water, particularly temporary streams in the south of its range. Frequently encountered near human habitation. Seems to prefer loose, sometimes sandy soils that are often of granitic or shaly origin. It digs burrows with its forelimbs and takes refuge in them during the day. Also found under stones and logs etc. and in the interstices of dry-stone walls. The Iberian Midwife Toad may aestivate in dry weather and is often the dominant amphibian in its favoured habitats. It is largely active at night and dusk, although occasionally seen by day in rainy weather. In areas where it occurs with the Common Midwife Toad, it tends to be replaced by this species at higher altitudes. Defensive behaviour is similar to the Common Midwife but Iberian Midwifes may produce a long squeaking cry when adopting the 'fright posture' and then emit copious skin secretion that smells like that of the Ribbed Newt.

Breeding is similar to that of Common Midwife Toad. Females often lay 30–60 eggs at a time and a male sometimes carries up to four clutches from more than one female, comprising a total of up to 180 eggs. He often retreats with these into a burrow where he controls humidity by adjusting its size and extent. Development takes around 3 weeks and eggs are deposited in slow, sometimes temporary streams, ponds, cisterns and other irrigation features. Broods from different parents often hatch simultaneously and the tadpoles usually mature in 3–5 months. In spite of its generally southern distribution, tadpoles of the Iberian Midwife Toad are very cold-resistant and can even survive beneath ice. Newly metamorphosed toads are 2–2.5cm long and sexual maturity occurs about 2 years after birth.

SOUTHERN MIDWIFE TOAD *Alytes dickhilleni* Pl. 8

Range South-eastern Spain: eastern Andalucía and adjoining areas south of the River Guadalquivir. A related species occurs in Morocco. Map 37.

Identification Up to 5.5cm; females larger than males. Very similar to the Common Midwife Toad. Like this species it has three tubercles on the palm but the skin tends to be rather smoother. Greyish, brownish or greenish above, but usually lighter on limbs and between the eyes. Small warts on the skin are not red or orange. Range should be taken into account when identifying this species.

Similar Species Common Midwife Toad (p. 64). Iberian Midwife Toad (opposite) has only two tubercles on palm.

Voice Apparently similar to the Common Midwife Toad.

Habits Mountainous areas from about 480m up to 2,300m. Often found in pine and oak forests, open rocky places and precipitous gullies where there is access to natural or artificial water. Behaviour and breeding is similar to the Common Midwife Toad. Newly metamorphosed frogs are 1.7–2.5cm long and males may mature at a length of 2.5cm.

MAJORCAN MIDWIFE TOAD *Alytes muletensis* Pl. 8

Range North-western mountains of Majorca (Serra de Tramuntana). Map 38.

Identification Up to 4cm; females rather bigger than males. Generally similar to other midwife toads but less robust with longer legs and a smooth skin. Like Common and Southern Midwife toads, it has three tubercles on the palm of the hand. Yellowish grey or yellowish above with dark greenish, blackish or brownish blotching that may sometimes be extensive, forming a marbling or being reduced to isolated blotches.

Variation Considerable variation in colouring.

Similar Species None within range.

Voice Males produce a melodious 'pi … pi … pi', like drops of water falling into a half-filled bucket. Heard mainly after dusk but animals sing from cover at any time of day. Can be confused with the call of the Scops owl. Females also sing just before egg laying, their calls being shorter and quieter than males.

Habits Now mainly confined to limestone mountains at around 300–400m where it occurs in narrow canyons often 30–60m deep, which have small streams running along their bottoms that persist as pools in summer. Usually found close to these in rock crevices and under stones, where it may be encountered in groups. Occurs less commonly by artificial water sources in more open country. Like other Midwife toads it is nocturnal and crepuscular, but it differs from them in not producing toxic skin secretions. Such loss of protective mechanisms is common on islands without

skin secretions. Such loss of protective mechanisms is common on islands without
natural predators.

Breeding males first grasp the females around the neck with their forelegs, later
gripping in the more usual position for Midwife toads around the loins. Females lay
7–20 eggs at a time. Unlike other Midwife toads, males rarely bear egg strings of
more than a single female, although this may sometimes occur in which case up to
34 eggs may be carried. Males retreat with the eggs into hiding places for about 2–3
weeks while development takes place. The tadpoles respond to the presence of
Viperine Snakes, which have been introduced into Majorca, by reducing their movements. They may take up to three years to metamorphose and have lived seven years
in captivity.

Other names *Baleophryne muletensis*.

Note This species was first described as a fossil in 1977 but discovered alive in 1979.
A related species occurred on Minorca in prehistoric times.

Spadefoots
(Family *Pelobatidae*)

A rather primitive family of about 100 species with representatives in Europe and
neighbouring areas of north-west Africa and Asia, in south-east Asia, and in North
America. Three rather similar species of spadefoot (*Pelobates*) are found in Europe.
These are relatively short-limbed animals that have a superficial resemblance to
Typical toads (p. 73), but differ in their very robust skull, vertical pupils, and very
prominent, flattened, sharp-edged 'spade' on the hind foot, which is in fact a modified metatarsal tubercle (Fig., p. 71).

Spadefoots are quite strictly nocturnal outside the breeding season. They are generally confined to areas with sandy soil, where they hide during the day, and in periods of drought, in deep often almost vertical burrows which may sometimes descend
as much as a metre. They dig these burrows themselves using the spades on their
hind feet alternately. In very light soils, spadefoots sink out of sight quickly and the
upper part of the burrow usually collapses, covering them completely. The toads use
their heavily ossified heads to push their way out of such hiding places.

Spadefoots breed in the spring, often in quite deep pools and at this time of the
year may be active during the day. There are no nuptial pads and males of all species
grip the females just in front of the hind limbs. The eggs are laid in thick bands that
smell of fish and are usually wound round the stems of water plants; the tadpoles
may grow very large (often up to 10 cm and occasionally to 17.5cm).

WESTERN SPADEFOOT *Pelobates cultripes*   **Pl. 9**

Range Spain except most of far north, Portugal, western and southern France. A
related species, Varaldii's Spadefoot (*Pelobates varaldii*) occurs in north Morocco.
Little, if any, actual overlap with Common Spadefoot. Map 39.

Identification Adults up to 11cm, larger than the Common Spadefoot. A big,
plump, smooth-skinned toad with large eyes, a vertical pupil and a black spade on
the hind foot; the top of the head is not obviously domed. Upperside greyish, yellowish, or whitish often with dark brown or greenish blotches, spots, or speckles that
rarely form longitudinal bands. Often greener than the Common Spadefoot, and less
frequently has orange spots. The eye is silvery gold or greenish. Sexual differences
are similar to those of the Common Spadefoot.

Similar Species Common Spadefoot (below) has a domed head and a pale spade. Midwife toads (p. 64) lack a spade as do Typical toads (p. 73) which also have a horizontal pupil.

Voice Breeding males produce a rapid quite deep 'co-co-co' under water which sounds rather like a clucking hen; females may also call.

Habits Generally like other spadefoots. Found in open areas usually with soft or sandy soils, mainly in the lowlands but up to 1,800m in Spain. Occurs in open woods, scrub, cultivation, dunes and sometimes marshes, even near the sea, and may be encountered close to human habitation. Often very common on sandy coasts. Largely nocturnal and occasionally to be seen in vast numbers after rain; abundant in some years but rare in others. Hides in burrows often about 20cm deep. When threatened may inflate body and mew like a kitten.

Breeds in often temporary water with thick vegetation, that may occasionally be brackish. Large numbers of males may appear quite suddenly at breeding sites, often during wet nights, the females coming later. The breeding period may last a month or only a few days and at this time the toads may be partly diurnal. Mating occurs both in the water and at its edge. Then females, which probably lay only once a year, produce bands of up to 7000 eggs which are sometimes up to 100cm long and are laid among vegetation or on the pond bottom. The eggs hatch in 1–2 weeks and the tadpoles take around 4–6 months to develop, unless they are killed by their ponds drying up, a frequent occurrence. Newly metamorphosed toads are 2–3.5cm long and take about 3 years to reach maturity. Western Spadefoots may live 10–15 years.

COMMON SPADEFOOT *Pelobates fuscus* Pl. 9

Range Much of lowland west, central and east Europe, from extreme north-east France eastwards, extending north to Denmark and Estonia and south to northern parts of Serbia, Bulgaria and Italy (Po valley in Piedmont and Lombardy and in coastal areas extending to the Slovenian border). A doubtful isolate in central France. Also reaches Ural Mountains, Kirghiz Steppes and Aral Sea. Map 40.

Identification Adults up to about 8cm but usually smaller, females bigger than males. A plump, smooth-skinned toad with large eyes and vertical pupils. A big, pale-coloured spade on the hind foot, which is more or less fully webbed, and a well marked dome on top of the head just behind the level of the eyes. Colour of upperside extremely variable: grey, pale brown, yellowish, or whitish, with darker brownish markings that may form blotches, marblings, speckles, or stripes. The sides and sometimes the back often have small orange spots. The eye is golden, orange, or coppery. This toad often smells strongly of garlic. Males tend to be smaller than females, and have a large, raised, oval gland on the upper arm, and pearly granules on the lower arm and hand during breeding season.

Variation Considerable individual variation in colour and pattern. North Italian animals have been named as *P.f. insubricus.*

Similar Species Western Spadefoot (opposite); Eastern Spadefoot (p. 70). Midwife toads (p. 64) lack a spade as do Typical toads (p. 73) which also have a horizontal pupil.

Voice Breeding males produce a repeated clicking 'c'lock-c'lock-c'lock' under water; females also call, grunting or making a more rasping 'tock-tock-tock'. If threatened, produces a shrill cry rather like that of a kitten or small baby.

Habits Generally similar to other European spadefoots. Usually found at low altitudes but can reach 675m. Occurs in flat open areas, dunes, heaths, open pinewoods, parks and cultivated areas, especially where asparagus and potatoes are grown. Although typically found in light soils it can sometimes be encountered in loam and clay regions and may live up to a kilometre from water. Nocturnal, usually not emerging until after sunset when it is particularly active in wet weather. Usually progresses in a series of short jumps. When attacked or alarmed has impressive but harmless 'threat' display in which it squeals repeatedly, inflates body, stands high on its legs, and may jump at its enemy with mouth open and sometimes even bite.

Partly diurnal during breeding season when it is found in deep pools, ditches, ponds and lakes, particularly nutrient-rich ones with a good growth of reeds and other plants at their edges, and sometimes even in rather brackish water; in Italy it sometimes breeds in rice fields. There may be 1,000–3,500 eggs in the bands produced by the female which are about 2cm thick and sometimes 50–100cm long. Tadpoles emerge after about 4–10 days and generally take 2–5 months to develop although some may overwinter. They occur in open areas of water, often being seen near the surface but diving down into mud if disturbed. Newly metamorphosed toads are about 2–4cm long and can mature the following year, although two years is more usual. Common Spadefoots may live 10 years.

EASTERN SPADEFOOT *Pelobates syriacus*   **Pl. 9**
Range Greece (including some neighbouring islands) and Turkey, north to Macedonia, east Serbia, south, east and north Bulgaria and south-west Romania. Also South-west Asia. Map 41.
Identification Adults up to 9cm. Similar to other spadefoots, but distinguished by combination of undomed skull, pale spade, and indented webbing on hind feet (Fig., opposite). The skin often has scattered, very small warts. Upperside is yellowish, greyish, or whitish, usually with large, often dark-edged greenish or brownish blotches or, more rarely, irregular stripes. Sides and back often have small yellow, pink or dark red spots. Males have a large gland on the upper arm.
Variation Some variation in colour pattern, but no obvious trends in Europe.
Similar Species Common Spadefoot (p. 69) has a domed skull and more fully webbed feet. Green Toad (p. 76) possesses a horizontal pupil, a fairly obvious eardrum, and no real spade, although its metatarsal tubercle is often fairly large.
Voice A fairly quick 'clock ... clock ... clock', rather like someone clicking their tongue in disapproval. Somewhat louder than the voice of the Common Spadefoot; usually produced under water and sometimes heard throughout the night.
Habits Generally similar to other Spadefoots. Mainly a lowland species but occurs up to 500m in places. Found in flat open habitats including cultivated areas, open woods and coastal dunes. Occurs in stony places as well as on more homogeneous light soils. Lives in deep burrows in typically moist soils, usually near its breeding areas which are generally in clear, deep, often temporary water, with relatively little vegetation.

Females lay 2,000–4,000 eggs which may hatch in about three days. Tadpoles usually take 3–4 months to develop in south of range and newly metamorphosed toads are about 2.5cm long. They mature at 5–6cm in two years for males and three for females. The Eastern Spadefoot may live 10 years.

western   common   eastern

Spadefoots, hind feet

Parsley Frogs
(Family *Pelodytidae*)

A group consisting of just three similar species, two of which are found in south-west Europe and the other in the Caucasus. Parsley frogs are related to spadefoots and sometimes placed in their family. However, they are much more frog-like, lack spades and almost all webbing on the hind feet, and do not dig much. Their name derives from their frequent appearance – speckled with green as if garnished with parsley. Like spadefoots, male Parsley frogs grip the females just in front of the hindlimbs and the eggs are produced in a gelatinous band that smells of fish. Parsley frogs differ further from spadefoots in having nuptial pads and in frequently breeding more than once a year.

PARSLEY FROG *Pelodytes punctatus*   **Pl. 8**

Range North-east and eastern Spain, France and extreme north-west Italy. Map 42.
Identification Rarely more than 5cm and usually less, females often bigger than males. A small, agile, slender, rather long-limbed frog, with rather flat head and prominent eyes, which have vertical pupils. Hind toes almost completely unwebbed, or with a thin flange of webbing on each toe, and skin of back is fairly warty, the larger warts often forming irregular lines. A glandular fold is present from behind the eye to at least the insertion of the forelimb. Ear-drum is indistinct but often visible. Usually pale greyish, yellowish, buff, or to light olive above, with small darker olive to bright 'parsley' green spots and often a light X-shaped mark on upper back. Can occasionally be mainly green and warts along the sides may be orange. Often smells of garlic. Male has shorter body, thicker forelimbs, and darker throat than the female. In the breeding season males often develop a blue to violet throat and nuptial pads consisting of rough dark brown or black patches under the upper and lower forelimbs and on the two inner fingers of each hand; tiny dark spines may also be present on the throat and belly.
Variation Some variation in colour, but not obvious geographical trends.
Similar Species See Iberian Parsley Frog (p. 72). Easily distinguished from Typical frogs (p. 80) and Painted frogs (p. 62) by vertical pupils and by almost complete lack of webbing on the hindfeet. Midwife toads (p. 64), are more robust and have shorter hind legs, the heels often not reaching the eye when they are turned forward.
Voice Mainly calls at night. Breeding males produce a sonorous two-tone 'co-ak … co-ak' under water, which may sound rather like a Corncrake or Little bustard.

Females respond to this with a rather feeble 'coo ... coo'. Call is sometimes audible for 300m but often sounds more distant than it really is. Out of water, males produces a medium pitched, weak 'cre-e-e-ek' - rather like a squeaky shoe or a low-pitched version of the noise produced by a cork being drawn from a bottle.

Habits Largely nocturnal, and terrestrial outside breeding season. Generally found in dry or slightly damp habitats up to 1,500m in south. Occurs in open woods, among bushes, in cultivated areas, in vegetation at the base of walls, and sometimes by small streams. Quite agile and moves by jumping. By day can be found under stones, in shallow burrows which it digs itself and in small caves, cracks and fissures in rocks and drystone walls. Frequents gypsum mines in the northwest of its Spanish range. Often encountered in calcareous or sandy areas. A good swimmer. Can climb smooth surfaces, maintaining position by broad contact of moist belly. Sometimes found sitting in bushes and on top of stones and rocks at night.

Often breeds in sunny open waters, usually with vegetation, such as temporary ponds, flooded ditches, pools in slow streams and brackish lagoons near the sea. Breeding is often stimulated by rain. Amplexus may last 5 hours. The female of a mating pair holds on to a stem of an aquatic plant with her forelegs and twists the egg band around it with her hind ones. The band is 3–4mm thick and often about 20cm long although it can sometimes reach 50cm; it contains 40–360 eggs. There may be several layings in a year and a female may produce a total of 1,600 eggs in a season. These hatch in around 3-19 days and tadpoles take about 3 months to develop in the south but up to 7–8 months elsewhere. Newly metamorphosed frogs are 1.5 – 2cm long and can mature in a year. Parsley frogs can live 15 years.

IBERIAN PARSLEY FROG *Pelodytes ibericus*
Range South and west Portugal and southern Spain. Map 43.
Identification Up to 4cm. Very like the Parsley Frog but adults are rather smaller with a somewhat broader head and shorter limbs. In males, the heel reaches at most to the front edge of the eye when the hind leg is turned forward, but extends anywhere from here to the snout tip in male Parsley Frogs. The tubercles underneath the hand at the base of each finger are conical, whereas they are only gently convex in the Parsley Frog (Fig., below). Skin is smooth to granular with small often dark warts. Upper parts are brownish grey, olive brown, or plain green or dark brown with green spots; sometimes a light X-shaped mark present on the back. Belly white to pale cream, throat of breeding males violet or dark grey. Nuptial pads similar to Parsley Frog.

Similar Species Parsley Frog (p. 71). See that species for distinction from other frogs and toads.

Voice Generally like Parsley Frog, but instead of consisting of two simply alternating notes, the first is followed by 2–5 repetitions of the second, this sequence usually being repeated a number of times. Unlike the Parsley Frog, Iberian Parsley Frogs do not usually sing under water.

Habits Similar to Parsley Frog and occurs up to 900m.

Breeding usually occurs after rain and animals

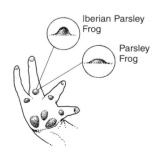

Iberian Parsley Frog

Parsley Frog

are then often found in sunny ponds occasionally up to 1.5m deep, and in seasonal streams and flooded areas. Females lay up to 350 eggs at a time and sometimes up to 1000 altogether. Newly metamorphosed frogs are 1-2cm long.

Typical Toads
(Family *Bufonidae*)

A large, widespread family found throughout much of the world but not most of the Australian region. The three European species all have warty skins, large paratoid glands (Fig., p. 74) and horizontal pupils. They are all largely nocturnal and terrestrial, but in the breeding season assemble in ponds and slow streams, often in large numbers. Males develop dark nuptial pads on the three inner fingers and call in chorus at night. In amplexus, females are grasped just behind the forelimbs. Each female may produce several thousand eggs which are laid in long strings. The tadpoles are among the smallest in Europe. The Natterack (*Bufo calamita*) has a more extended breeding season than the other species.

COMMON TOAD *Bufo bufo*   **Pl.10**

Range Found over most of Europe except Ireland, Corsica, Sardinia, the Balearics, Malta, Crete, and some other smaller islands. Also north-west Africa and western Asia. Related populations live in East Asia including Japan. Map 44.

Identification Adults up to 15cm long, but there is geographical variation in size and females are larger than males. The largest European toad; it is heavily built with a horizontal pupil, very warty skin and prominent paratoid glands that are slightly oblique (Fig., p. 74). Colour usually brownish, but varying from sandy to almost brick-red, rich dark brown, greyish or occasionally olive. Tends to be fairly uniformly coloured, but may have darker blotches and patches, especially in the south. Underside usually whitish or grey, often with darker marbling. Eye deep gold or copper-coloured. No external vocal sac.

Variation Animals from the Mediterranean area (*B. b. spinosus*) are often very large and have spinier skins; they grade imperceptibly into the more northern *B. b. bufo*. Toads from the Sierra de Gredos in central Spain with contrasting colouring and very large paratoid glands are named *B. b. gredosicola*. Very rarely, Common Toads hybridise with Natterjacks and Green Toads

Similar Species The two other Typical toads in Europe have parallel instead of oblique paratoid glands (Fig., p. 74) and lighter eyes. The Natterjack (below) also usually has a yellow stripe on the back while the Green Toad (p. 76) has distinctive pattern. See also Spadefoots (p. 68).

Voice Calls at night, the male 'release' call being most often heard. This is a not very powerful, rather high-pitched rough 'qwark-qwark-qwark', about 2–3 syllables a second, or more. The actual mating call is slower and the syllables are longer; it is rarely encountered.

Habits Found up to 2,500m in south of range. One of the most ubiquitous European amphibians, found in wide variety of often fairly dry habitats. Mainly nocturnal, hiding by day, usually in one particular spot, and emerging around dusk. Normally walks, but hops when alarmed. When approached by predators may assume characteristic posture with the head down, and the hindquarters raised. Can produce copious white skin secretion that is very distasteful to many predators, although some, like Grass Snakes and their relatives are unaffected. Eggs and tadpoles of the Common Toad are also distasteful. Like some other amphibians, Common Toads are often attacked by a flesh fly (*Lucilia bufonivora*) that lays its eggs on the skin where they hatch into maggots which enter the nostrils and eat the toad from within.

Migration towards the breeding ponds often begins in the autumn, but the final journey is made in the spring when large numbers of toads can be seen moving at night. Males usually arrive first and stop for several weeks while females visit just long enough to find a mate and lay their eggs. Males are very enthusiastic and will sometimes clasp fish or inanimate objects; several of them may attach to the same female forming a ball of up to a dozen animals. Males often fight for females rather than attracting them by calling. Reproduction in toads is very stressful and there is a high mortality among the participants. Large females lay 3,000–8,000 eggs, producing two egg strings simultaneously which swell to 5–10mm thick and are up to 5m long. They are often laid over several hours among the stems of aquatic vegetation, frequently in sunny situations. The eggs often develop in 2–3 weeks and the tadpoles swim in dense swarms that can be several metres long. Newly metamorphosed toads are very small, just 7–12mm, and are quite diurnal in their first months. In northern Europe animals may take 3–7 years to reach sexual maturity but the process is presumably more rapid elsewhere. Common Toads may live to about 10 years in the wild but a captive individual reached an age of 36.

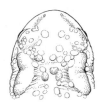

Common Toad oblique   Natterjack, Green Toad parallel

Typical toads, paratoid glands

NATTERJACK *Bufo calamita*   **Pl. 10**

Range West and central Europe, eastwards to the Baltic states, Belarus and west Ukraine; also Britain and south-west Ireland. Map 45.

Identification Up to 10cm but normally about 7 or 8cm; females larger than males. A robust, relatively short-limbed toad with prominent paratoid glands that are roughly parallel (Fig., above), horizontal pupil and typically a bright yellow stripe down the centre of the back (occasionally weak or absent). General colouring usually brownish, grey or greenish with darker markings that are rarely as distinct as in the Green Toad. Eye silvery gold. Tubercles under longest hind toe are usually paired (Fig., opposite). Throats of males have a bluish or purple tinge and there is a large external vocal sac under the chin.

Variation Some variation in colour and intensity of pattern. Iberian animals may grow particularly large and often lack a clear stripe on the back. Occasionally, hybridisation occurs with the Green Toad and very rarely, with the Common Toad.

Similar Species Common Toad (p. 73); Green Toad (p. 76); Spadefoots (p. 68). The Mauretanian Toad (*Bufo mauritanicus*) has been introduced close to Gibraltar.

It has a pattern of well defined blotches like a Green Toad but they are reddish or brownish.

Voice Sings mainly at night, usually in chorus, beginning shortly before sunset, but sometimes also heard in the day. Call is a loud rolling croak like a ratchet, which is repeated a number of times. Each croak begins and ends fairly abruptly and often lasts for about 1 or 2 seconds. On still, quiet evenings, a chorus can often be heard over 2km or more. Call can be mistaken for the song of a Nightjar and is sometimes stimulated by vaguely similar noises, such as passing trains.

Habits Very largely nocturnal. In north of range, mainly a lowland form, typically found in sunny open sandy areas, including heaths and dunes, often near the sea, and the open banks of rivers; also on moors and saltmarshes and disturbed sites like sand quarries. Elsewhere, can occur in a much wider range of habitats and in Spain may reach altitudes of 2,400m. Tends to run in short bursts like a mouse, instead of walking or hopping. Also climbs well on steep banks etc. Frequently digs shallow burrows in loose soil, using forelegs and, occasionally hind legs as well. It may use these burrows communally and hides in them and those of other animals during the day, but is also found under stones, logs etc. Often hunts in open areas and some animals may travel large distances (up to 2.5 km), partly using the earth's magnetic field to navigate. When alarmed, inflates body and raises rump, like Common Toad; also produces characteristic smell. Sometimes suddenly abandons established breeding ponds and migrates to a new area. Hibernates on land.

The breeding season is long in the north and breeding takes place in often shallow open sunny ponds which may be temporary and rather brackish. Most males attract females by calling while sitting in shallow water although a few save energy by remaining silent and depending on the calls of their follows. Females visit ponds briefly and lay 1,500–7,500 eggs arranged in a single row in strings of spawn that may be 1–2m long and about 5mm thick. Eggs hatch in less than a week and tadpoles metamorphose into tiny toads, usually under 1cm long. This can occur in less than a month, although it often takes substantially longer. Babies sometimes clump together to retain water in their dry habitat. Most animals in the north mature at 3 years when 5–6cm long , but 2 years is more usual in south. Has been known to live 17 years in the wild and longer in captivity.

Common Toad, Natterjack
tubercles paired

Green Toad
tubercles single

Typical Toads, undersides of hind toes

GREEN TOAD *Bufo viridis*   **Pl.10**

Range A mainly eastern species, extending as far north as Denmark, southernmost Sweden, the Baltic states and Russia, and as far west as Germany, extreme eastern France, and Italy; additionally Corsica, Sardinia and the Balearic Islands where it may have been introduced in the Bronze age. Also North Africa and south-west Asia eastwards to central Asia. Map 46.

Identification Up to l0cm but usually smaller; females larger than males. A robust toad with prominent paratoid glands that are roughly parallel and a horizontal pupil. Similar in general appearance to Natterjack but differs in following features: dorsal pattern usually much more contrasting especially in females (pale with bold, clearly defined greenish markings often with dark edges); nearly always no bright yellow stripe on back (although a weak stripe occasionally present); hind limbs rather longer; tubercles under longest hind toe unpaired. As in the Natterjack, an external vocal sac is present under the chin in males.

Variation Disposition and size of markings is very variable, but no obvious geographic trends. Hybrids with Natterjack occur occasionally in area of overlap. but hybrids with the Common Toad are very rare.

Similar Species Common Toad (p. 73); Natterjack (p. 74). Common and Eastern spadefoot (p. 69, 70) have vertical pupils and very prominent spades on heels.

Voice Males sing at night in chorus, animals sitting on the banks of breeding waters, in the shallows or among aquatic plants with their vocal sacs inflated as they call. Song a rather high pitched trilled liquid 'r-r-r-r-r-r ...' Starts quietly and often lasts up to about ten seconds and may be repeated about four times a minute. Not likely to be confused with other amphibians, but could possibly be mistaken for an insect song, for instance that of some Mediterranean crickets, or a Whimbrel.

Habits Mainly nocturnal, although sometimes active by day, especially in spring. In Europe, typically found in lowland areas but can reach 2,400m. Very resistant to aridity and high temperatures and often, although not invariably, found in dry and sandy habitats. Occurs in light woods, stony areas, coastal dunes, agricultural land, industrial and mining sites and, in northern Italy, rice fields. Frequently seen around human habitations and may be extremely numerous, entering villages in many areas and hunting insects around the base of street lamps. Often digs burrows in loose soil, or uses those of other animals, and hides under objects such as stones, logs etc. Hibernates on land. If disturbed may produce a poisonous white skin secretion with a characteristic smell.

In northern Europe breeds late in the spring when it makes use of a wide range of waters that are often shallow without much vegetation, including village ponds and temporary pools. Very tolerant of brackish or polluted conditions. Males spend quite long periods at the breeding ponds but females lay their eggs and depart. Strings of spawn 2–4m long are produced in which the eggs are arranged in 2–4 rows. The strings are deposited among aquatic plants or on the bottom of ponds and develop into tadpoles in about 3–6 days. Newly metamorphosed toads are about 1–1.7cm long and mature at about 3 years. Green Toads probably often live long lives like other European toads.

Tree frogs
(Family *Hylidae*)

A very large and successful family of about 750 species of frogs occurring in North, central and South America, the West Indies, Australia and New Guinea. One North American group of tree frogs, *Hyla*, extends into East Asia and westward to south-west Asia, Europe and north-west Africa. Like many of their American relatives, the European species are small, plump, smooth-skinned frogs with horizontal pupils, long legs, and disc-shaped adhesive pads on the tips of their fingers and toes. They are good climbers, usually being found in trees or other vegetation, and only rarely encountered on the ground except when going to water to breed. Tree frogs are the only European amphibians to climb very extensively (Parsley Frogs climb only occasionally). Colour can change quite rapidly and food consists substantially of flying insects which are captured with great agility. Breeding takes place at night in ponds, cisterns etc. Males have a large vocal sac under the chin and produce very loud calls (up to 87 decibels), They also have small nuptial pads on the thumbs and they embrace the female just behind the forelegs. The pale eggs are laid in clumps that swell to about the size of a wallnut. The tadpoles, which tend to be solitary, have a deep tail fin extending far forwards onto the back and swim with fast fish-like movements.

COMMON TREE FROG *Hyla arborea* **Pl. 11**
Range Most of Europe except the north, but absent from parts of southern France, much of the southern and eastern Iberian peninsula and the Balearic islands; replaced by similar species in Italy and Corsica and Sardinia. Also found east to Caspian sea and in Asiatic Turkey. Map 47.
Identification Adults up to 5cm but usually smaller; females tend to be bigger than males. Small, long-limbed and smooth-skinned with characteristic disc-like adhesive pads on tips of fingers and toes and a granular belly. Colour variable, usually bright uniform green, but may vary from yellow to blotchy dark brown. A dark stripe extends from the eye, over the ear-drum, and along the flank to the groin; just before this, it nearly always branches upwards and forwards. This stripe is often edged with white or cream, and there are prominent stripes on the limbs as well. Underside is whitish and green colour does not extend onto the throat. Males have a large, obvious yellowish or brownish vocal sac beneath chin, at least in the breeding season. This is spherical when blown up, and when deflated does not usually form longitudinal folds.
Variation Some regional variation in limb length: legs tend to be rather longer in south. Occasional individuals from south-east Europe have the upward branch of the flank stripe very narrow or even absent. Tree frogs from north-western Spain and northern Portugal (*H. a. molleri*) have distinctive biochemical features. Populations from Crete (*H. a. kretensus*) may have a reddish spot on the dark flank stripe and red spots on the back of the forelimbs.
Similar Species Other European Tree frogs.
Voice During breeding season, nearly always in evening and at night, a strident, rapid 'krak, krak, krak ', about 3–6 pulses a second. The call, which is repeated at intervals, tends to accelerate at the beginning and slow down at the end. Tree frogs form large dense assemblages and call from shallow water or when floating close to aquatic vegetation. Their vocal sac is blown up during bouts of singing but deflates

in between them. Choruses are very noisy and can be heard from over a kilometer away on still nights and, from a distance, can sound like quacking ducks. Outside the breeding season, may occasionally call briefly from trees or vegetation during the day, especially in late summer and in rainy weather. Can also be stimulated to call by sounds that have some of the qualities of its voice, for instance passing planes or staccato passages in string quartets such of those of Haydn.

Habits Mainly nocturnal and crepuscular but sometimes active by day, especially when conditions are warm and humid; at such times it may even be seen sitting fully exposed to the sun. Often spends the day resting on twigs and branches within the canopy of bushes and trees. Usually found in well-vegetated habitats, and shows a marked preference for areas with bushes, hedges, bramble patches, trees, or reed-beds, but occasionally occurs in more open places with low vegetation. Also found in gardens and parks. Reaches altitudes of up to 2,300m in south of range. Frequently occurs far off the ground, sometimes as high as 10m, but also often occurs low down in herbage, especially the young. Has been known to travel over 12km in a year. Hibernates on land among roots, dead leaves etc. The skin secretion helps deter some predators and will cause pain if accidentally rubbed in the eyes.

Breeds in a wide range of mainly still waters, typically with good exposure to sun and rich subaquatic vegetation, including even mountain pools and village ponds. Common Tree Frogs often start moving to their breeding sites after rain and the males may begin calling on their way to water, where they are very territorial. The loud call attracts potential mates but is very expensive in terms of energy and some males economise by not calling themselves, depending instead on the calls of others. Females lay 200–1,400 eggs in clumps containing 3–60 or more. Up to 50 separate clumps may be laid in a night. The spawn is usually attached to sub-aquatic vegetation and eggs take about 2–3 weeks to hatch. Newly metamorphosed young are about 1–2cm long; males may mature the next year and females the year after that. Individuals can live up to ten years in the wild and 22 years in captivity.

ITALIAN TREE FROG *Hyla intermedia*
Range Italy and Sicily; meets the Common Tree Frog in the region of the Italian-Slovenian border. Map 48.
Identification Recognised as a separate species on the basis of protein differences. Not distinguishable from the Common Tree Frog by appearance, although the ear is on average slightly larger.
Similar Species Common and other European tree frogs.
Habits As in the Common Tree Frog.
Other names Has previously been called *Hyla maculata, H. variegata* and *H. italica.*

TYRRHENIAN TREE FROG *Hyla sarda*   **Pl. 11**
Range Corsica, and Sardinia and their offshore islets; Also Elba, Capraia. Map 49.
Identification. Up to about 5cm but on average smaller than the Common Tree Frog. Differs from this species in its generally shorter snout, rather broader head and often more granular skin on the back. The dark flank stripe is poorly developed, breaking up on the hind body and always lacking an upward branch. Green, olive, yellow or grey above, often with distinct spots (dark green when the frog is light green, dark grey when the frog is light grey) and the hind legs may have cross-bars. Colour can change quite rapidly, especially when the frog moves onto a different background and with changes in temperature. Underside is whitish.

Similar Species None within range.

Voice Very like Common Tree Frog although call often slightly faster. Sings mainly at night but occasionally in day outside breeding season.

Habits Similar to Common Tree Frog in its behaviour and is like this species in being largely active at dusk and at night, although occasionally seen in the day in usually shady places. Differs in typically remaining close to water outside the breeding season, although it is occasionally encountered up to 500m from it. Often found in dense stands of reeds in ponds, in woods and vegetation close to pools and streams, in the vicinity of reservoirs, cisterns and springs, and in sometimes rather brackish pools near the sea. Quite heat-tolerant and often perches on rocks and dry-stone walls as well as vegetation. Most often encountered in lowlands and near the coast but reaches 1,750m in restricted areas of Corsica.

Females produce small clumps of spawn that are attached to plants or laid on the bottom of breeding waters; they take about 2 weeks to hatch.

Other names Previously regarded as a subspecies of the Common Tree Frog, *Hyla arborea sarda*.

STRIPELESS TREE FROG *Hyla meridionalis*   Pl. 11

Range Southern Portugal and Spain, north-east Spain, south-west and southern France, north-west Italy (Liguria); additionally Minorca, the Canary islands and Madeira, all localities where it is probably introduced. Also north-west Africa. Map 50.

Identification Adults up to about 6.5cm. Very similar to Common Tree Frog, but otherwise unmistakable; a small, smooth-skinned, long-limbed frog with obvious disc-shaped climbing pads on fingers and toes. Like the Common Tree Frog, most usually bright green, but may be yellowish or brownish, sometimes with small dark spots. At night perhaps more translucent than this species. Easily distinguished by lack of clear stripe along flanks. Other differences include larger average size, green pigment usually extending onto the sides of throat, stripes on limbs not very obvious, back of thighs often yellow, pale orange, or with dark mottling. Males possess a larger and broader vocal sac than the Common Tree Frog, which is longitudinally folded when deflated.

Variation Little geographical variation. Dark stripe may extend onto flank in young animals. Blue and yellow individuals are occasionally found. Hybrids with the Common Tree Frog are known to rarely occur.

Similar Species Common Tree Frog.

Voice A slow, quite deep, resonant 'cra-a-ar', repeated not much faster than once per second and typically more slowly. Sings mainly in the evening and at night in big choruses, both in the water and on the banks of pools.

Habits Very similar to Common Tree Frog but, where they occur together, the stripeless Tree Frog tends to be found at lower altitudes and rarely extends above 1,000m. Mainly active at night and in the morning. Found in orchards, vinyards, gardens and wooded and bushy areas. Lives in banana plantations in the Canary islands. Occasionally occurs far from water and sometimes in places with quite low rainfall. Breeding similar to Common Tree Frog, taking place in ponds, springs, irrigation ditches, temporary pools and flooded meadows. Females may lay up to 60 clumps of 10–30 eggs among water plants; these hatch in 2–3 days. Newly metamorphosed frogs are 1.5–2cm long.

Other names This frog was once called *Hyla barytonus* after the deep voice of the males.

Typical Frogs
(Family *Ranidae*)

A large group of over 700 species common over most of the world but not in much of Australasia, southern South America and the West Indies. All European species are slim-waisted, long-legged frogs with fairly smooth skins. With the exception of the American Bullfrog (*Rana catesbeiana*), which has been introduced, they can all be distinguished from other European frogs and toads by possessing both horizontal pupils and clear dorsolateral folds (Fig., p. 58). All are agile and progress on land by leaping, while in water they are often powerful swimmers. Native European species (all of which belong to the genus *Rana*) fall into two groups: the often aquatic, noisy Water or Green Frogs (13 species) and the frequently more terrestrial, quieter voiced Brown Frogs (eight species). Typical frogs often assemble in large numbers to breed. The males sing in chorus and develop prominent dark nuptial pads on the thumbs. They grasp the female around the body behind the arms. Eggs are laid in large clumps, some female Brown frogs producing up to 4,500 or more and some large female Green frogs as many as 16,000. The newly metamorphosed young are often small and are often difficult to identify.

COMMON FROG *Rana temporaria*   Pl. 12
Range Widely distributed throughout Europe, but absent from Portugal, most of Spain, much of Italy, and the south Balkans. Also eastwards, to Urals and adjoining western Siberia. Map 51.
Identification Adults up to 11cm, but usually smaller; males rather smaller than females. The most widespread Brown frog in Europe and, in many areas, the commonest species. Usually robustly built with relatively short hind limbs (heel only rarely extends beyond snout-tip), closely spaced dorsolateral folds, and weak metatarsal tubercles. Old specimens tend to be round-snouted, but younger animals vary considerably in head shape. Ear-drum usually fairly distinct, equal to or rather smaller than the eye.

Grey, brown, pink, olive, yellow or red above, usually with dark blotches (sometimes these are orange or red), and often a dark ^-shaped mark between the shoulders. Black spots frequently occur on the back and these may be extensive (more so than in any other Brown frog). Flanks are usually spotted or marbled. Underside white, yellow, or even orange, typically marbled, spotted, or speckled with darker pigment. Throat sometimes has light central stripe. Breeding males have very strong forelimbs, and are 'flabby' with a bluish tinge. They have black or dark brown nuptial pads on their thumbs and often a bluish throat. Breeding females develop pearly granules on their flanks and hind legs.
Variation Extremely variable in colour; albinos occasionally occur. The Common Frog shows considerable differences over its range. Populations from north-west Spain (*R. t. parvipalmata*) are relatively small with reduced webbing on the hind feet and a slightly slower call; some animals with reduced webbing also occur as far east as the Pyrenees. Some French lowland Pyrenees populations (Gasser's Frog) are very long-legged and in the past have been mistaken for Iberian (p. 87) and Agile Frogs (p. 83); similar animals occur in the Massif Centrale. At higher altitudes Pyrenean Common Frogs have been regarded as another subspecies, *R. t. canigonensis*. Long-legged populations in the Basses-Alpes (south-eastern France) have been

named *R. t. honnorati*, which is sometimes believed to extend into the Appenine mountains of Italy. All these forms need careful reassessment.

Similar Species The Common Frog is likely to be confused with almost all other Brown frogs, but especially the Moor Frog (p. 82). It usually has a less pointed snout than this species, is almost never striped (the Moor Frog often is), and has a small soft metatarsal tubercle (see Fig., p. 82). Most other Brown frogs tend to have longer legs (see these species for other differences). Individuals with relatively long legs (heel extending just beyond tip of snout) do occur in some areas, but are usually recognisable as Common Frogs by their close dorsolateral folds.

Voice A dull rasping 'grook … grook … grook', that is often produced by choruses of animals sitting in shallow water or floating on its surface but is occasionally also produced under water. Heard night and day during breeding, sometimes also in autumn.

Habits In the north may be largely terrestrial and is often only found in water during the breeding season or hibernation. In contrast in the south it may stop close to ponds and streams and is frequently confined to the mountains. Often found at high altitudes, even up to the snow line, and occurs up to nearly 3,000m in Pyrenees and to nearly 2,800m in southern Alps. Occurs in a very wide variety of habitats and may be encountered in almost any moist often shady place within its range, except in permanently frozen areas. Lives in woods, among dense vegetation, moors, cultivation and parks and gardens. Active by day and night. Hibernation takes place on land or in the water but may not occur in warm south-western localities. In north, Common Frogs begin activity early in year and are sometimes even seen moving about under ice.

An explosive breeder in early spring, large numbers of frogs travelling at night to the breeding waters, although they are also active in the day once they arrive. On rare occasions may cover up to 10km. Further south, breeding may extend over several months in the cooler part of the year. It takes place in a wide variety of sites: ponds in bogs, woods, meadows, arable land and farmyards, in ditches, slow stretches of rivers and streams, and even pools formed by melting snow. Females lay one or two clumps of spawn, containing a total of 700–4,500 eggs, which are deposited over vegetation in shallow water (usually no more than 30cm deep), and float at the surface, sometimes forming a continuous carpet with the spawn of other females. The spawn is largely fertilised by the female's mate, but other males may contribute by shedding sperm into the water nearby. The eggs often develop in 1–2 weeks but may take twice as long in cold conditions. Newly metamorphosed frogs are 1–1.5cm long and in the north many become sexually mature at 2–3 years at a length of about 5–6cm; small-bodied Iberian populations mature at much smaller sizes. Common Frogs may live 10 years or more.

Other names Grass Frog. *Rana aragonensis* in central Pyrenees.

PYRENEAN FROG *Rana pyrenaica*   **Pl. 12**

Range Western central Pyrenees mountains. In Spain from Roncal valley (Navarra) eastwards to Parque Nacional de Ordesa (Huesca); a few other localities immediately to the north in French Pyrenees. Map 52.

Identification Adults up to about 5cm; the smallest European Brown frog. Relatively delicately built, snout rounded with nostrils well separated (space between them wider than that between the eye bulges on top of the head, (Fig., p. 85), heel reaches snout or slightly beyond, dorsolateral folds quite closely spaced, ear drum small and often not visible, webbing on hind toes extensive and absent only from the tip of longest hind toe.

Light cream, pinkish, red-brown or olive-grey above, often with small dark marks that may be greenish; barring on hind legs is often weak. Generally, contrast in colouring is not strong. Upper lip usually conspicuously light. Throat whitish or weakly spotted or streaked with grey, especially at sides. Belly whitish, sometimes yellowish or pink especially under thighs. Nuptial pads of males pale.

Similar Species Common Frog (p. 80) is found in same areas but grows much larger. It also tends to have a more pointed snout when similar in size to the Pyrenean Frog and its ear drum is larger and often more pronounced. There are usually more dark markings above that contrast strongly with the ground colour and the throat is often generally darker with a light central band.

Habits Lives in and close to rocky mountain streams and torrents in which the water is cold, clear and well oxygenated; generally between 1,100 and 1,700m. Active by day and night. Young animals are less aquatic than adults. Swims well; timid and escapes by diving into water and hiding under stones and in crevices. Common Frog extends to higher altitudes, is more terrestrial and, when aquatic, is associated with slower moving and still water.

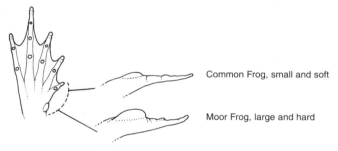

Common Frog, small and soft

Moor Frog, large and hard

Brown Frogs, metatarsal tubercle

MOOR FROG *Rana arvalis*   Pl. 12

Range From east France and Belgium eastwards; extends north to northern Sweden, Finland and Russia, and south to the Alpine region, north Croatia and north and central Romania. Also across Asia to the Lake Baikal region of Siberia. Map 53.

Identification Adults up to about 8cm. A robust frog that often has very short legs (heel does not extend to snout-tip over most of range), closely spaced dorsolateral folds, a pointed snout and a large, hard, sometimes sharp-edged metatarsal tubercle (Fig., above). Dorsal pattern often striped and belly usually almost unmarked. Eardrum well separated from eye.

Colour very variable: grey, yellowish or brownish. Sometimes uniform but usually with dark blotches and some black spots, which may be numerous. Many populations include animals that have a broad, pale vertebral stripe, often with a darker border. Sides usually blotched or marbled. Belly typically whitish ; often yellow on groin. Throat may be spotted or flecked and may have light central stripe. Breeding males have blackish nuptial pads on the thumbs and may be flabby with bright blue or violet coloration.

Variation Considerable variation in size, proportions and colour. Individuals with dark or spotted bellies are known. Animals from the south of the range (Austria, Slovenia, Croatia, Czech Republic, Slovakia, Hungary, south-east Poland, Romania, western Ukraine) tend to be larger and slimmer and have longer hind legs (heel may reach just beyond snout). These are often regarded as a separate subspecies, *R. a wolterstorffi.*

Similar Species Easily distinguished from the Common Frog by its usually more pointed snout, larger metatarsal tubercle (Fig., opposite), and frequently striped pattern. Long-legged animals in the south of the range can be confused with the Agile Frog but are more robust, usually shorter limbed with closer dorsolateral folds, and the ear-drum is smaller and further from the eye.

Voice Rather quiet, often described as like a blade being repeatedly and abruptly whipped through the air, or like air escaping from a submerged bottle. Difficult to transcribe but could be rendered as 'waup, waup, waup ...' At a distance, often sounds like a small dog barking. Males sing in the water or among water plants, often in chorus both at night and in the day, sometimes calling in the midday sun.

Habits Found at low altitudes (though up to 800m in south) in wet situations: lowland moors, sphagnum bogs, alder swamps, alluvial bogs, fens, damp fields and meadows, around lakes and on flood plains; also sometimes in open deciduous and coniferous woods. In tundra found in riverside wood and scrub. Common in agricultural habitats. Often in the same area as Common Frog, but shows a preference for wetter places, although may sometimes occur a kilometer from water. Quite shy and often takes refuge in water or in dense vegetation

Breeds explosively in the early spring, travelling in the evening and at night to its breeding places which are varied and include both permanent and temporary waters, preferably of low acidity, and artificial excavations such as peat cuttings. Males arrive first at the breeding assemblies which may contain a significant proportion of small frogs. Assemblies may last a month or so, the males being present much longer than individual females. These produce one or two clumps of spawn and a total of 2,000–4,000 eggs which may make up a third to a half of their pre-breeding weight. Spawn is laid in shallow water with rich aquatic plant growth. The eggs develop in 2–4 weeks. Newly metamorphosed frogs are about 1–1.5cm long and are active by day. They mature after 2 or 3 years, some males being only 3.5cm long at this time, and may live up to 10 years in the wild.

AGILE FROG *Rana dalmatina*   Pl. 13

Range Widely distributed east to Romania, absent from north Europe, Portugal and most of Spain (present only in Alava and Navarra in the Basque Country). Isolated colonies exist in north Germany, Denmark, extreme southern Sweden and possibly Sicily. Outside Europe, a single colony in north-west Anatolia. Map 54.

Identification Adults up to 9cm, but usually smaller; females rather bigger than males. An elegant, delicately coloured (sometimes slightly translucent) Brown frog. Hind limbs long (heel extends beyond snout tip in adults, often considerably), dorsolateral folds well separated, snout pointed, ear-drum large and nearly always close to eye (distance between them less than half the diameter of the ear-drum), and underside pale. Dark animals may have throat dark-stippled at sides but nearly always with a broad pale central area. Metatarsal tubercle quite large.

Colour not very variable or strongly contrasting; usually yellow-buff or pinkish brown above (often described as dead-leaf coloration) but can become darker. Faint vertebral stripe may be present, and back often marked with scattered darker blotches including a ^-shaped mark between the shoulders. Occasionally, a few jet black spots, especially on dorsolateral folds. Legs conspicuously banded. Sides often unmarked and groin frequently sulphur yellow. Nuptial pads greyish.

Variation No obvious geographical trends. Very rarely, throat may be rather dark with only a narrow light central stripe.

Similar Species Overlaps with virtually all other Brown frogs. Italian Agile Frog (below), Italian Stream Frog (p. 86), Balkan Stream Frog (opposite) and Iberian Frog (p. 87) are nearly always easily separated by their characteristic throat patterns and smaller ear-drums that are more widely separated from the eye. Common Frog (p. 80) is more robust, usually with shorter legs, closer dorsolateral folds, and a weak metatarsal tubercle. See also Moor Frog (p. 82).

Voice A rather quiet, fast 'quar-quar-quar-quar-quar-quar', (about 4–6 calls a second) that tends to increase in loudness. In breeding season calls sporadically under water, but there may also be large choruses of frogs singing together, even in the day, sitting in shallows or floating on water surface

Habits Typically found in light broad-leaved woods, with or without herbaceous vegetation, where colouring matches dead leaves well. Common in oak woods in central Europe but in many other tree associations elsewhere. Also in thickets and meadows. Often in fairly damp habitats but outside breeding season can occur in very dry parts of forests. Found up to 1,700m but mainly at low altitudes in north. Often crepuscular or nocturnal but sometimes also seen during day. An extremely agile frog capable of very long leaps. May sometimes take to water when pursued, but is rather poor swimmer. Usually hibernates on land but underwater in Greece.

Breeding takes place in still water of pools in fens and swamps, often in or close to woodland in which the frogs live at other times of year. In south of range especially, breeding sites are more varied and include brooks, ditches and ponds which may be quite shallow. An early breeder, sometimes even before the Common Frog. Clumps of spawn contain 450–1,800 eggs and are placed at some depth around twigs or stems, so that they look as if they have been impaled on these. The clumps later float up to the surface by which time they are often covered by green algae. The eggs take around three weeks to hatch. Newly metamorphosed frogs are about 1–2cm long and become sexually mature after two or three years; they probably live up to 10 years.

ITALIAN AGILE FROG *Rana latastei* Pl. 13

Range North Italy, and small adjoining areas of southern Switzerland, Slovenia and extreme north-west Croatia with an isolated population in Istria. Map 55.

Identification Adults up to about 7.5cm; males smaller than females. A graceful Brown frog with rather long hind legs (heel reaches beyond snout) and well separated dorsolateral folds. Throat consistently dark with a narrow light central stripe and frequently a pink flush on throat and limbs in males. The light stripe on hind upper lip stops abruptly under eye. Snout often fairly pointed, but may be quite blunt. Eardrum not very large, well separated from eye and sometimes not clearly defined.

Dorsal colouring rather variable: upperparts greyish or reddish brown, often with some darkish blotches, especially a bar between eyes and a ^-shaped mark between shoulders. No lichenous pattern on back or heavy blotches on flanks. Belly white, often marbled with grey at least anteriorly. Breeding males have dark brown nuptial pads on their thumbs, dark red-brown spots on the throat and orange to reddish colouring under the thighs.

Variation Some variation in colouring and snout shape, but no obvious geographical trends.

Similar Species Most likely to be confused with Agile Frog (p. 83) and Italian Stream Frog (p. 86). Differs from Agile Frog in throat pattern, smaller ear-drum well separated from eye, less prominent metatarsal tubercle, and sometimes reddish colouring on limbs and underparts (Agile Frog often has yellow on groin). For distinction from Italian Stream Frog see this species. Common Frog is more robust than Italian Agile Frog, with shorter legs, more closely spaced dorsolateral folds, and usually no red or orange beneath in breeding season.

Voice A long drawn-out thin 'mew', rather like that of a cat, with long intervals between calls which occur every 20-60 seconds. Usually produced under water, both by day and night. The 'release' call is quite different: a repeated 'kek ... kek ... kek'.

Habits A lowland species, but can occur up to 500m. Usually occurs in moist, partly inundated broad-leaved woods with swampy ground supporting a rich growth of herbaceous plants. Here it may be found on the borders of streams or lakes. Another typical habitat is poplar plantations with a thick understory. Occasionally occurs in meadows in Switzerland. Most active in late summer and early autumn, especially in the morning and around dusk. Generally similar to Agile Frog in behaviour and, like this species, capable of long and rapid leaps but more dependent on water. Hibernates on land when it may occur up to a kilometer from water.

Breeds in permanent and temporary water in wooded areas sometimes including slowly flowing parts of rivers. Active by night and day at this time. Females lay comparatively small clumps of spawn usually containing 300–400 eggs, although 90–900 have been recorded. As in the Agile Frog, the clumps are placed around twigs at some depth and then rise to the surface by the end of larval development which takes about 2–4 weeks. Newly metamorphosed frogs are about 1.5cm long. Appears short lived, most frogs dying in 2-3 years.

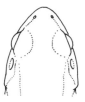

Italian Agile Frog

Pyrenean Frog
Stream Frogs

Brown frogs, heads

BALKAN STREAM FROG *Rana graeca*   Pl.13

Range Balkan peninsula from Bosnia, Montenegro, southern Serbia and south-east Bulgaria to southern Greece. Map 56.

Identification Adults up to 8cm, occasionally larger. A rather flattened, fairly long-legged Brown frog with often rounded snout, widely separated nostrils (Fig., above) and well-spaced dorsolateral folds. Throat dark with a light central stripe. Ear-drum small and not very distinct.

Colour variable; often greyish above, but can be brownish, coffee coloured, reddish, yellowish or olive. May have small scattered black spots, especially on dorsolateral folds, or obscure dark blotches including an ^-shaped mark between the shoulders. Frequently has lichenous pattern of soft-edged pale spots. Sides of snout often dark; flanks not normally marbled. Throat dark with pale, narrow central line and pale spots, especially at jaw edge. Belly usually pale, often a yellowish flush under hind legs. Breeding male has powerful forelimbs and blackish brown nuptial pads.

Variation Considerable variation in pattern. Northern animals tend to be larger.

Similar Species Has longer legs (though not invariably), less distinct ear-drum, and more widely separated dorsolateral folds than Common Frog (p. 80). Agile Frog (p. 83) is more delicately built and has a large prominent ear-drum and different throat pattern (Fig. Pl. 13).

Voice A very rapid low repeated 'geck-geck-geck-geck-geck', produced mainly at night. Calling frogs may sit completely out of the water, or with the foreparts resting on a rock and the hind limbs floating, or be completely submerged. Also produces a mewing call under water like Italian Agile Frog.

Habits Nearly always associated with cool running water, and most often found in mountain areas from 200 to 2,000m, being commonest at middle altitudes. Typically encountered in or near permanent mountain streams with rocky beds, usually without much aquatic vegetation although there may be dead tree leaves on the bottom. Springs, wet caves, and flowing irrigation channels are also occupied often in or near broad-leaved or mixed woodland. The species is usually seen on bank or rocks in streams from which it leaps into water when disturbed to swim strongly and hide beneath bank, in vegetation or under stones. Active by day and night; after dark, especially when rainy, may venture several metres away from water to forage.

Females lay relatively small clumps of spawn that are rather lumpy, like little bunches of grapes. These usually contain 600–800 (sometimes 200–2,000) eggs and are deposited under stones. They may possibly be guarded by the males and develop into tadpoles in 1–2 weeks, metamorphosed frogs maturing in about 3 years.

ITALIAN STREAM FROG *Rana italica*

Range Peninsular Italy but absent from much of east; isolated records from Monte Gargano area may really be based on Agile Frogs. Map 57.

Identification Up to 6cm. Very similar to Balkan Stream Frog (above) but largest adults smaller and there are some slight differences in body proportions. Best identified by range.

Variation Considerable variation in pattern.

Similar Species Agile Frog is more delicately built and has a large prominent ear-drum that is closer to the eye and a different throat pattern (Fig. Pl. 13). Italian Stream Frog may be confused with Italian Agile Frog (p. 84) in area south of river Po where their ranges meet, but Italian Stream Frog has the nostrils more widely separated (distance between nostrils is greater than distance between nostril and eye, Fig., p.85), a more extensive area of pearly granules on back of thighs, often little marbling on belly, and a less distinct ear-drum. Also the dark line running along outer side of hind leg is less clear and breeding males have no reddish colouring on throat and under hind legs.

Voice Similar to Balkan Stream Frog.

Habits A mainly montane species often found up to 1,000m and sometimes to 1,700m. Typically lives in small, fast-running streams with rocky bottoms and little

vegetation and it may sometimes follow these to quite low altitudes. Occasionally encountered in small ponds and mountain peat bogs. Females lay spawn under stones.
Other names Originally considered part of the Balkan Stream Frog but regarded as a separate species since 1985.

IBERIAN FROG *Rana iberica* Pl.13

Range Confined to Portugal and north-west and central Spain. Old records from some areas of the Pyrenees result from confusion with long-legged Common Frogs occurring there. Map 58.

Identification Adults up to about 7cm but usually 5.5cm or less; females are bigger than males. A rather small Brown frog with widely separated dorsolateral folds. Heel reaches or extends beyond snout tip; ear drum quite small and not very obvious being 1/3–3/5 diameter of eye; webbing on hind toes quite extensive.

Colouring very variable, upper parts often greyish brown, reddish-brown, olive, or sand coloured, with or without darker markings. Back sometimes has lichenous pattern of soft-edged pale blotches and back and flanks may be dark-spotted. Hind limbs with or without clear barring. Underside can be uniformly pale but often with dark marbling or spotting; if so, a light central line often present on dark throat. Nuptial pads on thumbs greyish in breeding males which may also have pale reddish undersides.

Variation Considerable variation in colour, including some between populations. Mountain animals often tend to be larger.

Similar Species Other Brown frogs, especially Common Frog (p. 80).

Easily distinguished from this species by sometimes longer hind legs, widely separated dorsolateral folds (Fig., p. 58), usually less obvious ear-drum and often fuller webbing on toes (frequently reduced in Common Frogs from north-west Spain, see Fig. below). In Basque Country (Alava) occurs close to Agile Frog (p. 83) which has longer hind legs, a larger and more prominent ear drum and distinctive colouring.

Voice A rapid 'cock-cock-cock-cock ...' or 'rao-rao-rao ...', about 3 calls a second. Sings mainly at night and under water.

Habits Usually a montane frog occurring up to 2,400m in south of range and found in areas with a rainfall of over 1,000mm. Typically encountered not far from cold, usually moving water, being found in or near streams, in shady places such as woods, but also in moors and meadows and areas of low scrub. Occurs also in bogs and around glacial pools and lakes. In north of distribution may occur down to sea level along slower-flowing rivers. Sometimes lives in polluted conditions. Active by day and night. Very agile, jumping into water and swimming strongly if disturbed.

Where climate permits may have a long breeding period, sometimes from November to April. Females lay 100–450 eggs in one or more clumps which are deposited in shallow water under stones or in aquatic vegetation. Newly metamorphosed frogs are about 1.3cm long. Males mature earlier than females, sometimes at a length of just 3cm.

Iberian Frog
webbing usually extensive

Common Frog
webbing often reduced

Brown Frogs in Northwest Spain.
Hind feet

Water Frogs

Water or Green frogs are often noisy, frequently aquatic and usually gregarious. They contrast with the quieter, more often terrestrial Brown frogs and can be distinguished from these by their more closely set eyes, the absence of a dark mask on each side of the head and the presence in males of external vocal sacs at the corners of the mouth. Their colouring is highly variable and adult females are bigger than males.

In Europe, there are no less than 13 kinds of Water frog. Many of them have been properly recognised only in the last 20 years, often on the basis of differences in call and in their proteins and DNA. Together they form one of the most complex groups among the European amphibians and reptiles, and their relationships are probably still not fully worked out. To understand their complexity, it is easiest to begin in the more northern areas where these frogs are found, outside the Mediterranean region. Here there are three well-defined forms: the Marsh Frog (*Rana ridibunda*), the Pool Frog (*Rana lessonae*) and the Edible Frog (*Rana* kl. *esculenta*). Marsh and Pool Frogs are essentially normal species except that they occasionally interbreed, the offspring so produced being Edible Frogs. Unlike the great majority of populations of hybrid origin, Edible Frogs are capable of breeding successfully with one or other of the parent species. Such matings would normally be expected to be unproductive or produce a range of intermediate forms. Instead they result in more Edible Frogs.

This odd situation, in which a form of hybrid origin can reproduce itself by breeding with one of its parent species, is connected with a peculiar chromosome mechanism. Like nearly all animals, first generation Edible Frogs have a set of chromosomes inherited from each parent, in this case one of Marsh Frog and one of Pool Frog origin. In nearly all other animals the two sets would become mixed up (by the chromosomes 'crossing over') when the eggs and sperm are formed, so each of these would contain a mixture of genes from the two parent species. In Edible Frogs one set of chromosomes is destroyed before this can happen, so their eggs and sperm contain the chromosomes and genes of only one of the two parent species. For instance, in Edible Frogs that breed with Marsh Frogs, the eggs and sperm usually have only Pool Frog chromosomes, the Marsh Frog set having been destroyed. So, the offspring produced from such matings will have one Pool Frog and one Marsh Frog set and are consequently Edible Frogs. Conversely in Edible Frogs that breed with Pool Frogs it is the Pool Frog set that disappears. This process goes on from generation to generation allowing Edible Frogs to persist indefinitely.

In fact the situation is even more complicated. Occasionally, Edible Frogs appear to be able to maintain themselves as a pure population without either of the parent species being present. In addition to normal *diploid* Edible Frogs with two sets of chromosomes, there are *triploid* ones with three, having an additional set of one or other of the parent species. In southern Europe there are more, normal species of Water frogs and some of these too may interbreed. In a couple of such cases this results in *hybridogenetic* forms like the Edible Frog, that arise by hybridisation and can persist indefinitely, so long as they can breed with one of the parent species.

Hybridogenetic forms deserve formal names so they can be referred to. They are not real species, as they are not independent of other forms, so it is inappropriate to give them a simple two part name, such as *Rana esculenta*. Instead, 'kl.' is inserted between the two words, thus: *Rana* kl. *esculenta*. The kl. stands for klepton, a word derived from the ancient Greek for thief. This refers, rather harshly, to the way hybridogenetic frogs 'steal' half their chromosomes in each generation from members of other species.

Identification Many of the different kinds of Water frog are very similar to each other but at the same time highly variable. This makes them the most difficult amphibian or reptile group in Europe to identify. Locality is very important in deciding what form you may be looking at. After this, the most useful features for identification, short of heading for a biochemical laboratory, are colour of the vocal sacs and the back of the thigh; hind leg length, size and shape of the metatarsal tubercle on the foot (Fig., p.91) and sometimes the distance of the ear-drum from the eye. Even these features will be insufficient in some cases, especially in southern France, north Italy and the southern Balkans.

WATER FROGS: North, central and eastern Europe

This area has two normal species of water frog, the Marsh and Pool Frogs, and single hybridogenetic form, the Edible Frog.

MARSH FROG *Rana ridibunda*   **Pl. 14**

Range Central and eastern Europe including the Balkan peninsula except much of Albania and Greece. Outside the Rhine area scattered colonies in France may be introductions . Introduced into southern England (first to the Romney Marsh area - hence the English name 'Marsh Frog'). Also east to central Asia. Map 59.

Identification Adults up to 15 cm (rarely 18cm); the largest native European frog. Big, robust and sometimes rather warty. Vocal sacs grey and back of thigh spotted whitish or grey (only rarely yellowish). Hind legs long (Fig. p. 90), the heel extending well beyond the eye; metatarsal tubercle small and usually low, about 25–40 per cent length of first toe (Fig., p. 91). Typically brown to grey above, often with a greenish or yellowish tinge; frequently some green or olive on back which has dark spots.

Variation Animals from central and eastern Europe (which includes the English populations as these originated from Hungarian stock) are rather sombre and often largely brownish with dark spots. Southern animals are more variable and often have bright green on the back.

Similar Species Most water frogs, but especially Edible and Pool Frogs (p. 92, p. 90). Easily separated from Brown frogs by closely set eyes and vocal sacs at corner of mouth in males.

Voice Sings by night and day. Most noisy during breeding season, but continues throughout summer. One of the most varied songs of the European amphibians: a wide variety of resonant croaks and chuckles - some of which can be rendered as and a vigorous, repeated 'bre-ke-ke-ke-ke-ke-kek'; also a metallic 'pink-pink' and 'croax-croax. The chorus often fluctuates erratically.

Habits Usually gregarious, diurnal, and very aquatic; but also active at night and animals are occasionally found some distance from water, especially in south. Over much of its range the Marsh Frog occurs in nearly all types of water including small ponds, ditches and streams, but where it overlaps with Edible and Pool frogs it tends to be confined to larger bodies of water, such as lakes and rivers. Frequently suns itself on bank, where it is often inconspicuous until it leaps into water with a characteristic 'plop'. Often seen on lily pads or floating among water weeds with only head exposed. Usually hibernates in water.

Breeds in late spring in north of range. Dominant males establish territories covering several square metres of water surface which they defend, often calling as they float with legs extended on the water surface. Females produce up to 16,000 eggs in a season, laying them in clumps of a few hundred in aquatic vegetation below the surface. They hatch in about a week, the tadpoles being rather solitary and living in deep water with vegetation and sometimes overwintering. Newly metamorphosed frogs are about 1.5–2.5cm long, males maturing at about 2 years old and females at 3. Some animals live 11 years.

Other names Lake Frog.

Marsh Frog   Pool Frog
Water frogs, differences in leg proportions

POOL FROG *Rana lessonae*   **Pl.14**

Range From France eastwards through central Europe to west Russia, north to Estonia with an isolate in southern Sweden and south to Po valley of northern Italy and to north Balkans. May possibly have been native to England but existing populations are all probably introductions. Map 60.

Identification Adults up to about 8cm but usually under 6.5cm. Generally similar to Marsh Frog but differs in following features: usually considerably smaller; vocal sacs of male white and back of thigh typically marbled yellow or orange and brown or black; hind legs quite short (Fig., above), the heel not reaching beyond the eye; metatarsal tubercle large, almost semicircular, hard and usually sharp-edged, often more than half the length of the first hind toe (Fig., opposite). Snout quite sharp. In many areas, most animals are grass green above while others are bluish green or brown, The ground colour is overlaid with well defined dark spots that tend to be black in females and brown in males and sometimes fuse to form irregular bands on the sides. A light stripe may be present along the centre of the back and the dorsolateral folds are often pale. Groin area often yellow and underside usually white, frequently with scattered grey spots. Breeding males may be a striking fairly uniform yellow or yellow green above with golden yellow eyes and have light grey nuptial pads.

Variation Great variation in colouring. Animals from north-west of range are more frequently brown and sometimes almost blackish, perhaps increasing the amount of heat they can absorb from the sun.

Similar Species Marsh Frog (p. 89) and especially Edible Frog (92). See also Italian and Albanian Pool Frogs (p. 94 and p. 95).

Voice 'Auwack ... auwack ... auwack ...', comprising a series of snoring croaks each up to 1.5 seconds long. Individual croaks are less resonant and more even and purring than Marsh Frog's and each ends strongly. Pool Frogs can sound like the quacking of Mallard ducks.

Habits May occur up to 1,550m in south of range but much lower in north. Typically found in small bodies of water: ponds in meadows and woods and at the edges of moors, also in flooded ditches and clay and gravel pits. Sun loving and often diurnal but active at night as well. Often forages nocturnally on land and babies especially may sometimes be found up to 500m from water. Occasionally lives in larger ponds and lakes and sometimes with the Marsh Frog although its most usual companion is the Edible Frog. In north of range confined to really open sunny ponds. Hibernation usually takes place on land.

Breeds in late spring, the males calling by day in dense associations with sometimes as many as 10 animals in a square metre of water. Females lay 600–3,000 eggs in a season, depositing the clumps on water plants near the surface. The eggs hatch in 1–2 weeks. Tadpoles are sociable and bask in shallows. Newly metamorphosed frogs are 2–3 cm long and may sometimes breed when a year old, but often take 2 years or more to mature in cooler areas. Pool Frogs can live 6–12 years.

Other names Early accounts of Edible Frog often refer partly or wholly to Pool Frogs.

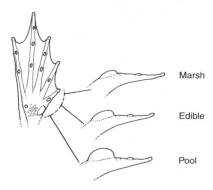

Marsh

Edible

Pool

Green frogs, hind foot showing typical sizes of metatarsal tubercle

EDIBLE FROG *Rana* kl. *esculenta*   **Pl. 14**

Range Apparently more or less the same as the Pool Frog but occurs alone in Denmark and extreme south Sweden. Some small colonies in southern England which were probably all introduced. Map 61.

Identification Up to 12cm. but usually smaller. Most similar to Pool Frog but bigger and many features intermediate between that species and the Marsh Frog. For instance Edible Frogs have rather longer legs than Pool Frogs. Vocal sacs white to dark grey and back of the thigh typically marbled brown or black and yellow or orange. Metatarsal tubercle often big, usually about 40–50 per cent as long as first toe with its highest point nearer to it. Most usually green with dark brown or black spotting. Generally bronze or brown animals also occur and often have green heads. Underside white, light grey or dark grey with darker spotting. Males often have light green or yellowish upper parts in the breeding season and grey nuptial pads.

Variation The above description is of the most usual diploid form of Edible Frog with two sets of chromosomes (one from the Pool Frog and the other from the Marsh Frog). Triploid animals with an extra set from the Pool Frog are generally similar but the metatarsal tubercle can be even bigger (43–55 per cent as long as the first toe) and has its highest point at the centre of the tubercle. Triploid animals with an extra set of Marsh Frog chromosomes have a smaller metatarsal tubercle (33–40 per cent as long as hind toe) and the back of the thigh may lack yellow colouring.

Similar Species Marsh Frog (p. 89); separation from this species is often quite easy. Pool Frog (p. 90) is usually distinguishable where it occurs in colonies with Edible Frogs and a number of animals can be compared, but certain identification of colonies made up of a single form can sometimes be more difficult.

Voice Tends to be intermediate between Marsh Frog and Pool Frog. A growling 'reh … reh … reh', each croak lasting about 1.5 secs.

Habits Occurs up to 1,550m. Usually very aquatic and can be found in a very wide range of waters, covering all the kinds inhabited by Marsh and Pool Frogs and brackish water too. Like other water frogs, it is frequently active by day and often basks in the sun. Sometimes found far from water, especially the young which often colonise new habitats. Has been known to travel 2.5km. Hibernates both on land and in water but more frequently on land. Most usually occurs in mixed colonies with Pool Frogs. These typically include diploid Edible Frogs of both sexes, but sometimes one or other sex is missing and triploid animals may also be present. Less commonly, Edible Frogs are found in colonies with Marsh Frogs (for instance in eastern Germany) and, rarely, they occur with both Pool and Marsh Frogs. Pure Edible Frog colonies are also sometimes encountered, for instance in north-east Germany and Denmark.

Females lay 3,000–10,000 eggs in a season. These vary greatly in size, from 1mm to 2.5mm across, the larger ones coming from triploid mothers. Tadpoles often bask in shallow in day but head for deeper water at night. Newly metamorphosed frogs are 2–3cm long and mature at about two years old in areas with warmer summers. Known to live 14 years in captivity.

Note In spite of its name, the Edible Frog is not peculiar in being edible, other Water frogs and Brown frogs being eaten in some areas.

WATER FROGS: Southern France, Iberian peninsula, Balearic and Canary Islands, Madeira.

Southern France may well have more forms of Water frogs than any other area. There are probably Pool and Edible Frogs, at least in the north, as well as scattered colonies of Marsh Frogs which may have been introduced. Also present are the Iberian Water Frog, and Graf's Hybrid Frog which has chromosomes from the Iberian Water Frog and the Marsh Frog. Graf's Frog probably originates from hybridisation between the Iberian Water Frog and the Edible Frog which carries Marsh Frog chromosomes. The Iberian Water Frog also extends over the whole of the Iberian peninsula and the rest of the region. In Spain it may possibly occur with Edible Frogs close to the Pyrenees and is found with Graf's Hybrid Frog in at least the Basque Country, Aragon and Catalonia. Elsewhere in the Iberian peninsula the Iberian Water Frog seems to be the only native member of its group, although the similar North African Water Frog (*Rana saharica*) and Marsh Frog, are known to have been introduced on a limited scale in places. Only the Iberian Water Frog occurs in the Balearic and Canary Islands and on Madiera.

IBERIAN WATER FROG *Rana perezi*

Range Southern France, Spain, and Portugal. Introduced into Madeira, the central and west Canary islands and the Balearic islands. Map 62.

Identification Usually up to about 8.5cm but occasionally to 10cm. Vocal sacs of male grey and back of the thigh and groin not yellow. Hind legs quite long with heel reaching beyond eye; metatarsal tubercle small and flat less than 40 per cent of the length of the first toe. A relatively slender Water frog with a pointed snout. Green, grey or brown with a pattern of darker brown or blackish irregular spots. Often a light stripe on midback and, in green animals, dorsolateral folds are often brown. Underside whitish sometimes with a grey reticulation. Breeding males have grey nuptial pads.

Similar Species Other water frogs in area.

Voice Like Edible Frog.

Habits Highly aquatic and often abundant, occurring in a wide range of waters up to 2,400m in the south of its range. These include, ponds, reservoirs, canals, slow rivers and streams, mountain lakes, marshes, and moors, rice fields, flooded areas, ditches, cisterns and other irrigation structures in farm land, and even cattle troughs. Can survive in some polluted and brackish water. Active by night and day, often conspicuous as it sunbathes in the shallows. Flees into water when disturbed although sometimes seen quite far from it, especially young animals. Hibernates in water or on land.

Extended breeding period. Females lay 800–10,000 eggs in a season. Tadpoles develop in 2–4 months and are very tough, surviving high temperatures and low levels of oxygen in the water. They metamorphose faster in water with poor resources that is likely to dry up, and then emerge at a relatively small size. Newly metamorphosed froglets are usually 1.5–3cm long. Males mature in 1–2 years and females in 2–3 years. Tadpoles may over winter and then can reach 11cm. Maturity occurs at about 5.5cm and frogs may sometimes live to an age of 10 years, but usually less.

GRAF'S HYBRID FROG *Rana* kl. *grafi*

Range Southern France (north to area of Lyon), north-east Spain (Catalonia, Aragon and Basque country). Map 63.

Identification Up to about 11cm and generally similar to the Iberian Water Frog. Differs in its larger size, rather more fully webbed feet and more prominent metatarsal tubercle

Similar Species Distinguishable from the Marsh Frog (p. 89) by having the eardrum further from the eye (separated by only half its width in the Marsh Frog).

Voice Rather like Marsh Frog. The pauses between the croaks are longer than in this species.

Habits Occurs in mixed populations with the Iberian Water Frog and survives by reproducing with it.

WATER FROGS: Italian area

This region has Pool and Edible frogs in the extreme north, in the Po valley, and possibly some introduced Marsh Frogs in the north-west (e.g. around Imperia). Further south, it has two species of its own. The Italian Pool Frog, which is closely related to the Pool Frog, occurs in Italy itself, and on Sicily and Corsica. The hybridogenetic Italian Hybrid Frog also covers this area except for Corsica. It depends on the Italian Pool Frog for its reproduction and has chromosomes from this species and from the Marsh Frog. It may have arisen through hybridisation of the Italian Pool Frog with either Marsh Frogs, or Edible Frogs which carry Marsh Frog chromosomes.

ITALIAN POOL FROG *Rana bergeri*

Range Peninsular Italy approximately south of a line joining Genoa and Rimini; also Sicily and Corsica. Map 64.

Identification Up to about 8cm, but usually smaller. Very similar to Pool Frog in colouring but rather longer legged like an Edible Frog. Vocal sacs whitish and back of the thigh with yellow or orange spots. Metatarsal tubercle relatively long and high, often over 40 per cent the length of the first toe. Frequently green above with black spots, and often a paler green stripe along the mid-back. Underside white, yellow or light grey, frequently marbled or spotted with darker grey. Males have grey nuptial pads in the breeding season.

Variation This species may sometimes intergrade with the Pool Frog where their ranges meet. In the Abruzzi mountains of central Italy, between 1,000m and 1,500m, there are large water frogs up to 11cm long with heavily spotted bellies; their status and relationships are presently unknown.

Similar Species Italian Hybrid Frog (below); Pool Frog (p. 90).

Voice Like Pool Frog.

Habits A typical water frog. It matures at a smaller size than the Italian Hybrid Water Frog.

Other names *Rana maritima* may be available for this or the next form.

Note Has sometimes regarded as a subspecies of the Pool Frog as *Rana lessonae bergeri*.

ITALIAN HYBRID FROG *Rana* kl. *hispanica*

Range Italy except north, including Sicily; possibly also north-west Croatia and Slovenia. Map 65.

Identification Up to about 10cm. Vocal sacs dark grey and back of the thigh with small spots that are usually white but sometimes light yellowish or light grey. Hind legs moderately long; metatarsal tubercle lower and smaller than in Italian Pool Frog, usually less than 40 per cent the length of the first toe. Back green to brown with black spots, sometimes with a green stripe along the mid-back. Nuptial pads of breeding males dark grey.

Similar Species Pool Frog (p. 90) and Italian Pool Frog (opposite). Italian Hybrid Frog differs in its darker vocal sacs, lack of strong yellow or orange colouring on back of thigh, rather longer limbs, smaller metatarsal tubercle and rather different song.

Voice Like Edible Frog.

Habits Usually lives in mixed colonies with the Italian Pool Frog.

WATER FROGS: Southern Balkans

The area north to Montenegro and north-east Greece has Marsh Frogs in the north and a further six species of Water frogs confined to the region. Range is helpful in identifying all of them. Only the Epirus and Albanian Water Frogs can be identified with any confidence by their appearance. Here and elsewhere song may also be helpful.

ALBANIAN POOL FROG *Rana shqiperica*

Range Coastal areas of Albania and extreme south-western Montenegro. Map 66.

Identification Grows on average to about 7.5cm. Similar to Pool Frog. Vocal sacs olive to grey and back of thigh with yellow pigment. Hindlegs moderately long; metatarsal tubercle usually large with a high arched edge, its highest point towards the leg rather than the first toe, about 40 per cent of the length of this. Males are greenish or brownish and females light brown to olive. Both sexes have darker brown or black spots which may be chocolate coloured in females. A light green stripe may be present along back. Underside is cream with weak dark spotting and the groin is often yellow, this colour sometimes extending to the belly and the underside of the hind legs and feet. In breeding season, back of males is grass-green to yellow olive.

Variation Hybridises quite regularly with the Greek Marsh Frog and hybrids can breed with parent species.

Similar Species Occurs with the Greek Marsh Frog (p. 96), and perhaps with the Epirus Water Frog (below) in a very small area of south Albania. Differs from both of these in usually having yellow on back of thigh and a large metatarsal tubercle

Voice Like Pool Frog.

Habits Occurs in lowlands within 60km of the sea, around the shores of Lake Scutari and in swamps, ditches, marshes and along the edges of slow-flowing rivers. Prefers water with heavily vegetated margins, reeds and floating water weeds.

Note Sometimes regarded as an isolated population of the Pool Frog.

EPIRUS WATER FROG *Rana epeirotica*

Range Western Greece including Corfu; also extreme southern Albania. Map 67.

Identification Up to about 8.5cm. Vocal sacs dark grey to black in breeding season, otherwise slightly olive. The back of the thigh is marked with curved dark stripes running from the greenish upper surface to the whitish lower one. Hind legs relatively short; metatarsal tubercle very small and triangular, about a third of the length of the first toe. Most usually uniform olive green above with irregular dark green spots and often a light green stripe along the midback. Groin yellowish and dorsolateral folds less prominent than in Marsh Frog.

Variation Often hybridises with Greek Marsh Frog and up to a sixth of animals in a mixed colony may be hybrids. Hybrids may also sometimes breed with parent species.

Similar Species Other water frogs. Of these, only the Greek Marsh Frog (below) certainly occurs in the same area. In this species the vocal sacs are always grey to black, the back of the thigh is not markedly striped, and there is usually no yellow on groin, the hind legs are rather longer and the metatarsal, tubercle is larger.

Voice A rattling call consisting of a series of up to 30 croaks each about half a second long and separated by about half to one second. The call tends to increase in intensity. May call earlier in the day than Greek Marsh Frog and, unlike in this species, the call is usually delivered from among clumps of water plants, the body angled in the water so just the head projects above the surface. Hybrids between the Epirus Water Frog and the Balkan Marsh Frog have voices like Edible Frogs.

Habits A mainly lowland species occurring up to 500m and found in lakes, marshes, slow rivers and canals, which all often have rich vegetation at their edges. When disturbed it often flees on to land, while the Greek Marsh Frog which occurs with it often flees into water. Greek Marsh Frogs are usually in the minority in shared waters but are rather more abundant in the smaller ones.

GREEK MARSH FROG *Rana balcanica*

Range Albania and Greece except north-east; also Thasos and Zakinthos islands. Map 68.

Identification Up to 10cm. Very similar to Marsh Frog with grey vocal sacs, no bright yellow colouring on back of thigh, relatively long legs and small metatarsal tubercle. Often brown with a green back, frequently with a lighter central stripe. Irregular darker green or brown spots usually present. Distinguished from Marsh Frog by subtleties of the song and to a lesser extent by molecular features. In the field it is best separated by range, as it only contacts the Marsh Frog in north-east Greece.

Similar Species Marsh Frog (p. 89). Occurs with the Albanian Pool Frog (p. 95) in western Albania and neighbouring Montenegro, and with the Epirus Water Frog (p. 95) in western Greece and extreme southern Albania.

Voice Generally like Marsh Frog; individual croaks last around 0.5 seconds. Often calls floating horizontally on the surface of open water.

Habits Generally found in more open water than Epirus Frog and can also occur at higher altitudes.

Other names *Rana kurtmuelleri*

Remarks Sometimes not regarded as distinct from Marsh Frog.

KARPATHOS WATER FROG *Rana cerigensis*

Range Karpathos island, in the Aegean Sea between Rhodes and Crete. Map 69.

Identification Up to about 7cm. Vocal sacs dark grey. Long-legged; metatarsal tubercle moderately long and high, about 40 per cent of first toe length. Grey to olive above with or without brown spots. Underparts cream with dark grey mottling. Best identified by range.

Similar Species None within distribution.

Habits Aquatic like other water frogs.

CRETAN WATER FROG *Rana cretensis*

Range Crete. Map 70.

CRETAN WATER FROG *Rana cretensis*

Range Crete. Map 70.

Identification Up to about 8cm. Vocal sacs dark grey and yellow pigment present on hind legs and groin. Hind legs moderately long; metatarsal tubercle quite well developed, about 40 per cent length of first toe. Usually light grey to brown above; occasionally grass green with brown or olive-grey spots. Best identified by range.

Similar Species None within distribution.

Habits Aquatic like other water frogs.

LEVANT WATER FROG *Rana bedriagae*

Range European islands close to the coast of Asiatic Turkey; also Asiatic Turkey itself, Syria, Israel, Jordan and Egypt. Map 71.

Identification A typical Water frog generally like the Marsh Frog but distinguished by molecular features and song. Probably the only Water frog in its European range and best identified by distribution.

Similar Species Other Water frogs.

Voice Like Iberian Water Frog and Edible Frog.

Habits Mainly found in relatively warm areas with mild winters.

Other names *Rana levantina*

AMERICAN BULLFROG *Rana catesbeiana*   Pl. 14

Range Native to eastern North America, but now introduced into many other places including parts of Europe. Established longest in the Po valley of northern Italy (now in Lombardy, Veneto and Emilia Romana); also found in Tuscany and near Rome, in western Spain (Cáceres), France (around Bordeaux) and the Netherlands (Breda); has also bred in southern England. Map 72.

Identification Adults up to 20cm. A very large aquatic frog unlikely to be confused with any native species. Distinguished from European Water frogs by its lack of dorsolateral folds, its very large ear-drum (as large as the eye in females and up to twice as large in males) and by the vocal sac being beneath the chin and not at sides of mouth. The American Bullfrog tends to be more uniformly coloured than European Water frogs, being usually green, olive-green, or brown above, sometimes mottled with grey or darker brown and without a light stripe along the mid-back; the head is often light green and the legs usually banded. The belly is whitish, mottled or spotted with grey, and the throat is often cream in females and with a yellowish flush in males.

Similar Species European Water frogs (pp. 88–97).

Voice Quite different from native European species. A very deep loud groaning call: a slow 'br-wum,' sometimes described as 'jug-o-rum'.

Habits Very aquatic. Found at low altitudes in Europe (usually under 50m but sometimes to over 300m), in lakes, ponds, cisterns, rice fields and small rivers. Prefers bodies of water with good growth of vegetation at the edges. Active by day and night, but more often heard calling after dark. Adults include other frogs in their diet, as well as water snakes and chicks of water birds.

Males are territorial in the breeding season when they have a dark nuptial pad on the thumb. Some 10,000–25,000 eggs are laid in a flat floating mass.

TORTOISES, TERRAPINS AND SEA TURTLES
(Testudines)

There are some 300 species in this group which is characterised by its members possessing a bony shell that encloses the body and is covered by horny plates or, less commonly, tough skin. In Europe there are three native land species and three that are semi-aquatic in fresh or brackish water; five marine species have also been recorded. In other European reptiles, their chromosomes determine the sex of individuals, but in shelled reptiles temperature is responsible. Eggs incubated at relatively high temperatures become females while those in rather cooler conditions become males.

In Britain, no one vernacular name exists that covers all members of the group. *Tortoise* is used for the strictly land-dwelling forms, *terrapin* for the semi-aquatic ones and *turtle* for the marine species and sometimes for one or two other very aquatic groups. In the United States and Canada, tortoise is used more or less as in Britain, terrapin for edible, partly aquatic, hard-shelled forms, but turtle can cover all aquatic and semi-aquatic species and may sometimes be used for all shelled reptiles.

Key to Tortoises, Terrapins and Sea Turtles

Note *Carapace* is the domed, upper part of the shell; *plastron* the flatter, lower section, and the area on each side joining these two regions is the *bridge*.

1. Carapace strongly domed; no webbing between toes; terrestrial   **2**
   Carapace rather low; webbing present between toes; semi-aquatic in fresh or brackish water   **3**
   Carapace low; forelimbs form very flat, paddle-shaped flippers; marine   **6**

2. **TORTOISES**
   A large, well-defined horny scale on tail-tip; no spurs on backs of thighs; usually two supracaudal plates present (Fig., p. 101)
   Hermann's Tortoise, *Testudo hermanni*, (p. 100, Pl. 15)
   No large, well-defined scale on tail-tip; spurs often present on backs of thighs; usually only one supracaudal plate present (Fig, p. 101)
   Spur-thighed Tortoise, *Testudo graeca*, (p. 101, Pl. 15)
   Marginated Tortoise, *Testudo marginata*, (p. 102, Pl. 15)

3. **TERRAPINS**
   Neck usually with light spots and streaks but no stripes; virtually inguinal plates (Fig., p. 104); plastron hinged in adults (front part moves slightly)
   European Pond Terrapin, *Emys orbicularis*, (p. 103, Pl. 16)
   Neck with distinct stripes; clear inguinal plates present (Fig, p. 104); plastron not hinged in adults.   **4**

4. Bright red patches on side of head

Red-eared Terrapin, *Trachemys scripta* (p. 105; Pl. 16)
No bright red patch on side of head   5

5. South-east Europe. Plastron often predominantly dark; eye dark
Balkan Terrapin, *Mauremys rivulata*, (p. 105, Pl. 16)
Spain and Portugal. Plastron often with light areas; eye light.
Spanish Terrapin, *Mauremys leprosa* (p. 104, Pl. 16)

6. **SEA TURTLES**
Carapace covered by horny plates, with at most three ridges   7
Carapace covered by skin, with five or seven prominent ridges.
Leathery Turtle, *Dermochelys coriacea*, (p. 110, Pl. 18)
7. Five costal plates on each side; nuchal scale in contact with first of these
8
Four costal plates on each side; nuchal scale separated from first of these 9

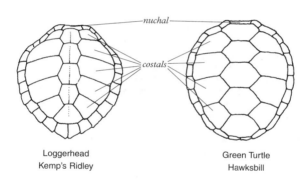

Loggerhead
Kemp's Ridley

Green Turtle
Hawksbill

Sea Turtle carapaces

8. Carapace clearly longer than broad; bridge with three inframarginal plates
on bridge (if more, then numbers on each side tend to be different); no pores
in inframarginal plates (Fig., Pl. 17)
Loggerhead Turtle, *Caretta caretta*, (p. 107, Pl. 17)
Carapace often almost as broad as long, or broader; bridge with four (rarely
three) pairs of inframarginal plates, each with a pore (Fig., Pl. 17)
Kemp's Ridley, *Lepidochelys kempii*, (p. 107, Pl. 17)

9. Plates on carapace do not overlap; one pair of prefrontal scales on snout
(Fig., p. 109)
Green Turtle, *Chelonia mydas* (p. 108 Pl. 18)
Plates on carapace usually do overlap; two pairs of prefrontal scales on snout
(Fig., p. 109)
Hawksbill Turtle, *Eretmochelys imbricata*, (p. 109, Pl. 18)

TORTOISES
(Family *Testudinidae*)

This group contains nearly all the mainly herbivorous shelled-reptiles that are modified for life in dry or fairly dry habitats. There are about 40 species distributed throughout most of the warm parts of the world except the Australasian region. Only three species occur in Europe and these are restricted to the south.

European tortoises occur in a wide variety of habitats varying from meadowland to almost barren hillsides. They can often be detected by the constant rustling they make as they push through the undergrowth (quite unlike the intermittent noise made by lizards or the more even, continuous sound of a moving snake). They raise their temperature by basking and, in the summer, tend to be active in the morning and evenings, spending the remainder of the day resting. Although mainly herbivorous, tortoises eat varying amounts of carrion, faeces and invertebrates.

The tail of adult male European tortoises is distinctly longer than that of the female, and the plastron is often more concave. Courtship consists of the male chasing the female, butting and tapping her shell, and biting at her limbs. Mating is rather precarious, the male often falling off the smooth shell of his partner. Copulating males produce a regular series of high-pitched cries, sometimes likened to the whine of a puppy or the miaowing of a cat. Females lay white, hard-shelled eggs that are spherical or oval. The clutch size varies with the size of the tortoise, being often 2–5 but sometimes up to 12. Eggs are usually laid in a hole the female digs herself in sunny, loose soil and more than one clutch may be produced in a year. Hatchling tortoises are basically similar to the adults, but have a more rounded shell and sometimes more clearly defined markings.

Mediterranean populations have suffered severely from over-collecting, either for the pet-trade or, less commonly, for food. Tortoises are also frequently killed on roads and in scrub and forest fires, especially in late summer.

Tortoises are usually quite easy to identify but confusion may arise because people have often transported European species to places outside their normal ranges and other forms are commonly imported into Europe and frequently escape.

HERMANN'S TORTOISE *Testudo hermanni*   Pl. 15
Range Balkan peninsula (mainly south of Danube), Ionian islands, south and west Italy, Sicily, Elba, Pianosa, Corsica, Sardinia, Balearic islands (Majorca and Minorca and perhaps Formentera), south-east France and north-eastern Spain. Introductions have also been made elsewhere. Map 73.
Identification Shell length of adults usually up to about 20cm; males smaller than females. Upper shell strongly domed, sometimes rather lumpy. Shell yellowish, orange, brownish or greenish, usually overlaid with varying amounts of dark pigment. Differs from other European tortoises in having following combination of features: large scale on tail tip, no spurs on thighs, usually two supracaudal plates above the tail (Fig., opposite), scaling on front of fore limbs tends not to be very coarse.
Variation Considerable variation in colour and shape. Old animals may sometimes be very dark and are often scarred, producing a gnarled appearance. Western animals (*Testudo* h. h*ermanni*), from Spain, France, Balearic islands, Corsica, Sardinia and

Italy, tend to be relatively small (females to 17cm), with a higher shell and brighter colouring including a light spot below and behind the eye. Animals in the rest of the range (*Testudo h. boettgeri*) are larger (females to 20cm), with a lower shell and duller colouring. Other more local variations also exist.

Similar Species Spur-thighed Tortoise (below); Marginated Tortoise (p. 102).

Habits Restricted to areas with hot summers. Found in a variety of habitats: lush meadows, cultivated land, scrub-covered hillsides, light woodland, stabilised dune areas and even rubbish dumps. Occurs up to 600m in west of range and up to 1,500m in south-eastern Europe. Males may have home ranges of about 2 hectares and females half this. Animals can occur at densities of 10 per hectare in east of range. In some areas diet includes a high proportion of leguminous plants (wild peas, lupins, beans etc) but many others are also taken, including composites, labiates, grasses as well as fruits.

Copulation may last 1–2 hours, the male using the horny tip of his tail to stimulate the female's cloaca. She lays one or two clutches of 3–12 eggs (average about 3 in west and 5 in east), which are 30–45mm by 20–30mm. The eggs hatch in 2–3 months, the babies being about 3.5cm in shell length. Males mature in 8–12 years at a shell length of 12–13cm and females in 11–13 years at 15cm.

Other names Greek Tortoise (see next species). Eastern animals were once called *T. h. hermanni* and western ones *T. h. robertmertensi*.

Hermann's Tortoise
often two supracaudals, large scale on tip
of tail

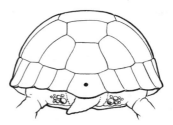

Spur-thighed Tortoise
often one supracaudal, spurs on thighs

Tortoises, back views

SPUR-THIGHED TORTOISE *Testudo graeca*   Pl. 15.

Range Southern Spain (between Huelva and Jerez, and the Murcia-Almería area). Balkans south of Danube to Macedonia, European Turkey and north-eastern mainland Greece; present on Aegean islands of Thasos, Samothraki, Limnos, Lesbos, Chios, Samos, Levos, Kos, Symi, Milos, Salamis, and, doubtfully, Euboea. Also found on Majorca, perhaps Ibiza and Formentera; in south France, Sicily, Italy, Sardinia and additional parts of Spain, Many of these localities are the result of introductions and breeding colonies may not be established. Outside Europe occurs in North Africa, Asia Minor and the Middle East to Iran. Map 74.

Identification Carapace length of adults sometimes up to 25 cm or even larger. Very like Hermann's Tortoise although shell rarely lumpy; differs in having obvious spurs on thighs, typically a single supracaudal plate above tail (Fig., above) and frequently coarser scaling on the front of the fore-legs; there is no large scale on the tail-tip.

Variation Considerable variation in colour and shape. Spanish and north-west African animals (*Testudo g. graeca*) tend to have a cream, yellow or reddish ground colour, the underside of the shell has well defined often symmetrical dark blotches, and the head has a black and yellow pattern. South-east European animals (*Testudo g. ibera*) have a horn to brownish ground colour, the underside of the shell has a less conspicuous, more irregular pattern and the head is more uniform; they also tend to have a broader, flatter shell than those in west, but this difference is not very constant. Old animals may have the back of the shell slightly flared, but not as much as in the Marginated Tortoise.

Similar Species Hermann's Tortoise (p. 100); Marginated Tortoise (below).

Habits Generally similar to Hermann's Tortoise. Often found in fairly dry grassland and scrub, sand dunes, open woods and edges. Can occur up to 1,300m in eastern Europe. In breeding season animals move considerable distances.

May mate in spring and autumn. Females lay 1–4 clutches of 1–7 (often 3–4) eggs, 30–40mm by 20–30mm. These hatch in 2.5–4 months, the babies being about 3–3.5cm in shell length. Males mature in 7–8 years and females in 9–10 years. Has lived about 100 years in captivity.

Other names Moorish Tortoise. *Testudo terrestris* has been used for eastern populations. In the past, there has been considerable confusion of the vernacular names of the two more widespread species of tortoise in Europe. Both *Testudo graeca* and *T. hermanni* have been known as the 'Greek' Tortoise: *T. graeca* on account of its scientific name, even though it has a very restricted distribution in Greece, and *T. hermanni* because it is the most widespread species in that country.

MARGINATED TORTOISE *Testudo marginata*   Pl. 15.

Range Much of Greece except the extreme north and the north-east; southern Albania near Greek border, some Aegean islands including Skyros, Poros and Kyra Panagia in the Northern Sporades. Known from recent fossils on Crete where now extinct. Introduced into Italy (western coast and south of Po delta) and also occurs on Sardinia (mainly northeast). Map 75.

Identification Shell length of adults up to about 38cm but usually smaller; males larger than females. Mature animals very distinctive: upper shell long and strongly flared at rear, mainly black except for a characteristic light orange or yellow patch on each large plate. Hatchlings look very like young of other species, as they are rounded and lack flaring and distinctive colouring. They can usually be separated from Hermann's Tortoise because there is only one supracaudal plate above the tail in most cases, sometimes weak spurs on the thighs, and scaling on the front of forelimbs is coarse. Separation from Spur-thighed Tortoise is more difficult and range is often the best clue; also spurs may sometimes be absent in the Marginated Tortoise and, if present, are rather small. Half-grown Marginated Tortoises may lack flaring, but the shell tends to be longer than in similarly sized animals of other species and often has the distinctive Marginated Tortoise pattern.

Variation Some old animals are entirely black. Populations from a 40km coastal strip west of the Taygetos mountains, southern Peloponnese, between Kalamata and Areopolis, have been named as a separate species, Weissinger's Tortoise (*Testudo weissingeri*). In this area maximum shell length is around 27cm and tortoises frequently have features of subadult Marginated Tortoises, for instance the upper shell is often not strongly flared. The shell is brown or blackish with lighter often greyish yellow patches flecked with grey, or even entirely yellowish. These populations sometimes appear to interbreed with more normal Marginated Tortoises where the two meet and the separate status of Weissinger's Tortoise is uncertain. It is presently treated as a subspecies of the Marginates Tortoise as *t. m. weissingeri*.

Similar Species Spur-thighed Tortoise (p. 101); Hermann's Tortoise (p. 101).

Habits Generally like other European tortoises. Often found on scrub-covered, rocky hillsides but may spend some time in neighbouring olive groves and grassy vegetation. Often encountered in hilly country and reaches altitudes of 1,600m. In areas where it occurs with Hermann's Tortoise (p. 100) tends to be more montane. At least west of Taygetos mountains, this species may dig 3m long burrows.

Male combat and courtship are particularly vigorous in Marginated Tortoises. Females lay clutches of 3–11 eggs, 30–40mm by 25–35mm, which hatch in 7–9 weeks producing babies with shell lengths of about 3.5–4cm.

Note Sardinian populations have been said to result from a tradition in Roman times of Sardinian sailors bringing tortoises from Greece as presents to their proposed wives but other origins have also been suggested.

Terrapins
(Families *Emydidae and Bataguridae*)

These are two generally similar groups of terrapins. Emydids have 40 species only one of which, the European Pond Terrapin, naturally occurs outside the Americas. Another, the Red-eared Terrapin has been introduced from the United States. The 63 species of Batagurids in contrast are largely Asian with just one group in tropical America and two species in Europe. Both families have long histories, Emydids occurring back to the Cretaceous and Batagurids, which are related to tortoises, to the Paleocene. Unlike tortoises, terrapins are aquatic or semi-aquatic and mainly carnivorous, feeding on a wide variety of aquatic animals (especially fish, amphibians and large invertebrates), and carrion. Differences between the sexes and mating behaviour are generally similar to those of tortoises. The elongate eggs are laid in soft ground and young animals are more brightly marked than adults and also longer tailed and rounder in outline when viewed from above.

EUROPEAN POND TERRAPIN *Emys orbicularis* Pl. 16.

Range Most of Europe except north and parts of centre. Found on Majorca, Minorca, Corsica, Sardinia and Sicily; absent from Aegean except Thases and Samo Hiraki. Introductions beyond present natural range are fairly common and may include the populations in Denmark, Germany and surrounding areas. Also west Asia and north-west Africa. Map 76.

Identification Carapace of adults usually up to 20 cm but occasionally to 30cm. Black, blackish or brownish, usually with pattern of light, often yellow, spots and streaks. Underside yellow with varying amounts of black. Readily distinguished from tortoises by more flattened shell. Only likely to be confused with Spanish and Balkan Terrapins but differs in colouring and lacks distinct stripes on neck and usually inguinal plates (Fig., p. 104, occasionally present but poorly developed). Shell more or less oval in outline, slightly wider behind, with a central keel in young but not in old specimens. Plastron rather flexible in young; more rigid in adults but a transverse hinge allows the front section to move up and down slightly. Males have thicker tails than females with the vent closer to the tip, and curved claws.

Variation Considerable geographical variation in size, shell proportions and colouring of shell, head, eyes and limbs. Many subspecies have been described.

Similar Species Spanish and Balkan Terrapins (below).

Habits The most northerly chelonian. Usually found in still or slow moving water with a good growth of aquatic plants and overhanging vegetation, including ponds, lakes, rivers, canals, bogs, ditches and brackish areas. Reaches 1,400m in Sicily. Typically seen when basking on stones or logs at water's edge, sometimes with only the head and neck projecting above surface. Rather timid and dives when disturbed. Feeds largely under water on a wide range of invertebrates, amphibians including tadpoles, fish and some vegetation.

Said to produce short piping calls in mating season. Females lay 3–18 (usually 9–10) leathery-shelled eggs, 30–40mm by 18–20mm. These are placed in a cavity that the female digs herself in a sunny place often some considerable distance from water. Favoured areas may contain the nests of several females. In north of range needs hot summers for successful breeding and this may occur only every 4–5 years. Eggs usually hatch in 2–4 months producing babies 2–2.5cm in shell length. Males mature at 6–13 years and females at 18–20 years, when they reach a shell length of about 12cm. This terrapin may live 30–40 years in natural conditions.

Note After the last Ice age, European Pond Terrapins also extended naturally into Britain, the Netherlands and south Scandinavia but became extinct in these areas later as a result of climatic cooling.

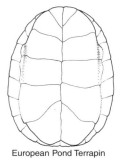

European Pond Terrapin

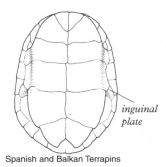

inguinal plate

Spanish and Balkan Terrapins

Terrapins, undersides of shells

SPANISH TERRAPIN *Mauremys leprosa*   Pl. 16.

Range Iberian peninsula, except parts of north; possibly introduced elsewhere. Also north-west Africa from Morocco to north-east Libya. Map 77.

Identification Carapace of adults usually up to about 20cm, occasionally longer; females larger than males. Only likely to be confused with European Pond Terrapin (above). Differs in having usually less dark colouring, often stripes on neck, and inguinal plates (Fig., above). Typically grey-brown or greenish. Young animals are frequently warm brown with clear red or yellowish markings on each plate of the carapace, and neck striping is also clear, often together with a light spot between the ear drum and the eye. Plastron is yellowish, largely obscured by dark colouring. Two dark spots or a streak on the bridges joining the plastron to the carapace. Eye is light-coloured. Shell has distinct central keel (best developed in young but visible at least at rear of shell in adults). Plastron rather flexible in young but rigid in older animals, without a hinge. Many individuals have algal infection of shell which makes it flaky and often results in the horny plates sloughing off.

Variation Considerable variation in colouring.

Similar Species European Pond Terrapin (p. 103); Balkan Terrapin (below) is outside range.

Habits Very like European Pond Terrapin but sometimes found in larger, more open waters, such as lakes, reservoirs and big rivers. Also present in small and shallow pools and streams, tending to aestivate by burrowing into mud if these dry up. Very tolerant of brackish or polluted water. Occurs up to 1,000m. Often quite abundant. May produce musky odour when disturbed. Diet is similar to European Pond Terrapin.

Females lay 2–3 clutches of 1–13 (usually 5–10) hard-shelled eggs, 25–40mm by 18–35mm, which hatch after 4–16 weeks into babies 2–3cm long. Males mature in about 7 years and females in about 10, when they have shell lengths of 14–16cm. Can live about 20 years, at least in captivity.

Other names. Once considered a subspecies of the Caspian Terrapin as *Mauremys caspica leprosa*.

BALKAN TERRAPIN *Mauremys rivulata* Pl. 16

Range Balkan peninsula north to Montenegro and extreme southern Croatia, Macedonia, southern Bulgaria and European Turkey; many Aegean islands including Crete. Also parts of Asiatic Turkey to Israel. Map 78.

Identification Up to 25cm; females larger than males. Very similar to Spanish Terrapin. Adults typically grey-brown or greenish above, younger animals have reticulations on the carapace. Plastron and bridges connecting it to the carapace usually totally blackish but this colour is sometimes limited to edges of the plate in adults. Eye is dark coloured. Head is narrower than in Spanish Terrapin and any stripe along the centre of the neck is narrower than those at the sides.

Variation Some in colouring.

Similar Species Only likely to be confused with European Pond Terrapin (p. 103); Spanish Terrapin (above) is outside range.

Habits Similar to Spanish Terrapin. Mainly a lowland form but occurs up to 800m. Found in a wide range of fresh and brackish water including, lakes, ponds, marshes and irrigation channels.

Females lay 4–6 hard-shelled eggs, 35–40mm by 20–30mm, which hatch into babies 3–3.5cm long.

Other names. Until recently was considered a subspecies of the Caspian Terrapin as *Mauremys caspica rivulata*.

RED-EARED TERRAPIN *Trachemys scripta* Pl. 16.

Range Widely introduced in Europe (for instance in Britain, parts of France, Germany, west, south and north-east Spain, and Italy). Original range is Mississippi drainage in U.SA but now established in such places as Mexico, Central America, northern Brazil, Bahrain (Arabian Gulf) and South-east Asia.

Identification Up to 28cm; females larger than males. Characterised by a strongly striped head with a bright red patch on its side behind the eye. Babies have strong and complex markings on shell but large animals tend to become dark and more uniform. Males have very long curved claws.

Variation The form introduced into Europe and elsewhere is *T. scripta elegans*.

Similar Species Spanish and Caspian Terrapins.

Habits A highly adaptable species found in a wide range of habitats, including lakes, ponds, canals and slow-moving rivers. Basks on logs and rocks. Said not to breed in northern parts of its introduced range but does so in south. Large numbers of elegantly coloured babies have been regularly imported into Europe and many other areas as pets. Many of these, and more grown animals, have been released into the wild. This terrapin is very vigorous and sometimes impinges on native terrapins. For instance, on Bahrain, it appears to have displaced the Caspian Terrapin (*Mauremys caspica*).

A courting male faces the female and vibrates his long claws against the sides of her head before copulating with her. In the original range of the species, females lay one or more clutches of 2–19 (often 6–11) eggs. These are placed in a hole on land that she digs with her hind legs. Hatchlings are 2–3.5cm in shell length.

Note The related Painted Terrapin (*Pseudemys picta*) is introduced around Barcelona. It is dark with a yellow spot behind the eye and is widespread in the U.S.A.

Sea Turtles
(Families *Cheloniidae* and *Dermochelyidae*)

Most Sea turtles belong to the family Cheloniidae and have a shell made up of large bones covered by horny plates, like other shelled reptiles. There are six living species of which four have been recorded in European waters. A single species, the Leathery Turtle, makes up the related family Dermochelyidae in which the shell consists of many small bones and is covered by tough skin. Sea turtles have long histories, probably both groups occurred at least as far back as the Upper Cretaceous period, over 65 million years ago, and were more diverse in the past than they are now. Sea turtles are superb swimmers and spend virtually all their lives in water, usually only the females coming ashore occasionally to lay their eggs. Sea turtles may, however, be thrown up on beaches when sick, damaged or dead. All species have streamlined shells, heads that cannot be retracted into these and flat paddle-like fore limbs (flippers). When swimming, the long forelimbs are flapped slowly up and down like the wings of a bird, so that the turtle essentially flies through the water. Sea turtles are mainly warm-water animals but some kinds reach cooler seas and may be found in the north-east Atlantic ocean, as well as the Mediterranean. There is some evidence that many of the turtles encountered off north-west Europe swim or drift from the west Atlantic with the Gulf Stream, but the Leathery Turtle migrates actively. Female Sea turtles produce large clutches of rounded, soft-shelled eggs which are usually buried at night on sandy beaches of warm seas, the young entering the water as soon as they hatch. More than one species may use the same beach.

Turtles suffer considerably from exploitation and other human activities. They and their eggs are eaten in many areas, the horny plates of mainly one species provide 'tortoiseshell', and others are often butchered to make souvenirs, soup etc. The opening up of originally deserted sandy beaches for tourism can have a catastrophic effect on the breeding success of turtles. Other casualties arise from collisions with boats and from eating rubbish thrown in the sea.

When stranded, or egg-laying, turtles should present no real problems of identification, even though they may sometimes be encrusted with sea-weed and barnacles. At sea, where clear views are much harder to obtain, turtles can be difficult to identify. In the following descriptions, the shell lengths given are taken over the curve of the upper surface.

LOGGERHEAD TURTLE *Caretta caretta*   Pl. 17

Range Atlantic (occasionally north to Norway and beyond), Black Sea and Mediterranean. Breeds, in Greece (on Zakynthos, Cephalonia, Peloponnese and north-west Crete); and in Sicily, Linosa and Lampedusa. Previous breeding records exist for Corsica, south Italy and Gozo. Common in the Canary islands. Also Pacific and Indian oceans.

Identification Carapace up to about 123cm in adults, but more usually 85–100cm. Often weighs 100–150 kg but may even exceed 450 kg. A large horny-shelled turtle, with an oval, often rather long carapace, five costal plates (Fig., p. 99) and usually three inframarginal plates on each side, which always lack pores (Fig. Pl. 17). Inframarginals may exceed three but numbers on each side then tend to be different. Young have three keels along the carapace, the central one involving a backward projection on each vertebral plate, giving the animal a 'saw-backed' appearance. Adults are red-brown above, sometimes tinged olive, and yellow to cream below. Shells of babies are brown to greyish black above and sometimes streaked.

Variation Some in colour.

Similar Species Juveniles may look superficially like Kemp's Ridley (below). Adults have some resemblance to Green Turtle (p. 108). See also Hawksbill (p. 109).

Habits Occurs in deep water but often found relatively near shore, entering bays, lagoons, creeks and river estuaries. In north-east Atlantic (French coast northwards) most animals encountered are immature. The commonest turtle in the Mediterranean. May migrate long distances to feed. Atlantic animals appear to travel on the great circular current in the north Atlantic ocean, going from the West Indies north-east to Madeira and the Canary islands and then returning south and westwards. East Atlantic animals may migrate into the west Mediterranean and breeding populations from the east of this sea migrate to Tunis and other areas. Omnivorous, feeding especially in rocky places and around reefs, taking hard prey such as crabs, sea urchins, molluscs and other invertebrates which it crushes with its massive head and jaws. Also eats sea weed, sea grasses etc. In deep water tends to feed especially on salps and jellyfish.

Females lay 23–198 eggs, 3.5–5.5cm across, and produce up to 1–6 clutches in a season. They usually reproduce every 2–4 years. The eggs hatch in 7–11 weeks producing babies about 3.5–5.5cm long. These mature in 12–30 years at a length of 75–100cm. Has lived 60 years in captivity.

KEMP'S RIDLEY *Lepidochelys kempii*   Pl. 17

Range Atlantic coasts (especially British Isles and France, but also Netherlands, Spain, Madeira and Azores) but generally rare in European waters; only a single record for the Mediterranean. Breeds almost entirely around the Gulf of Mexico, mainly on one beach in Tamaulipas, north-east Mexico. Individuals reaching Europe drift across the Atlantic from this area, presumably being mainly carried by the Gulf Stream.

Identification Carapace of adults up to 75 cm, but nearly all specimens recorded from European waters are young. A small, very broad shelled turtle. Young have three keels along the carapace, the central one involving a backward projection on each vertebral plate, giving the animal a 'saw-backed' appearance. Rather similar to young Loggerhead Turtles, but shell distinctly wider; usually four inframarginal plates on each side, and these have pores (Pl. 17). Colour above is not reddish but varies in young from blackish to brownish, grey, or olive, often with weak streaking; adults are grey to yellowish with an immaculate white underside.

Similar Species Loggerhead (p. 107) is superficially similar. Further south in the Atlantic, and in the Pacific and Indian oceans, the similar Pacific Ridley (*Lepidochelys olivacea*) occurs. This can nearly always be recognised by having 6–9 costal plates (Fig., p. 99) on each side instead of 5. It has not been recorded from European waters, except for a single individual off Madeira.

Habits Appears to be an essentially sedentary shallow-water species that is largely a bottom forager. Eats principally crabs, especially when young, but also takes shrimps, barnacles, molluscs, echinoderms, other invertebrates and some marine plants.

Females are exceptional among sea turtles in visiting their nesting beach during the day, having mated near by. Previously, very large numbers were involved, up to 40,000 being recorded in a single day. They each lay 50–180 eggs, 3.5–4.5.cm across, and may produce up to 4 clutches in a season. The eggs hatch in 7–10 weeks producing babies 3.5–5cm long. Sexual maturity is attained after about 6 years at a length of 60cm.

Note The most endangered of the Sea turtles, only a few hundred now nesting each year.

GREEN TURTLE *Chelonia mydas*   **Pl. 18**

Range Very rare in European Atlantic waters, although recorded from British Isles, Netherlands, Belgium, France, Spain, Portugal and Madeira; however it is common around the Canary islands. Also rare in most of Mediterranean, although a few breed in the north-eastern parts of that sea, especially in Asiatic Turkey and Cyprus; Green Turtles were once much more abundant in this area. Perhaps only one record from the Black Sea. Occurs elsewhere in the Atlantic, Indian and Pacific oceans.

Identification Carapace length of adults up to 150cm. Rarely, weighs almost 400kg. A large turtle with an oval shell; only four costal plates on each side (Fig., p. 99) and one pair of prefrontal scales on snout (Fig, opposite). Shell usually brown or olive above, often with darker mottling, streaks or blotches; occasionally black. Underside immaculate white or yellow. Light edges of scales on top of head are conspicuous. Babies are dark green to brown above and may be mottled.

Similar Species Loggerhead (p. 107) is sometimes mistaken for Green Turtle, but has five (not four) costal plates on each side. See also Hawksbill Turtle (below).

Habits A mainly tropical turtle living largely in warm, shallow water with a good growth of submerged sea plants, but capable of long trans-oceanic migrations of some thousands of kilometres. Sometimes comes ashore to sunbathe. Adults more vegetarian than other turtles, eating sea grasses, sea weeds and other plants but also invertebrates; babies are largely carnivorous, perhaps eating jellyfish.

Females nest communally, having mated offshore. They lay about 50–240 eggs, 3.5–6cm across, and produce up to 8 clutches in a season. Females usually breed every 2–5 years. Eggs hatch in 7–9 weeks producing babies 5.5–6cm long which mature in 18–27 years in the wild.

Note Has been heavily exploited as food. The name 'Green Turtle' refers not to its external appearance but to the colour of its fat.

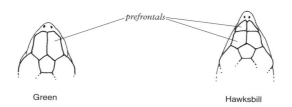

Green

Hawksbill

Sea Turtles, snouts

HAWKSBILL TURTLE *Eretmochelys imbricata* Pl. 18

Range Extremely rare in European Atlantic and Mediterranean waters. Found mainly in tropical areas of Atlantic, Pacific and Indian oceans.

Identification Shell of adults up to 110cm but usually smaller. May reach a weight of 125kg. A moderately small, sea turtle, usually easily distinguished from all others by the distinctly overlapping horny plates on its carapace. Old adults may lack this feature but can be separated from Green Turtles by having two pairs (not one) of prefrontal scales on the snout, and a nuchal scale of which the front border is only about half the length of the hind one (Fig, p. 99). Shell dark greenish brown above in adults and yellow below. Younger animals have a warm yellowish brown upper shell with conspicuous dark mottling.

Variation Considerable variation in shape: adults have narrower shells than young animals.

Similar Species Old adults can be confused with Green Turtles (opposite).

Habits An essentially sedentary tropical turtle, although it is sometimes known to travel long distances, especially when young. Found typically in shallow water around rocky areas and reefs but in other coastal situations as well. Omnivorous, but adults take mainly invertebrate food, especially sponges; hatchlings appear to be herbivorous. The horny plates of the Hawksbill Turtle are the main source of commercial tortoiseshell, an adult producing 4–5 kg. Usually, animals are killed to obtain this, but sometimes the horny plates are removed using heat and the turtles released; some of them may possibly survive this process and grow new plates. Hawksbills are eaten but their meat is sometimes poisonous if they have taken up toxins from the animals on which they feed.

Breeding usually takes place away from other species on sheltered beaches often within coral reefs, mating occurring offshore from these. Females often nest alone, laying 50–200 eggs, 3.5–4cm across, and producing up to 4 clutches in a season; they usually reproduce every 2–6 years. The eggs hatch in 8–9 weeks producing babies about 4.5cm long.

LEATHERY TURTLE *Dermochelys coriacea*   **Pl. 18**

Range Atlantic ocean, from southern Spain to Norway, with many records from around the British Isles and particularly the area between the estuaries of the Loire and Gironde rivers in France. Rare but regular in the Mediterranean and once occasionally bred there (including Sicily), although there are no recent records; breeds occasionally in the Canary islands. Also Pacific and Indian oceans. The only Sea turtle regularly occurring in cool waters, even below 6°C, and the most widely distributed reptile.

Identification The largest surviving shelled reptile with a shell up to 290cm long, but usually less than 200cm. The biggest adults may weigh more than 900k. A huge dark turtle with seven (sometimes five) very prominent ridges along the upper shell, and four below. The shell is covered by thick leathery skin instead of horny plates and the ridges are sometimes notched. Skin black or dark brown above, usually with lighter, white to pink flecks, especially on head and neck; pale below. Upper jaw has two distinct tooth-like points at front.

Variation Some in colour.

Similar Species Unlikely to be confused with any other turtle if a clear sighting is obtained. Sometimes sticks head and neck right out of water, when it may look quite unturtle-like.

Habits Pelagic, usually living in very deep waters but comes inshore at times. A regular summer visitor to the Atlantic off Europe and a strong swimmer regularly covering 30km in a day and sometimes travelling 6,000km from its breeding beaches. Frequently dives to 60m and sometimes to 1,200m. Ability to maintain a body temperature up to 18°C above the surrounding water enables it to survive in the cool areas it regularly reaches. Heat generated from muscle activity is conserved by the relatively small surface area associated with large size and by layers of fat. There are also heat exchange mechanisms at the base of the limbs that keep warmth in the body instead of dissipating it through the flippers. Mainly carnivorous, feeding principally on floating jellyfish and salps (relatives of sea squirts); adults may take 50 per cent of their weight of these in a day and young turtles even more; other animals and floating vegetation are also sometimes eaten. The gullet is lined with horny, backwardly pointed spines that apparently help this turtle retain soft-bodied prey. Leathery Turtles may swallow floating plastic bags, presumably mistaking them for jellyfish, sometimes with fatal results.

Copulation appears to occur far from land and nesting takes place on exposed tropical beaches on both sides of the Atlantic. Females lay 80–128 eggs, 5–6.5 cm in diameter, and produce up to 12 clutches a season; they usually reproduce every 1–4 years. Fertile eggs are mixed with smaller egg-like 'spacers' and hatch in 6–9 weeks producing babies 5–7cm long. Males take 5–6 years to reach sexual maturity and females 13–14 years. There are perhaps only 40,000 Leathery Turtles left in the world.

LIZARDS

(*Sauria*)

More than 3,000 lizard species exist of which 79 occur in Europe. All these have scaly skins, relatively long tails and in most cases, closeable eyelids and two pairs of legs. However, six more or less limbless species are found in the area covered by this book (for separation from snakes, see p. 195). Most lizards feed on invertebrates, especially arthropods although a minority take a significant amount of vegetable food. Unlike mammals, they have joints within the skull that often permit the upper jaw to move as well as the lower one, probably making grabbing agile prey easier. Such movement is often lost in forms that do not appear likely to benefit from it, such as some herbivores, and chameleons that catch prey almost entirely with their long sticky tongues. Lizards shed the whole of their skin at intervals, instead of it flaking steadily away as it does in tortoises and mammals. Male lizards each have two organs used for introducing sperm into females, the hemipenes. Probably the ancestors of lizards originally had a single penis like that of mammals which was later lost and the hemipenes represent a different structure with the same function that was 'invented' later. Most of the time the hemipenes lie in the tail base like separate, backwardly directed fingers of a glove. When in use, they turn inside out and project from the cloaca and have a groove on the outside which carries the semen. Only one hemipenis is used at a time, and they are often employed alternately. The great majority of lizards lay eggs although a few species, mainly skinks and slow worms, give birth to fully formed young.

Identification Some of the features important for identifying lizards are shown in Figs., p. 112, 113 and in the keys. As with other groups, body proportions, colour and pattern are also helpful in confirming identifications. A hand lens (8x or 10x) is very useful for checking details of lizard scaling.

Counting Body Scales Where these are all roughly the same size, a count is made round the body, midway between the fore and hind legs (or halfway between the head and vent in limbless species). If belly scales are distinctly and abruptly larger than the others (as in lacertid lizards), they are counted separately from the remaining, dorsal scales.

Back (dorsal) scales. A count is made, in as nearly a straight transverse line as possible across the back from the outermost belly scale on one side to the outermost belly scale on the other.

Belly (ventral) scales. The number of longitudinal rows of enlarged scales at midbody is counted. At the sides, any scale that is as long as the other belly scales is included, even if it is narrower than the rest.

Pattern Some common features of lizard dorsal patterns, especially of lacertid lizards, are shown in Fig., right.

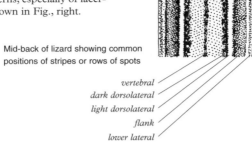

Mid-back of lizard showing common positions of stripes or rows of spots

vertebral
dark dorsolateral
light dorsolateral
flank
lower lateral

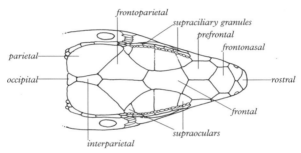

frontoparietal
supraciliary granules
prefrontal
frontonasal
parietal
occipital
rostral
frontal
supraoculars
interparietal

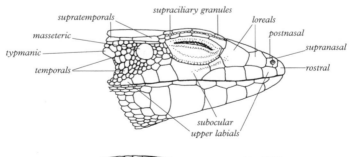

supratemporals
supraciliary granules
loreals
masseteric
postnasal
typmanic
supranasal
temporals
rostral
subocular
upper labials

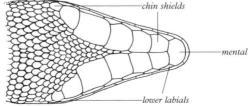

chin shields
mental
lower labials

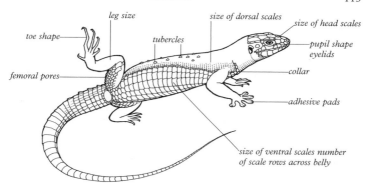

Key to Lizards and Ampbisbaenians

1. Worm-like, with regular ring-like grooves on body   **2**
   Not worm-like; if legless, then without ring-like grooves   **3**

2. Spain and Portugal; snout does not overhang lower jaw
           Iberian Worm Lizard *Blanus cinereus* (p. 194, Pl. 41)
   Islands close to Asiatic Turkey; snout overhangs lower jaw
           Anatolian Worm Lizard, *Blanus strauchi* (p. 194, Pl. 41)

3. Pupil a vertical slit in bright light   **4**
   Pupil not slit-shaped in bright light   **6**

4. GECKOS

   Toes with flat adhesive pads   **5**
   Toes without adhesive pads and strongly kinked (Fig., below)
           Kotschy's Gecko, *Cyrtopodion kotschyi*, (p. 125, Pl. 20)

5. Toes with flat adhesive pads only at tips (Fig., below); no tubercles on back
           European Leaf-toed Gecko, *Euleptes europaea* (p. 124, Pl. 20)
   Toes with undivided, flat adhesive pads along their whole length (Fig., below); obvious claws only on third and fourth toes of each foot
           Tarentola Geckos (p. 121, Pl. 20)
   Toes with flat adhesive pads divided on lower surface, and not extending to toe tips (Fig., below)

           Turkish Gecko, *Hemidactylus turcicus* (p. 124 Pl. 20)

Kotschy's (side)   European Leaf-toed (underside)   Tarentola (underside)   Turkish (underside, side)

Gecko toes

6. Body strongly flattened from side to side; eyes very bulging   **7**
   Body not strongly flattened from side to side, eyes not very bulging **8**

7. Flaps on each side of back of head, no spurs on hind feet
                Common Chameleon, *Chamaeleo chamaeleon* (p. 119, Pl. 19)
   No flaps at back of head; males have spurs on hind feet.
                African Chameleon, *Chamaeleo africanus* (p. 120; Pl. 19)

8. South Balkans, Aegean islands and Malta only. Scales on head all small; a
   large, flattened, rather spiny lizard
                Starred Agama, *Laudakia stellio* (p. 118, Pl. 19)
   Scales on head relatively large   **9**

9. Scales on back small or, if large, then keeled and often pointed; legs and
   femoral pores present (Fig., below)   **10**
   Scales on back large, often more or less smooth and shiny; legs sometimes
   absent or reduced; no femoral pores (Fig., below)   **22**

femoral pores

Lacertid   Skink

Lizard thighs from beneath

10. Well defined collar present (Fig., below)   **11**
    No collar, or at most a poorly defined one; dorsal body, scales relatively large
    and keeled   **20**

collar

collar present   collar absent

Lizard necks from beneath

11. Dorsal body scales large, larger than those on tail   **12**
    Dorsal body scales small, smaller than those on tail   **15**

12. **ALGYROIDES**

Scales on back blunt, much larger than flank scales   **13**
Scales on back pointed, about the same size as flank scales   **14**

13. East Adriatic area. Back scales strongly keeled
> Dalmatian Algyroides, *Algyroides nigropunctatus*, (p. 131, Pl. 22)

Spain. Back scales flattish and feebly keeled
> Spanish Algyroides, *Algyroides marchi* (p. 133, Pl. 22)

14. Corsica and Sardinia. Very small; about 13–23 scales across mid-back.
> Pygmy Algyroides, *Algyroides fitzingeri* (p. 132, Pl. 22)

South Greece and Ionian islands. Not very small; 18-25 scales across mid-back
> Greek Algyroides *Algyroides moreoticus* (p. 132, Pl. 22)

15. Iberian peninsula, East Romania to southern Ukraine. No occipital scale (Fig., p.112),
subocular scale does not reach lip, or only very narrowly (Fig., below)   **16**
Occipital scale nearly always present, subocular reaches lip widely   **17**

16. East Romania to southern Ukraine. First upper labial scale well separated from nostril (Fig., below); 12–20 large scales across mid-belly
> Steppe Runner, *Eremias arguta* (p. 130, Pl. 21)

Iberian peninsula. First upper labial scale borders or is close to nostril (Fig., below); ten (rarely eight) large scales across mid-belly.
> Spiny-footed Lizard, *Acanthodactylus erythrurus* (p 129, Pl. 21)

subocular   first upper labial

Steppe Runner   Spiny-footed Lizard

17. Canary islands. Usually 10 or more large scales across mid-belly; one postnasal scale
> Canary Island Lizards, *Gallotia* (p. 180, Pls. 37 and 38)

Other areas. Not usually this combination of features   **18**

18. Collar more or less smooth-edged (Fig., below)   Small lacertas (p. 142)
Collar strongly serrated (notched)   **19**

collar smooth-edged   collar serrated (notched)

Lacertid necks from beneath

19. Large (7–20 cm or more from snout to vent), robust, belly scales very strongly overlapping. Often green, at least on back or flanks. Often two postnasals, and supratemporals large

Green Lizards (adults, p. 134)

As above, but small; often not green; frequently with characteristic pattern (Pl. 26). Typical baby lizard shape: very large rounded head and big eyes

Green Lizards (juveniles, p. 134)

Small (up to 9 cm from snout to vent but usually smaller), often brown, with darker flanks, frequently one postnasal scale. If green, then supratemporal scales shallow (Fig., below)

Small lacertas (p. 142)

20. South-east Balkan area, east Aegean. Eye without obvious eyelids; first upper labial scale separated from nostril (Fig., p. 112)

Snake-eyed Lacertid, *Ophisops elegans* (p. 130, Pl. 21)

South-west Europe. Eye with normal eyelids; first upper labial scale reaches nostril   **21**

21. **PSAMMODROMUS**
    Adults less than 5.5cm snout to vent, weak collar present, scales on side of neck small and granular (Fig., p. 128),

Spanish Psammodromus, *Psammodromus hispanicus* (p. 128, Pl. 21)

Adults up to 9 cm snout to vent, no collar, scales on side of neck strongly overlapping and keeled (Fig., p. 128)

Large Psammodromus, *Psammodromus algirus* (p. 127, Pl. 21)

22. Legs absent, or only a minute hind pair   **23**
    Two pairs of legs present, although these may be very small   **26**

23. Balkans only. A prominent groove on either side of the body
    European Glass Lizard, *Ophisaurus apodus* (p. 193, Pl. 41)
    No groove on either side of the body   **24**

24. Head blunt (viewed from above); 23 or more rows of scales round mid-body; over 20cm when adult   **25**
    Head pointed (viewed from above); 18 (occasionally 20) rows of scales round mid-body; rarely over 18cm.
    Limbless Skink, *Ophiomorus punctatissimus* (p. 191, Pl. 41)

25. Usually 24–30 scales round mid-body; if sides of forebody dark, their upper border is not wavy

> Slow Worm, *Anguis fragilis* (p. 192, Pl. 41)

South Greece and Ionian islands only. Often 34–36 scales around mid-body; sides of forebody dark with a wavy upper border

> Peloponnese Slow Worm *Anguis cephallonica* (p. 193)

26. South-east Europe only. Diminutive; head and body less than 6cm long. Eye without obvious eyelids and will not close if gently touched

> Snake-eyed Skink, *Ablepharus kitaibellii;* (p. 186, Pl. 39)

Head and body of adults over 6cm long. Eye with normal lids   **27**

27. Rhodes etc. only. Rostral scale separated from nostril; back scales often with three feeble keels; frequently two rows of dark bars along back and a pair of pale stripes on each flank

> Levant Skink, *Mabuya aurata* (p. 191, Pl. 40)

Rostral scale contacts nostril; back scales smooth; pattern different   **28**

28. **CHALCIDES**

Snake-like; legs minute with only three toes on feet; 20–26 scales around body **29**
Thick-bodied; limbs not extremely small, with five toes; 24–38 scales around body   **30**

29. South-west Europe including north-west Italy. Pattern of 9–13 dark lines; middle hind toe about equal to outer one

> Western Three-toed Skink, *Chalcides striatus* (p. 188, Pl. 39)

Italy, Sicily, Elba and Sardinia. Pattern variable but usually with six or fewer dark lines; middle hind toe longer than outer one

> Italian Three-toed Skink, *Chalcides chalcides* (p. 189, Pl. 39)

30. Europe including some Mediterranean islands   **31**
Canary islands   **32**

31. Spain and Portugal only. Loreal scale borders second labial scale (Fig., p. 187); 24–28 scales around mid-body

> Bedriaga's Skink, *Chalcides bedriagai* (p. 188, Pl. 39)

Sardinia, Malta, Sicily, Naples area, Greece, some Aegean islands including Crete. Loreal scale borders second and third labial scales (Fig., p. 187); 28–38 scales around mid-body

> Ocellated Skink, *Chalcides ocellatus* (p. 187, Pl. 39)

32. East Canary islands. 30–32 scales around mid-body; rows of small pale spots on back

> East Canary Skink, *Chalcides simonyi* (p. 189, Pl. 40)

Gran Canaria. 25–38 scales around mid-body; pattern very variable but belly not very dark.

> Gran Canaria Skink, *Chalcides sexlineatus* (p. 190, Pl. 40)

West Canary islands. 28–32 scales around mid-body; belly dark

> West Canary Skink, *Chalcides viridianus* (p. 190, Pl. 40)

Agamas
(Family *Agamidae*)

Agamid lizards are mainly confined to the warmer regions of Africa, Asia and Australia, being replaced in America, Madagascar and Fiji by the closely related Iguanids. There are about 300 species in the family, only one of which reaches Europe. This belongs to the genus *Laudakia* which is widely distributed in central and south-west Asia. Agamids, together with chameleons, have distinctive teeth that are pointed, flattened and bladelike and fused to the jaws, making these look like pinking shears. Agamids are capable of some colour change, like geckos often becoming lighter, when warm and darker when cold. Food consists mainly of invertebrates but some vegetable matter is also taken and there are specialised agamids that concentrate on plants and others that take a very high proportion of ants. The great majority of species lay eggs.

STARRED AGAMA *Laudakia stellio*   Pl. 19.

Range Small colonies in Greece and the Greek islands, including the area round Thessaloniki, the Cyclades (Mykonos, Rhineia, Delos, Paros, Despotiko, Antiparos and Naxos) and islands close to Asiatic Turkey (Lesbos, Chios, Samos, Ikaria, Rhodes etc); introduced to Corfu around 1915, and to Malta. Also south-west Asia and north-Egypt. Map 79.

Identification Adults up to about 30cm, including tail which is about as long as the body. Unmistakable; a large, robust lizard with a conspicuously flattened, relatively short body, rather flat triangular head, well-defined neck and long legs and tail. Noticeably spiny, particularly on neck, with rows of tubercles across the body that are especially well developed on flanks. Scales on head and belly are small. Colour is variable and can change quite quickly; usually light or dark brown or grey, with a series of roughly diamond-shaped, often yellowish markings along middle of back. Tail may be conspicuously barred. Throat often dark flecked. Dominant males tend to be particularly brightly coloured.

Variation Considerable variation in colouring. Animals from Mykonos island usually have pale yellowish heads and unspotted throats.

Similar Species Not likely to be confused with any other European species.

Habits A diurnal, sun-loving lizard found in a variety of dry habitats. Frequents walls, rocky hillsides and olive groves, where it climbs well both on trees and rocks. Hides in holes and crevices and characteristically bobs head vertically. Often quite abundant where it occurs. Feeds on a wide variety of invertebrates but will also take small lizards and some plant material, particularly fruits and flowers.

Females lay about 6–10 eggs at a time, about 20mm x 10mm, and may produce 2 clutches a year; babies are approximately 3.5cm from snout to vent.

Other names. Hardun, *Agama stellio, Stellio stellio.*

Chameleons
(Family *Chamaeleonidae*)

This family contains about 85 species, all rather similar to each other. Chameleons are slow-moving lizards most of which, climb in bushes, trees and other vegetation. The great majority of species are found in Africa and Madagascar, but a few occur on islands in the Indian Ocean and in south-west Asia and two just reach Europe. Chameleons can change colour quite rapidly and are able to move each eye independently of the other, something they share with the related agamids. The body is flattened and leaf-shaped which, with ability to change colour, gives good camouflage. The toes are arranged so that the feet can grip small branches and stems firmly and the tail is strongly prehensile and readily twisted around twigs. The mainly insect prey is caught on the sticky tongue which can be shot forwards for more than the length of the body. Accurate aim of the tongue depends on good sight and the eyes are large, bulging and capable of being turned forwards to give good binocular vision. A few species give birth to living young, but the majority produce eggs that are laid in the ground.

MEDITERRANEAN CHAMELEON *Chamaeleo chamaeleon*   Pl. 19.
Range Extreme southern Portugal and Spain, Sicily, Malta, Crete, and Chios and Samos in the east Aegean sea; introductions have been made in some of these areas, the Canary islands and possibly elsewhere. Also found in south-west Asia and North Africa. Map 80.

Identification Adults up to about 30cm, including tail (which is rather shorter than the body) but usually smaller. Slow movements, laterally flattened body, prehensile tail, and bulbous eyes that can be moved independently make this species almost unique within its European range. There is a flap-like lobe extending backwards on each side of the head. Colour is very variable and capable of quite rapid changes, associated with camouflage, controlling body temperature and social interactions. Often green but can be blackish, brown, grey or whitish frequently with a varied pattern which can include dark, white or orange spots and an often broken white line along the side. Frequently especially pale at night.

Variation No geographical trends recorded in Europe.

Similar Species African Chameleon (p. 120).

Habits Nearly always found climbing in bushes, often in quite dry habitats, which include, open pine and eucalyptus plantations, orchards and gardens, and less commonly olive groves and vineyards. Occurs up to 800m in Spain. Usually stalks prey slowly and catches it with the sticky extensible tongue. Diet consists largely of flying insects which are usually caught as they rest on vegetation, but small lizards and even fledgling birds may also be taken. Only rarely descends to ground, for instance when depositing eggs. When disturbed, inflates itself with air, opens mouth and often becomes very dark.

Males defend a territory very vigorously and guard particular females, being more assiduous in the case of large ones. During this time, which may last 2 weeks or so, they copulate repeatedly, each mating taking 5–10 minutes, but females often have other partners so their offspring may have several fathers. Females produce a single clutch a year, digging a gallery, 20–50cm long, in which they deposit 5–35 eggs that may constitute over half their body weight at this time. Incubation takes 8–12 weeks and babies are about 5.5–7cm long, some becoming sexually mature in less than a year. The strain of reproduction is severe and 30–40 per cent of females may expire afterwards. Not surprisingly most chameleons die before their third birthday. This short life-span means that chameleon populations often fluctuate greatly in response to environmental changes that temporarily affect breeding success.

AFRICAN CHAMELEON *Chamaeleo africanus*   Pl. 19.

Distribution A small area of the south-western Peloponnese, in southern Greece where the species may be a long-term introduction from Egypt. Map 81.

Identification Up to 46cm including tail. Very similar to the Mediterranean Chameleon but grows larger and lacks flap-like lobes at the back of the head. The males also have a spur on the back of each hind foot.

Variation The African chamaeleon is a complex of forms widespread along the southern edge of the Sahara but which also occurs in the Nile Valley with a colony as far north as Alexandria; Greek animals resemble these.

Habits Found climbing in vegetation such as tamarisk and reeds (*Phragmites*). Also sometimes occurs on the ground, especially females when they go to lay their eggs. This takes place in sandy areas near beaches.

Geckos
(*Gekkota*)

There are about 900 species of geckos distributed throughout the warmer parts of the world. Only eight reach our area and these are all members of the family Gekkonidae. Typically geckos are small, plump lizards with large heads and eyes and a soft, granular skin that may have scattered tubercles. In Europe geckos differ from other lizards in possessing a cat-like vertical pupil which is most obvious in good light. Most geckos are basically nocturnal, although some of them may be at least partly active by day, especially in temperate regions such as Europe. Like many other night-active animals, they have especially large eyes. These are usually covered by a transparent 'spectacle', as in snakes and the Snake-eyed Lacertid, and cannot be closed. In fact, geckos have eyelids but in most species they are fused together and the spectacle develops as a very large window in the lower lid. Nearly all geckos differ from most other lizards in having a voice which varies from a thin squeak in some small species to a strident bark in the 35cm long Tokay (*Gekko gecko*) of south-east Asia. Many geckos are superlative climbers and, in addition to claws, may have sophisticated adhesive pads under the toes. These are covered beneath by a mass of hair-like structures (setae) that frequently branch several times. The ends of the setae are brought into intimate contact with the surfaces on which geckos climb, generating weak molecular forces that enable them to adhere and permit many of them to scale even extremely smooth surfaces such as panes of glass.

Many species enter buildings and hunt on the walls and even on ceilings. Like agamas and chameleons, geckos can change colour quite rapidly, although this is mainly confined to substantial lightening or darkening of the basic pattern. Change is related to temperature (hot animals get lighter and cold ones darker) and the general colour of the environment where these lizards live. European species lay hard-shelled eggs that are often attached to the walls of crevices etc. Usually there are two eggs in a clutch, but sometimes only one. Because they enter buildings, geckos are often accidentally transported in cargo and may turn up a long way from their original range, especially in port areas.

Tarentola Geckos
(*Tarentola*)

One mainland European gecko and the four native species on the Canary islands are quite closely related and all belong to the genus *Tarentola*, a group of about 24 species. They are robust, plump climbing geckos with flattened bodies and heads and have broad, adhesive pads that extend along the length of the toes and are widest near the tips (Fig., Pl. 20). The toes can be curled upwards to protect the adhesive pads when these are not in use. Well developed claws are confined to the third and fourth digits of each hand and foot. Females and sub-adult males however usually have tiny claws on the other digits. Tarentolas climb with great agility, even on overhanging rock faces, the body being kept close to the surface and the legs projecting sideways. Like most other geckos, tarentolas have a vertical slit-shaped pupil in bright light.

Males are rather bigger than females when adult and are usually territorial in the breeding season, when they are quite noisy producing a variety of calls. Females may return these before and during copulation, which often lasts less than a minute. The male often holds the female by the head when he mates and may insert one of his hemipenes (see Glossary) and then the other. Eggs may be laid in crevices or buried in litter, sand or earth. Most Tarentolas prefer relatively dry open areas including artificial habitats such as houses and drystone walls. They are largely active at night but may also be seen by day when they may bask in the sun.

There are several species of Tarentolas in North Africa, one of which, the Moorish Gecko, reaches southern Europe and has recently been introduced to Madeira by human activity. Tarentolas are also good travellers in their own right and over the past few million years have made several long journeys across the sea from North Africa, presumably accidentally travelling on natural rafts of vegetation carried by long-distance currents. They have travelled an estimated 6,000km to colonise Cuba and the Bahamas and have invaded three groups of the Canary islands separately (the eastern islands; the Selvages, Gran Canaria and Hierro; and the other western islands). A form from the western Canary islands then went on to colonise the Cape Verde archipelago, 1,500km to the south, and evolved there into another 9 species.

MOORISH GECKO *Tarentola mauritanica*   Pl. 20.

Range Mediterranean area, especially west: Iberian peninsula, Balearic islands, southern coast of France, Corsica, Sardinia, Italy, Sicily, Malta, Adriatic coastal area (south Slovenia, Croatia, Ionian islands, north and west Peloponnese), Crete; also many small islands across range including Pantellaria and Lampedusa. Additionally in North Africa and Israel with small introductions around the Atlantic including Madeira. Map 82.

Identification Usually up to about 8cm from snout to vent, although sometimes even larger; tail equal or slightly larger. A typical Tarentola gecko. Body and tail have prominent keeled tubercles giving a rather spiny appearance. Colour variable but usually brownish or brownish-grey or whitish, often with dark bands that are best developed on the tail and in young animals; the eye is usually grey and the underside whitish or yellowish. May have minute bright red spots on toes which are actually parasitic gecko mites (*Geckobia*). Regenerated tail is uniform and lacks tubercles.

Similar Species Other geckos within range.

Habits Found mainly in warm, dry, lowland coastal areas but extends inland in some regions, especially the Iberian peninsula. Usually found under 400m but extends to 1,400m in southern Spain. Often occurs on dry-stone walls, ruins, boulders, cliffs, outcrops, wood piles and the outside walls and tiled roofs of buildings, which it enters; occasionally found on trees. Usually climbs but on islands like the Balearics is often also ground-dwelling. May be seen near lights, catching insects attracted to them. Most usually active at night, if temperature is above 15°C, but also basks in sun, particularly in cooler parts of year.

Females lay 2–3 clutches of 1 or 2 oval eggs, 12–15mm x 9–12mm, which may sometimes be laid communally and are placed under stones, in cracks, hollow trees etc. Hatching time is said to vary greatly, from a few days to 14 weeks or more; babies are 2–2.5cm from snout to vent (4–6cm in total length). Has lived 8 years in captivity.

EAST CANARY GECKO *Tarentola angustimentalis*

Range East Canary islands of Fuertaventura, Lanzarote, Lobos and the small islands north of Lanzarote. Map 83.

Identification Up to 8cm from snout to vent, intact tail about the same length or rather longer. Similar to Moorish Gecko. Tubercles not large and often with more than one keel. Colouring very variable; a narrow, light, often broken stripe usually present along mid-back and there may about five broad dark cross bands on body, frequently with lighter hind edges; the eye is usually brown or gold. Underside whitish, sometimes with yellow markings, especially around the vent. Colouring is often more contrasting in babies.

Variation Considerable variation in general colouring.

Similar Species None within range. Differs from Moorish Gecko (p. 121) in eye colour, and in having the nostril separated from the rostral scale and sometimes extra keels on the body tubercles.

Habits Very common; found in most habitats up to about 800m, including stony and rocky places, gullies, lava fields (except old ones which are usually very dry) and drystone walls. Also on dunes and salt flats with vegetation (even if stones are absent), in cultivated areas and human habitation.

Females lay probably repeated clutches of 2 eggs or sometimes of a single one, about 10–14mm x 7–12mm. Babies are approx. 2–3cm from snout to vent. Females first breed when about 6cm long. Has been known to live over 17 years in captivity.

GRAN CANARIA GECKO *Tarentola boettgeri*

Range Gran Canaria and El Hierro in the Canary islands, and the Selvages group. Map 84.

Identification Up to 6cm from snout to vent, intact tail rather longer. Similar to other Canary geckos but appears smoother and has smaller, flatter tubercles on the back and tail. Colouring is very variable, but a narrow light stripe sometimes present along mid-back; eye usually grey or grey-blue. Underside whitish to grey, sometimes with dark speckling. Babies are said to have a light grey or whitish tail tip.

Variation Gran Canaria animals (*T. b. boettgeri*) are often tinged yellow, orange or pink and usually have smooth, unkeeled tubercles. Populations from the north and south of Gran Canaria are genetically different from each other. The Selvages population (*T. boettgeri bischoffi*) is greyish with a large head, clearly keeled tubercles and brown or grey-brown eye. It is sometimes treated as a separate species, the Selvages Gecko,

Tarentola bischoffi. El Hierro animals (*T. boettgeri hierrensis*) are small (less than 5.5 cm from snout to vent), lack bright colouring like Selvages animals, and the tubercles are lightly keeled.

Similar Species None within range. Small tubercles often with reduced keeling distinguish this species from other Canary and European Tarentola geckos.

Habits Often very common, particularly near coast. Found especially in rocky places and frequently encountered on the ground under stones; also sometimes present in damp places and occasionally in houses. Common up to 1,000m on Gran Canaria and sometimes to 1,500m; probably to 500m on El Hierro. On the Selvage islands found under stones and in rock crevices and perhaps, the burrows of petrels.

Males are aggressive and females produce presumably repeated clutches, each usually consisting of a single egg about 13–14mm x 10–11mm, which is large for the size of the lizard. The eggs hatch in perhaps 2–3 months and babies are 2.5–3cm from snout to vent. On Gran Canaria, females become sexually mature at around 5cm from snout to vent.

TENERIFE GECKO *Tarentola delalandii*

Range Tenerife and La Palma in the Canary islands. Map 85.

Identification Up to 8cm from snout to vent, intact tail equal or slightly longer. Body tubercles lightly keeled. Colouring highly variable; usually no light stripe along back; lip scales dark spotted; eye gold brown to dark grey. Underside whitish or yellowish and intense yellow pigment may be present around the vent, especially on La Palma. Babies have a light tail tip

Variation Animals from north and south Tenerife show some minor differences.

Similar Species None within range.

Habits Very common, especially near coast, but extends up to 1,500m and even 2,300m on Tenerife. Found frequently in open rather dry, rocky areas and around buildings, sometimes entering houses. Avoids dense woods including native laurel and pine forest. Like other Canary geckos, often clings to undersides of stones and runs around them if they are turned over, so keeping hidden on their lower surface; in this situation, it rarely drops to the ground and flees.

Females produce several clutches of 1–2 eggs, 11–14mm x 9–11mm, which hatch in about 8 weeks, the babies being around 2cm from snout to vent (4–4.5cm total length).

LA GOMERA GECKO *Tarentola gomerensis*

Range La Gomera in the west Canary islands. Map 86.

Identification Up to 7.5 cm from snout to vent, intact tail equal or slightly longer. Generally similar to Tenerife Gecko but fairly uniform above with scattered small light spots centred on the tubercles and no other larger light areas; any dark bands are usually continuous across the back; the eye is usually coppery or orange, and the underside is whitish. The toes are rather broader than in other Canary Geckos and babies are more contrasting than adults.

Similar Species None within range.

Habits Found up to 1,150m, often in humid habitats including disturbed areas, plantations and human structures, although it rarely actually enters houses. Frequently found on drystone walls and under stones; it avoids dense native forest.

Females lay repeated clutches of 1–2 eggs, sometimes communally; they apparently hatch in 2–3 weeks.

TURKISH GECKO *Hemidactylus turcicus*   **Pl. 20**

Range Mainly close to Mediterranean sea and present on numerous islands including many of those in the Aegean area; extends inland in southern Spain and parts of Italy. Also North Africa and Levant. Introduced into the Canary islands (Las Palmas on Gran Canaria and Santa Cruz on Tenerife) and parts the southern U.S.A, Mexico and central America. Map 87.

Identification Usually up to about 10cm, sometimes rather larger. A quite slender gecko with tubercles on the back and tail and flat adhesive pads that do not extend right to the tips of the toes (Pl. 20). Often a pale and rather translucent pinkish or buff, rarely so dark and opaque as other European geckos. Back usually marked with irregular dark blotches but intact tail increasingly banded towards its tip, especially in young. Males have 'femoral' pores in front of the vent.

Variation Some variation in colour; animals on very small islands are sometimes dark. A striped population occurs on the islet of Addaya Grande, near Minorca.

Similar Species Other geckos within range.

Habits Mainly in warm, coastal areas including some very small islands; sometimes spreads inland along river valleys. Usually at low altitudes but up to 300m and exceptionally 1,100m. Found on dry-stone walls, cliffs, caves, on boulders, rocky hillsides, tree boles, between the leaves of agave plants, and among their fallen remains, among rubbish and in both empty and occupied houses. Sometimes penetrates into large towns. Where it occurs with the Moorish Gecko it tends to be found lower down on walls etc. than this species. Often catches insects attracted to lights. Largely crepuscular and nocturnal but sometimes active by day at cool times of year, even sunning itself. Very fast and agile and an excellent climber, although it sometimes occurs on the ground. Has a range of calls used in different social situations.

Males tend to be territorial in breeding season when they call and may fight. Females lay 2–3 clutches a year of 1–2 eggs, 9–12mm x 8–10mm, which are placed under stones, in crevices, under dead vegetation or buried in the ground. They hatch in 6–12 weeks producing babies 4–5cm long that can mature in as little as 6 months. Has lived 7 years in captivity.

EUROPEAN LEAF-TOED GECKO *Euleptes europaea*   **Pl.20**

Range Corsica and Sardinia and their offshore islets, the Tuscan archipelago east of Corsica, and offshore islets of Liguria and of south-east France (Gulf of Marseilles to Isle d'Or, including the Iles d'Hyères). Known from a few mainland localities close to the sea in Tuscany (including Monte Argentario), Liguria and south-east France. Also islands off northern Tunisia. Map 88.

Identification Adults up to about 8cm including tail, but usually about 6cm; the smallest European gecko. Head broad and flat, body rather long so the limbs appear fairly short; no tubercles on back; intact tail may sometimes be quite thick, especially in females, and is often very swollen when regenerated. Toes have flattened adhesive pads only at their tips (Pl. 20). Colour rather variable: often brownish or greyish with yellowish marbling or spotting, or predominantly yellowish.

Variation Colouring and size varies considerably between small islands.

Similar Species Other European geckos all have tubercles on the back.

Habits A secretive, mainly nocturnal gecko that is largely restricted to rock surfaces, being found on rock exposures, drystone walls and boulders, especially granite ones. Occurs up to 1,500m but is less common at high altitudes and usually confined there to areas that are close enough to the sea not to get too cold. Avoids cultivated regions, high maquis, woods and the floors of narrow valleys. Not often associated with people although sometimes found in outbuildings and abandoned houses and, more rarely, occupied ones. Spends a lot of time in narrow crevices and can be very abundant in these. On offshore islands, densities of 200 animals per square metre have been recorded under loose flakes on granite boulders. When in crevices, the geckos benefit on cool nights from the way rock retains heat from the sun long after sunset.

Occasionally found under loose bark of dead trees and logs in degraded scrub habitats, olive groves etc. Such animals may move only temporarily into these situations when the nearby rocks they usually inhabit get extremely hot at the height of summer. Call sounds like 'tsi, tsi, tsi'. Very agile and can make long leaps for its size. Often stalks prey very slowly, like a cat, before making a final sprint to catch it

Mature females lay two (occasionally one) eggs at a time which are about 8mm across. 2–3 clutches a year may be produced in lowland areas but only one in the highlands. The eggs are placed in crevices and cemented to the rock. Laying may be communal in traditional sites, where there are often deep deposits of old empty egg shells. Development takes 8–13 weeks and hatchlings are about 3cm long, maturing in about 2–3 years. Has lived 22 years in captivity.

Other names. *Phyllodactylus europaeus*.

Note The largely island distribution, with just a few tiny sites on the European mainland, suggests that this gecko has undergone a relatively recent contraction in range there. See also the Tyrrhenian Painted Frog (p. 63).

KOTSCHY'S GECKO *Cyrtopodion kotschyi* **Pl. 20**

Range South and east Balkans north to Albania, central and east Greece, Macedonia, south and east Bulgaria, Turkey, the Ionian and many Aegean islands including Crete, the south Crimea and south-east Italy (Puglia, Basilicata). Also Asiatic Turkey to Israel, and Cyprus. Map 89.

Identification Up to 10cm including tail; females rather bigger than males. Has characteristic gecko shape but is rather slender with fairly slim tail and limbs and, at a distance, can look almost like a small lacertid lizard. Tubercles are present on back, and tail (if not regenerated); toes lack adhesive pads but have a characteristic 'kinked' shape (Pl. 20). Colour variable: back often grey or grey-brown with dark, pale-edged, V-shaped cross bands, but ground colour can be dark brown, blackish reddish, yellow or orange. Some animals are more or less uniform. Underside often yellow or orange. Males may have 'femoral' pores in front of vent.

Variation Some regional variation in proportion, colour and details of scaling (distribution, number, size and shape of tubercles, number and size of belly scales, and presence or absence of pores in front of vent in males.). At least 17 subspecies have been recognised on this basis, mostly on the Aegean islands.

Similar Species None really similar but has superficial resemblance to other geckos and to some small lacertas.

Habits Typically found in rather dry rocky or stony places: on ground among stones in low dense scrub (phrygana), on dry-stone walls, outsides of buildings, cliffs and sometimes tree-boles. Prefers areas with shale, schist, granite, marble or volcanic rocks. Usually at low altitudes but can reach 1,400m. Does not enter houses very often but tends to be associated with people in north of range. Climbs less, and less high, than Moorish and Turkish Geckos, but very agile in spite of having no adhesive pads. When disturbed, may run onto underside of rocks and cling on upside down, or may retreat into holes, or bases of bushes. Often at least partly nocturnal in summer, but regularly active by day especially in the cooler parts of the year, particularly in the morning and evening. Often abundant where it occurs, and frequently encountered alongside Wall lizards. Common call is 'chick' repeated many times and males and females may call to each other in courtship.

Females lay 2 eggs (occasionally 1), 8–11mm x 7–9mm, that are placed in cracks or under stones. They take about 11–18 weeks to hatch and babies are about 2cm from snout to vent and mature in 2–3 years. Has lived 9 years in captivity.

Other names *Cyrtodactylus kotschyi*, *Tenuidactylus kotschyi*.

Lacertid Lizards
(Family *Lacertidae*)

There are about 240 species of lacertid lizards which occur throughout much of Europe, Africa and Asia. All members of this family are alert, active, diurnal lizards most of which measure less than 8cm from snout to vent, although a few species, including the European Green Lizards (p. 134) and the Canary Lizards (p. 180), grow much larger. Lacertids are relatively long-bodied with well defined heads, long tails and well-developed legs. The top of the head and the belly are covered with large scales and femoral pores are present.

Lacertids are a very important and characteristic part of the European reptile fauna. They make up about 70 per cent of the lizard species in the region (54 out of 79) and up to seven may be found at a single locality. All are active hunters, preying mainly on invertebrates, although some take varying amounts of vegetable food as well. Competition between species occurring together is limited in various ways. For example, adult Green lizards take much larger food than the other lacertids in the same area, and many species tend to hunt in different places. Thus, at one locality, some forms may be ground-dwellers in open areas, others may live in dense vegetation, and some may be climbers on stone piles and rock faces. Humidity may also be important. For instance, in southern Bosnia and neighbouring regions, there are two species climbing on rocky surfaces, the Sharp-snouted Rock Lizard (*Lacerta oxycephala*) and the Mosor Rock Lizard (*Lacerta mosorensis*); both live on cliffs, boulders, screes and rock-pavements, but the Mosor Rock Lizard is often restricted to cooler, moister places than the Sharp-snouted Rock Lizard. Such habitat differences can sometimes be helpful in confirming identifications. Many of the characteristic features of lacertid species are connected with where they live and hunt. For instance, rock species that usually hide in crevices tend to be very flattened, and forms that hunt in vegetation are frequently green.

Male lacertids often fight, or at least display towards each other in the breeding season. Most species have a characteristic threat posture with head tilted down, throat pushed out and body flattened from side to side. This shows off the often brightly coloured flanks and underparts, which may vary between species and can sometimes be important as a means of recognition both for lizards and for lizard watchers. All lacertids lay eggs except most populations of the Viviparous Lizard (*Lacerta vivipara*) which gives birth to live young over most of its range. Clutch size varies. In small species it is often 1–4 eggs but in larger forms the number is greater (up to 23 in some large Green Lizards). Some smaller lacertid species are also relatively large-brooded, the Viviparous Lizard produces up to 11 young and clutches can reach this number, or more in the Iberian Rock Lizard (*Lacerta monticola*), the Large Psammodromus (*Psammodromus algirus*) and the Steppe Runner (*Eremias arguta*). The time taken to hatch is rather variable but may be about six weeks in many small species. The bigger lacertids are often long-lived: the Ocellated Lizard (*Lacerta lepida*) has survived 20 years in captivity and the giant lizards of the Canary islands are capable of living considerably longer.

In general, male lacertids have bigger heads than females, shorter bodies and better developed femoral pores (Fig., below). Also, the base of the tail is often swollen in the breeding season. Young animals have relatively larger, more rounded heads, relatively larger eyes and shorter tails than their parents.

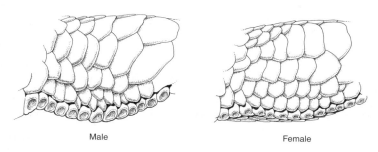

Male   Female

Lacertid lizards,
underside of thighs showing sexual difference in femoral pore size

LARGE PSAMMODROMUS *Psammodromus algirus*   Pl. 21

Range Iberian peninsula (except much of north Atlantic coast) and a small adjoining area of Mediterranean France as far east as the Rhone valley. Also on Conigli island near Lampedusa and in north-west Africa. Map 90.

Identification Adults usually up to 7.5cm from snout to vent but sometimes to 9cm; tail two to three times body length. Relative size of males and females varies from place to place. A thick-necked lizard with a long thin, rather stiff, tail and no collar. Scales on back and flanks are large, flat and pointed with a prominent keel. They overlap strongly, as do the belly scales. Colour fairly constant: usually metallic brownish with two conspicuous white or yellowish stripes on each side, the upper ones bordered above by dark dorsolateral stripes. Flanks are often dark and there may be vague dark stripes on back. Some animals, especially old males, are almost uniform. Males often have one or more blue spots in the shoulder region. Underparts slightly iridescent-whitish or even tinged green. Breeding males have throat and sides of head orange, red or yellow, and flanks and chest may also be yellow; females occasionally show similar head colouring. Babies are like adults, but pale stripes are less obvious and tail is often more orange.

Variation Some minor variation in pattern.

Similar Species Spanish Psammodromus (p. 128), is much smaller, has a weak collar and usually different colouring; also, scales on the sides of the neck are granular or only weakly keeled (Fig., p. 128) Spanish Algyroides (p. 133), which is restricted to a small area of south-east Spain, is smaller and flatter with no light stripes or reddish tail; also large scales on back are blunt and scales on flanks are small.

Habits Occurs up to 2,600m in southern parts of range but most abundant at much lower altitudes where it may reach densities of more than 100 per hectare in favourable habitats. A typical inhabitant of very dense bushy places, although it sometimes occurs in more open areas. Often found in open or degraded woodland, in undergrowth in pine and eucalyptus forest, and among very dense spiny shrubs, dwarf oak, heather, gorse, brambles and even prickly pear (*Opuntia*). Spends most of time around the base of these plants hunting in leaf litter etc., but may climb in bushes and sometimes makes excursions across more open areas. The most abundant lizard species in many parts of the Iberian peninsula, but well-camouflaged and not always conspicuous. Sometimes squeaks, especially when picked up but also at other times. The Large Psammodromus tends to be replaced by the Spiny-footed Lizard and Spanish Psammodrous in more open areas. It has a pocket in the skin on each side of neck where chiggers (the red larvae of trombiculid mites) often accumulate. These mites feed on the body fluids of lizards for some weeks and often attach around the eyes or ears, reducing efficiency of these sense organs and damaging delicate tissues. The pockets are believed to reduce this problem by luring mites into areas where they do less harm. Usual diet is arthropods, with occasional small lizards and vegetation.

Male may hold neck of female in his jaws during mating (most lacertids hold side of body); eggs are laid 2–4 weeks after this. Females lay up to 2–3 clutches a year of 2–11 eggs, 12–14mm x 7–9mm, which hatch in 5–6 weeks producing babies about 2.5–3cm from snout to vent. These mature in 1 or 2 years and may sometimes live 5–7 years.

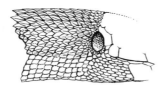

Large Psammodromus
scales keeled

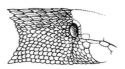

Spanish Psammodromus
scales granular

Psammodromus, scales on sides of necks

SPANISH PSAMMODROMUS *Psammodromus hispanicus*   Pl. 21
Range Iberian peninsula (not much of extreme north), French Mediterranean coast as far east as Var. One or two similar species occur in north-west Africa. Map 91.
Identification Adults usually less than 5cm from snout to vent; tail 1.4–2 times body length. A diminutive unflattened lizard with rather large, keeled and overlapping scales, and a weak collar. Colouring very variable: grey, metallic brown, olive or ochre above. Often a pair of whitish, yellow or greenish lines on each side and other pale stripes on mid back that are often broken up into rows of short streaks with blackish bars or spots between the rows. Some animals quite uniform. Belly whitish or reddish. In breeding males the pale lines on the sides may be intense yellow or green.

Variation Individuals from eastern Spain and France (*P. h. edwarsianus*) often have finer body scaling than those from the central and west Iberian peninsula (*P. h. hispanicus*), the hind feet also tend to be larger, the subocular scale is usually separated from the lip, and the scales on the throat are coarser.

Similar Species Large Psammodromus (see p. 127). Because the Spanish Psammodromus is so small, the large, overlapping scales are not always obvious and it is possible to mistake it for a young small lacerta but the weak collar, only well developed at the sides of the neck, will identify it. If a lens is available, overlapping body-scales and keeling on underside of toes can also be seen.

Habits A lowland ground lizard of mainly Mediterranean habitats found up to 1,600m in south but usually lower. Occurs mainly in dry open situations on firm or sandy soils, especially in areas with a patchy covering of low (often under 30cm high) dense, bushy plants. Here it is inconspicuous and usually glimpsed dodging from one clump to another. It also occurs in more barren habitats such as sand flats and gravel plains with very sparse vegetation, and in this environment often runs quite long distances at high speed. Takes refuge in the twiggy base of plants or buries itself among their roots, or sometimes under stones. Squeaks when picked up, and also more or less spontaneously in breeding season. Eats small arthropods, females and babies taking more flying prey than males.

Copulation is brief, often less than a minute, the male holding the female by the neck as well as the body. Two clutches are often laid, of 2–6 eggs, 9–13mm x 6–8mm, which hatch in around 7–9 weeks. Babies are about 2–2.5cm from snout to vent and mature very quickly, in 8–9 months. Most do not live longer than 2–3 years and many die after their first reproductive season.

SPINY-FOOTED LIZARD *Acanthodactylus erythrurus*   Pl. 21

Range Iberian peninsula, but not much of north or parts of south-west. Related populations occur in north-west Africa. About 32 other species of *Acanthodactylus* are found across the arid and desert regions of north Africa and the Middle East to north-west India; many of these have fringes of long spiny scales on the toes which act like snow shoes when the lizards cross loose sand. These are not obvious in the European species. Map 92.

Identification Adults up to 8cm, from snout to vent; tail often twice as long as body. A moderate-sized, ground dwelling lacertid with an alert, upright stance. Head rather large, with pointed snout; tail very slender, but strongly swollen at base in adult males. Adults very variable in pattern, but usually grey, brown, or coppery with up to ten pale, often weak, streaks, or rows of spots, separated by dark bars or blotches, especially on flanks. Some animals almost uniform. Underparts white, although the base of the tail and parts of the hind legs may be red in females carrying eggs. Juveniles are very distinctive being black with white or yellowish stripes, a bright red tail and sometimes thighs. The red may persist in older animals. Easily identified at close quarters by lack of an occipital scale at the back of the head, only two large supraocular scales over each eye (Fig., right), and an obvious groove on top of snout.

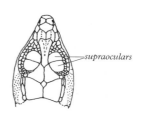

supraoculars

Spiny-footed Lizard

Variation Some variation in dorsal pattern, both within populations and from place to place.

Similar Species Unlikely to be confused with any other European lizard.

Habits A ground-dwelling lizard often found in open sandy areas with sparse, shrubby vegetation, or occasionally in completely bare places, such as beaches or even rocky plains. Occurs up to 1,400 m in south of range but usually under 400m. Can occur at densities of 200 animals per hectare and individual ranges may be about 600 sq. m. Not particularly shy, but when disturbed may run long distances in a series of straight bursts with the tail raised in a gentle curve. Often skulks around bases of open bushes but, when really pressed, retreats into spiny vegetation or into an often short burrow. Has characteristic resting position with forequarters raised. May lift limbs to cool them and reduce heat intake when sitting in hot sun. Juveniles often wave tail slowly when basking in the open. Eats arthropods, especially ants, beetles and bugs; also a small amount of vegetation

Copulating males do not maintain their bite on the body of the female as most other lacertid lizards do; eggs are laid around 20 days afterwards. In coastal areas females may produce two clutches in a year but usually one elsewhere. Each consists of 1–8 eggs, about 15mm x 9mm, which hatch in around 8 weeks, producing babies about 3cm from snout to vent. These mature in 18 months or less at around 6–6.5cm from snout to vent. In the wild, quite short-lived often not surviving its second year.

STEPPE RUNNER *Eremias arguta* **Pl. 21**

Range Eastern Romania (Danube Delta area) to southern Ukraine including Crimea. Also eastwards to Mongolia and China. About 25 other species of *Eremias* are found in central and south-west Asia. Map 93.

Identification Up to about 7.5cm snout to vent, tail as long or a little longer. A quite plump lizard with a pointed snout and rather prominent nostrils. Distinguished from all other east European lacertids by high number of large belly scales (14–20 across mid-belly), and by the subocular scale which does not reach the lip (Fig., p. 112). Usually grey or greyish brown, often with a pattern of ocelli or light, often broken stripes and irregular dark markings; sides may be darker than back.

Variation Considerable variation in pattern. Form in Europe is *E. a. deserti*.

Similar Species None in Europe.

Habits Typically found in dry open places with some low bushy vegetation, including sandy beaches, coastal dunes and sparsely vegetated river plains, where it occurs up to 300m. Often quite common, actively hunting for its mainly insect prey. When disturbed it may run very fast for long distances between bushes. Sometimes takes refuge by sheltering under spiny shrubs but also uses burrows about 10–25cm long that it digs itself, often at the base of bushes. May also hide under stones and in rodent burrows and dive into loose sand when pursued.

Females may lay about 2 clutches of 1–12 (usually 3–4) eggs, 10–20mm x 6–10mm, which produce babies 2.5–3cm from snout to vent; these probably become sexually mature after their second spring.

SNAKE-EYED LACERTID *Ophisops elegans* **Pl. 21**

Range North-east Greek mainland, adjoining southeast Bulgaria, European Turkey, Thasos, and Aegean islands close to coast of Asiatic Turkey (Lesbos, Ikaria, Samos, Patmos, Kalymnos, Rhodes and Karpathos). Outside Europe extends east to Iran and to parts of North Africa. Seven related species occur in dry steppe country of North Africa, the Middle East, and the Indian subcontinent. Map 94.

Identification Usually under 5.5cm from snout to vent; tail about twice body length. A small ground-dwelling lacertid lizard easily identified in the hand by its eyes which lack obvious eyelids. Instead, the eye is covered by a transparent spectacle as in snakes, and cannot be closed, although the spectacle may sometimes be pulled briefly downwards. Also distinguished from other south-east European lacertids by lack of a collar beneath neck. Body scales relatively large, keeled, pointed and overlapping. Colouring variable; often brown, frequently tinged greenish above with two pale, narrow stripes on each side, the upper one stronger, frequently green and often bordered with black lines, spots or blotches. Flanks may be mottled and sometimes greenish; underside usually pale and throat sometimes yellow; tail often reddish. At close quarters has staring 'expression' produced by large eyes and overhanging ridges above them.

Snake-eyed Lacertid

Variation No obvious variation in Europe where the local form is *O. elegans ehrenbergii*.

Similar Species None within European range.

Habits A ground-dwelling lizard usually found on open arid plains, fields, and stony hillsides with sparse grass, crops, or low scrub. May occasionally occur on almost bare ground. Frequently not especially fast but often dodges from plant to plant when pursued and may take refuge in dense vegetation, in crevices in the ground or under stones. Often buries itself in soft soil or sand at night, a slow process that takes place under cover. Pulling the spectacle downwards is a means of cleansing its outer surface of dust. This is done by the muscle that opens the lower eyelid in other lizards and the spectacle is wiped clean against the upper edge of the suborbital scale.

Females lay 2 or 3 clutches of 1–6 (usually 3–4) eggs, 10mm x 6mm which swell to 14–25mm x 8–9mm. Babies are about 2cm from snout to vent (5cm total length) and are usually sexually mature by their second spring. Appears to be relatively short-lived.

DALMATIAN ALGYROIDES *Algyroides nigropunctatus* **Pl. 22**

Range East Adriatic coastal region: extreme north-east Italy and adjoining western Slovenia, west and south Croatia, west Bosnia, Albania and adjoining areas of Serbia and Macedonia, north-west Greece as far as the Gulf of Corinth and the Ionian islands (possibly not Zakynthos). Map 95.

Identification Adults up to 7cm snout to vent; tail usually about twice as long; females rather smaller than males. A small, dark lizard about the size and build of a wall lizard but easily distinguished from these and other Small lacertas by rough appearance of back scales and sombre colouring above. Back scales are large, blunt and strongly keeled while those on the flanks are much smaller (a total of 20–29 across the mid-body). Typically dark grey-brown to reddish-brown above, often with scattered black spots. Adult males have an intense blue throat and eye, and the belly is orange to red, the colour often extending onto flanks.

Variation Animals from Cephaloria and Ithaca have been named as *A. n. kephallithacius*. Males have lemon-yellow bellies and steel-blue throats in the breeding season, and greenish throats at other times. Similar animals are found on Lefkada.

Similar Species Greek Algyroides (below) is only likely to be confused in Ionian islands; back and flank scales are all large, pointed and keeled.

Habits Seen in a wide variety of habitats, but usually in open woods, degraded scrub, on hedges, walls, and bushes between fields and olive groves etc. Found up to 700m and exceptionally to 1,200m in the south of its range. A good climber most conspicuous when perched on pale walls, boulders etc, but also occurs commonly on tree-trunks, such as those of olive trees, and in bushes where it clambers among the leafy twigs, a frequent behaviour in summer. Tends to prefer shady or partly shaded areas and may be rather secretive. Often occurs at higher densities in situations associated with people. On Corfu, where there is no climbing small lacerta to do so, it enters towns and villages and is abundant there.

Copulation is short, around 40 seconds, but males grip their mates with their jaws for up to 20 minutes afterwards. Females lay more than one clutch of 2–6 (often 3–4) eggs, about 11mm x 7mm swelling to 16mm x 12mm. These hatch in about 5–7 weeks, producing babies 2–2.5cm from snout to vent (about 6cm in total length).

GREEK ALGYROIDES *Algyroides moreoticus*   **Pl.22**

Range Southern Greece (Peloponnese) and Ionian islands (Cephalonia, Ithaca, and Zakynthos); Strofades islands. Map 96.

Identification Adults up to 5cm from snout to vent, tail about l.5–2.4 times as long. A small lizard easily distinguished from small lacertas in its range by the large, pointed and keeled scales on its back and flanks (18–25 across the mid-body). Typically dark brown or reddish-brown above; females more or less uniform, but males usually have darker flanks with light spots and often a pale dorsolateral streak on each side. Belly is whitish, sometimes flushed yellowish-green and often with a few black spots; underside of tail and hind legs may be rust red or yellow.

Variation No obvious geographical trends recorded.

Similar Species Dalmatian Algyroides (p. 131).

Habits A rather secretive, inconspicuous lizards and, like other Algyroides, usually prefers semi-shaded, sometimes damp situations, although it basks more obviously in cooler seasons. Quite often encountered on north-facing slopes and near water, frequently in open woodland and clearings. Found up to 1,200m. Occurs in piles of brushwood, in hedge bottoms, among leaf litter at base of bushes etc. May occasionally be seen on tree-trunks and walls where it can be conspicuous, and in cultivated land. Females lay few eggs in each clutch (sometimes 2); babies are 2–2.5cm (around 6cm in total length).

PYGMY ALGYROIDES *Algyroides fitzingeri*   **Pl. 22**

Range Corsica and Sardinia. Map 97.

Identification Adults usually less than 4cm from snout to vent; tail about twice as long. A very small lizard, easily distinguished from young Small lacertas occurring on Corsica and Sardinia by its uniform, often dark, colouring and coarse scaling on back and flanks, the dorsal scales being large, keeled, and pointed (13–23 across mid-body). Head and body rather flattened; tail long and often fairly stout. Normally dull brown, olive brown, grey, or even blackish above, sometimes with a vertebral streak or scattered black spots. Belly of both sexes may be yellowish or orange and the throat whitish; babies may be brown beneath.

Variation No obvious geographical trends recorded.

Similar Species Not likely to be confused with any other species in range.

Habits An inconspicuous species found in a range of habitats, particularly Mediterranean ones at moderate altitudes, although it may occur up to 1,830m on Sardinia and 1,390m on Corsica. Present in forest situations and places with rich vegetation but also in degraded scrub environments and ones with sparse grass, especially on rocky slopes and in mixed habitats including small-scale cultivation. May also occur on dry-stone walls, and sometimes climbs in brushwood or on tree boles (particularly bases of oak and olive). Hides under stones, loose bark etc. and is like other Algyroides in preferring semi-shaded places. It is often seen relatively close to water.

Females lay clutches of 2–4 oval eggs about 8mm long, which hatch in about 11 weeks.

SPANISH ALGYROIDES *Algyroides marchi* Pl. 22
Range South-east Spain in the mountain ranges of Cazorla, Segura and Alcaraz. Map 98.

Identification Adults usually less than 5cm from snout to vent; tail about twice body-length. A rather small, flattened and lightly built lizard with large, weakly-keeled scales on back and small flank scales (a total of 24–31 dorsal scales across the mid-back). Usually coffee-brown, occasionally with a greenish tinge, with very dark flanks and dark sides to the head and tail; back may be uniform or with dark markings that may form a vertebral streak. Throat usually whitish, occasionally yellowish in males; breeding males have a yellow belly, sometimes with a greenish tinge and this colour may occur in females or their belly may be white. Babies tend to be darker than adults, especially their tails.

Variation The throat of males in some populations is blue.

Similar Species Iberian Wall Lizard (p. 151) has small scales on back. Psammodromus species (p. 127) have only a weak collar or none at all, flank scales that are about equal in size to back scales and different colouring.

Habits Typically found in or near woodland, from 700–1,700m, often occurring in quite shady places along small fast-flowing streams that may be seasonal. Also occurs on the overgrown banks of forest tracks and occasionally in more open situations. A shy but agile lizard that climbs well, clambering among boulders, tree-stumps, and fallen branches, hunting and hiding under loose bark and in cracks and even running on the undersides of branches and rocks.

Males may nod rapidly at rivals. Females lay clutches of 1–5 (usually 2–3) eggs, about 11mm x 7mm, which take about 3–6 weeks to hatch, producing babies about 2–2.5cm from snout to vent.

Note Another algyroides, *Algyroides hidalgoi* was described from the Sierra de Guadarrama in central Spain, but has never been recorded since. The description suggests something similar to the Pygmy Algyroides (opposite). Its existence was discounted until the discovery of the Spanish Algyroides in 1958, by which time the original specimen had been lost. As time passes, it becomes increasingly unlikely that an Algyroides really exists in central Spain, but the possibility should still be borne in mind.

Green Lizards
(*Lacerta*)

Green lizards occur in Europe, where they form a characteristic and spectacular part of the European fauna. The Ocellated Lizard is significantly different from the others, but it is convenient to include it here. All are moderate to large lizards, mature animals being at least 7cm from snout to vent and, in one species, sometimes over 20cm. Adults, especially males, are often brilliant green above, or at least on the flanks, and frequently have a greenish or yellow belly (never strong red, orange or blue as in at least the males of many Small lacertas). Babies tend to be very characteristically patterned (see Pl 26) with whitish bellies.

Green Lizards are all very robust lacerteds with a deep, powerful head (usually much larger in males than females), the collar is deeply serrated and the belly scales overlap strongly and have angled sides (Fig., below). The small dorsal scales are strongly keeled (except in the Ocellated Lizard), the supratemporal scales are quite deep and there are usually two postnasal scales (some exceptions, especially in the Sand Lizard). Green Lizards tend to live in and around dense vegetation, but also occasionally in other habitats. At any locality only one, two or very rarely three species are likely to be present. Like many Small lacertas, Green lizards tend to be highly variable but they are usually quite easily identified, except in parts of the Balkans.

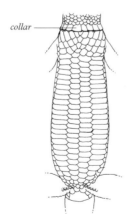

collar

Green Lizard
(collar serrated; belly scales
overlapping, with angled sides)

Sharp-snouted Rock Lizard
(collar smooth-edged; belly scales
scarcely overlapping, with straight sides)

Undersides of lacertid lizards, contrasting types

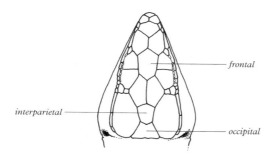

Ocellated Lizard

KEY TO GREEN LIZARDS

1. Iberian peninsula, southern France, and north-west Italy only. Occipital scale very big, nearly always wider than hind margin of frontal (Fig., above); dorsal body scales smooth or feebly keeled, belly scales in eight or ten rows. Often obvious blue spots on flanks of adults. Very large, may grow to over 20cm from snout to vent
    Ocellated Lizard, *Lacerta lepida* (p. 136, Pls 23 and 26)
Occipital scale much narrower than frontal; dorsal body scales strongly keeled, belly scales in six or eight rows. Rarely much over 16cm snout to vent, usually less   **2**

2. Usually under 9cm (rarely to 10cm) from snout to vent. A stocky lizard with relatively short legs and tail, and relatively stubby feet. If green pigment is present, it is usually confined to flanks and only occasionally extends to mid-line. Usually a band of distinctly narrow scales along centre of back. Rostral scale small and does not normally enter nostril; often only one postnasal scale
    Sand Lizard, *Lacerta agilis* (p. 140, Pls 25 and 26)
Adults over 9cm from snout to vent, often considerably so. More elegant than Sand Lizard, usually with longer tail and longer, narrower toes. If green pigment present, then usually on back, and may cover whole body. Scales on centre of back not, or only slightly, narrowed. Rostral enters nostril in many cases; often two postnasal scales   **3**

3. West, north-west and central Iberian peninsula only. Usually eight rows of belly scales; occipital scale often distinctly broader than interparietal (especially in adults). Belly spotted with black in males and some females; no obvious light spots on top of head. Never well-defined, narrow light stripes in pattern. Babies have light, black-edged ocelli or bars, especially on flanks
    Schreiber's Green Lizard, *Lacerta schreiberi* (p. 137, Pls 23 and 26)
Many regions, but only the north of Iberian peninsula. In this area separated from Schreiber's Green Lizard by following characters: usually six rows of belly-scales (rarely eight); occipital scale often narrower than interparietal; belly rarely spotted with black; light spots often present on top of head.

Sometimes well defined narrow light stripes in pattern. Babies without
black-edged ocelli or bars

Most areas: Green Lizard, *Lacerta viridis* (p. 138, Pls. 24 and 26)
Balkan peninsula: Green Lizard, *Lacerta viridis* or Balkan Green
                                 Lizard, *Lacerta trilineata* (p. 139, Pls 24 and 26)

OCELLATED LIZARD *Lacerta lepida*   **Pls 23 and 26**

Range Iberian peninsula, south France, extreme north-west Italy. Similar, related
species occur in north-west Africa. Map 99.

Identification Usually to 20cm from snout to vent but sometimes to 26cm; tail 1.5
to over twice body length and huge animals with total lengths of over 80cm have
been reported. The largest European lacertid, and size alone will identify most
adults; males also have characteristic massive, broad head. Frequently, prominent
blue spots present on flanks. Dorsal ground colour typically green, but sometimes
grey or brownish, especially on head and tail. Ground colour usually overlaid with
black stippling that may often form a bold pattern of interconnected rosettes. Belly
and throat usually yellowish or greenish. Females tend to be less brightly coloured
than males and the blue spots are fewer or absent.

Subadults might be confused with adult Schreiber's Green Lizard or Green
Lizard, but tend to be flatter-bodied with different patterning. Also the occipital scale
is much wider (Fig., p. 135), dorsal scales are smaller (over 55 across mid-body and
sometimes as many as 90) and not strongly keeled, and there are often ten rows of
belly scales. Babies are usually green, grey or brown with characteristic pattern all
over the body of yellow or whitish ocelli that in most cases have black edges sometimes fusing into cross-bands; light spots may also occur on limbs and on the tail.
Uncommon juvenile animals without light spots or ocelli can be recognised by scale
characters.

Variation Animals from south-east Spain (*L. l. nevadensis*). are less richly coloured
than the more widespread *L. l. lepida*, being more uniformly greyish without much
black stippling, the blue spots on the flanks are reduced and the belly is whitish.
Babies in this area may also be fairly uniform. The head tends to be more pointed
than elsewhere and the average number of large scales across the mid-belly is lower.
These populations are probably a separate species, *L. nevadensis*. Populations from
west Galicia and extreme north-west Portugal are sometimes distinguished as *L. l.
iberica*. Animals from the north-west of the Iberian peninsula have relatively small
even teeth and these become steadily bigger and more irregular towards the south-
east. This trend appears to be related to the problems of eating increasingly more
robust beetles.

Similar Species Schreiber's Green Lizard (opposite); Green Lizard (p. 138).

Habits Found in wide variety of uncultivated and arable habitats from sea-level up
to 1,000m in Alps and Pyrenees, and 2,100m in south Spain, although less common
at higher altitudes. Prefers rather dry, bushy places with lots of refuges, such as open
woodland and scrub, old olive groves and vineyards, road banks etc., but sometimes
found on more open rocky or sandy areas. Occasionally seen basking on the edges
of roads. Often very noisy as it crashes through undergrowth. Largely a ground
lizard, but climbs well on rocks and even in trees. Usually takes refuge in bushes
(often thorns), stone piles, dry-stone walls, rabbit burrows and holes it digs itself.
When cornered opens mouth and hisses and may jump towards enemies. Feeds
mainly on large insects especially beetles, but also robs birds' nests in spring and

occasionally takes vertebrates such as Worm lizards and other lizards, small snakes, frogs, and small mammals including even baby rabbits. It also eats fruits and other vegetable matter, especially in dry areas where prey is less abundant.

Males are territorial in the spring and fight with rivals. Females lay 2.5–3.5 months after copulation producing clutches of 5–22 eggs, about 20mm x 14mm, which are deposited under stones and logs or in leaf litter or loose damp soil. The number of clutches a year varies geographically, for instance there is usually one in some areas and two in others. Fewer larger eggs are laid in dry areas such as the south-east of the Iberian peninsula. Eggs hatch in 8–14 weeks and babies are 4–4.5cm from snout to vent on hatching and can become sexually mature at the beginning of their third year. The Ocellated Lizard has been known to live 20 years in captivity.

Other names. Eyed Lizard. *Timon lepidus*

SCHREIBER'S GREEN LIZARD *Lacerta schreiberi*   Pls 23 and 26

Range North-west, west and central areas of the Iberian peninsula and isolated areas further south. Map 100.

Identification Usually less than 12cm, from snout to vent but can reach 13.5cm; tail about twice body length; males rather smaller than females. Has typical green lizard shape and males are predominantly green with small black spots which tend to be larger on the back than on the flanks. Females more variable: usually brownish or greenish, or with greenish flanks or back; often a band of irregular usually large dark spots along mid-back, and flanks also heavily spotted. Belly heavily spotted with black in males; not or less so in many females. Throat often white or cream but blue in breeding males and some females and this colour may extend to whole head. Young are strikingly coloured: dark brown or blackish with white or yellow, black-edged spots or bars, especially on sides; tail often orange or yellow. The pale spots may be retained in subadults.

Variation Females especially are quite variable. No subspecies recognised, but there are minor differences between populations and western and northern animals are genetically distinct from more central ones.

Similar Species Ocellated Lizard (opposite). Green Lizard (p. 138), from which Schreiber's Lizard can be distinguished in north-west Spain by following characters: distinctive range of patterns; never clearly defined, narrow, white stripes on back; males without light spots on top of head; often dark-spotted belly; eight (rarely 6 or 10) rows of belly scales; occipital scale often distinctly wider than interparietal.

Habits Found mainly in comparatively moist, hilly areas and may reach altitudes of 2,100m, but not in north of range. Lives in typical green lizard habitats often with lots of bushy vegetation, such as wood edges, open especially broad-leaved woods and clearings in them, overgrown roadside walls and banks, bramble patches and even pasture; regularly occurs close to water. Where Schreiber's Lizard is found with the Ocellated Lizard this is generally in drier more open areas. In north-west Spain, the Green Lizard sometimes replaces Schreiber's Lizard at higher altitudes. Diet consists largely of arthropods and includes significant numbers of flying insects such as mosquitos and flies; beetles and grasshoppers are also taken as well as occasional small reptiles.

Females produce 11–18 (occasionally 7–24) eggs, 12–18mm x 8–12mm, which are laid in sunny soil or sometimes mammal burrows and hatch in 8–13 weeks. Babies are 2.5–3.5cm from snout to vent. Males mature at beginning of third year and females in their fourth or fifth year.

GREEN LIZARD *Lacerta viridis*   **Pls 24 and 26**

Range Much of the southern half of Europe extending north to most of France, the Channel Islands, west and south Switzerland, south and east Austria, parts of the Czech Republic and Slovakia and southern Ukraine including the Dneiper Valley. Also isolated populations in the Rhine valley. Extends south to north Spain, Sicily, and north and central Greece. Not known from many Mediterranean islands, but present on Elba, Corfu, Euboa, Thasos, Skiathos and Samothraki. Populations in extreme east Germany and adjoining Poland are now extinct. Map 101.

Identification Adults up to about 13 cm, from snout to vent; tail often twice body length or more. A large, elegant lizard with a rather short, deep head, especially in males. Males usually almost entirely green with a fine black stippling above and a darker, light-spotted head. Females very variable: sometimes uniform green or brown, or with blotches; frequently two or four narrow light stripes on body, which may be edged with black lines or spots. More rarely, both sexes have a bold irregular pattern of interconnected black blotches. Belly yellowish, nearly always without black spots. Throat blue in mature males and some females. Young often beige, uniform or with a few light spots (without black edge) on flanks, or else with two or four narrow light lines.

Variation Populations vary in body size, colouring and minor details of scaling and a number of subspecies have been recognised. In Sicily and extreme southern Italy (*L. v. chloronota*) breeding males have blue on the sides of head and further subspecies have been described in other areas of the Italian peninsula. In populations from the Black sea area of Romania, Bulgaria and European Turkey (*L. v. meridionalis*) males are often uniform green, without black stippling, and in this region the hind legs and tail may often be uniform brownish. Further north in Europe, eastern animals are assigned to *L. v. viridis* while those from the Rhine area, France, northern Italy and Spain are named *L. v. bilineata*. This last unit is often treated as a full species, the Western Green Lizard, (*L. bilineata*), because it is apparently not fully fertile with *L. v. viridis*. All-black specimens of the Green Lizard are known.

Similar Species All other members of the Green Lizard group, but especially the Balkan Green Lizard (opposite) and Schreiber's Green Lizard (p. 137). Overlaps extensively with the Sand Lizard (p. 140) but usually easily distinguished.

Habits Typically found in and around dense bushy vegetation with good exposure to sun, for instance in open woods, hedgerows, wood and field edges, bramble thickets, on overgrown embankments etc. In south of range, often confined to damp situations or to highland areas, where it may occur up to 2,200m. In north sometimes found in heath areas, provided bushes etc. are present. Often quite common and can occur at densities of 200 per hectare. The Green Lizard hunts and climbs in dense vegetation but comes out to bask, especially in morning and evening. When pressed it takes refuge in bushes, rodent burrows, crevices etc. Food consists mainly of invertebrates, but fruit and eggs and nestlings of small birds are also eaten at times.

Females lay 6–23 eggs in a clutch, 13–20mm x 8–12mm, which hatch in about 7–15 weeks, producing babies 3–4cm from snout to vent (7–9cm total length). Animals may become sexually mature in their second spring when females are about 8cm from snout to vent.

BALKAN GREEN LIZARD *Lacerta trilineata* **Pls.24 and 26**

Range Balkan peninsula north to north-west Croatia, southern Bosnia, Albania, Macedonia, Bulgaria and south-east Romania; present on the Ionian and many Aegean islands including Crete. Also Asiatic Turkey. Map 102.

Identification Adults up to 16cm or more, from snout to vent; tail twice body length or more. A large Green lizard with the typical 'jizz' of the group. Often like a big version of the Green Lizard. Adults fairly uniform bright green (more rarely yellow or brownish) generally with fine black stippling on back. Babies and subadults usually brown (babies frequently dark), often with three or five narrow light stripes or a few light spots on flanks, or dark spots on midback. The Balkan Green Lizard and Green Lizard can be very difficult to distinguish at times, especially as both are very variable. The Table on p. 140, together with remarks under **Variation**, should identify most individuals.

In places where identification of adults is difficult, it helps to look out for striped babies or subadults; usually animals with three or five stripes (including one in the centre of back, which may be faint) are Balkan Green Lizards, those with two or four are Green Lizards.

Variation In east Adriatic coastal areas (*L. t. major*) animals have all the distinctive features of the Balkan Green Lizard. Those from south-east Romania and coastal Bulgaria (*L. t. dobrogica*) differ in being small with heads shaped like Green Lizards, and the throat and sometimes sides of the neck blue in males. In central and southern Greece (*L. t. trilineata*) the species is quite variable and sometimes very like the Green Lizard: adults may have a blue throat, and sometimes six rows of belly scales and rather few temporal scales, while both striped and unstriped young may occur within the same locality; the presence of at least some young with three or five stripes should identify most populations in the north of this area where the Green Lizard occurs. In the Greek islands, Balkan Green Lizards are quite variable. Those on Andros and Tinos (*L. t. citrovittata*) have a light green back, grey-green flanks with yellow stripes, a bluish throat and whitish belly often with dark spots. On Milos, Kimolos, Sifnos and Serifos (*L. t. hansschweizeri*) lizards lack bright green colouring, being brown or yellow brown. Animals from Crete (*L. t. polylepidota*) have a blue throat and are small with fine scaling. On islands close to Asiatic Turkey (*L. t. cariensis*) lizards are larger but again blue-throated with fine scaling, the last feature being especially developed on Rhodes (*L. t. diplochondroides*). There is usually no identification problem in the southern Ionian and most Aegean islands as the Green Lizard is absent.

It is probable that the Balkan Green and Green Lizards may occasionally hybridise. All-black Balkan Green Lizards occur occasionally.

Similar Species Green Lizard (opposite).

Habits Similar to Green Lizard but tends to be found in warmer, drier places and is largely confined to areas with a Mediterranean climate. A lowland form in the north of its range but extends up to 1,500m in the south. Where the two species occur close together, the Balkan Green Lizard is often replaced by the Green Lizard in moister habitats and at higher altitudes. However, the Balkan Green Lizard sometimes lives near water (for instance in east Romania and parts of southern Greece) and may even be seen swimming across shallow streams and ditches. Like the Green Lizard it is often found in bushy places, although the young may live in clumps of low herbage. Also occurs on overgrown sand-dunes, dry-stone walls, and in ruins on arable land.

Females lay 6–20 (usually 7–14, rarely up to 30) eggs, 14–22mm x 9–14mm, which hatch in about 11–14 weeks producing babies 3.5cm–5cm from snout to vent.

	L. trilineata	*L. viridis*
Rows of belly scales	Usually eight	Usually six
Head of male	Snout typically narrow, but back of head broad. Top of head often with light vermiculations	Snout typically blunter, back of head not very wide. Top of head often with small light spots
Supracillary granules	Often a continuous line of granules	Often few, or even none
Temporal scales	Numerous (often over 20, sometimes less)	Fewer (usually less than 20, often less than 15)
Rostral scale reaches nostril	Usually	Often not
Throat in males	Often yellow	Often blue

Green Lizards in south-east Europe, features to check. Temporal scales are dotted.

SAND LIZARD *Lacerta agilis*   **Pls 25 and 26**

Range Most of Europe north to southern and north-west England and southern Scandinavia, but rare or absent in much of west and south-east France, and from Italy, European Turkey, most of Greece and nearly all of the Iberian peninsula. Also eastwards to central Asia and Mongolia. Map 103.

Identification Most adults up to about 9cm, from snout to vent, but usually smaller and rarely larger; tail about 1.3–1.7 times body length. A short-legged, stocky lizard with a very short deep head (especially in males) and usually a distinct band of narrowed scales along the mid-back. Not very like other Green Lizards in build and, with experience, easily distinguished from them. Colouring extremely variable; usually a dark band or series of marks along centre of back that is often complex and may contain darker blotches, a light central streak (often broken up) and light spots. In most animals ocelli, dark spots, or mottlings are present on sides but these are separated from back band by an unmarked area. Males have green, yellow-green, or greenish flanks (very intense when breeding) and some may be almost entirely green, except for a dullish central band. Less commonly they have large ocelli all over the back. Females are grey or brown, rarely with green, and the dark central band tends to be broken up. Some specimens of both sexes have the entire back brown or reddish. Underparts whitish, greenish or yellow, often with dark spots which are usually very numerous in males.

Young often have a weaker version of adult markings (but no green) and frequently prominent ocelli as well, especially on flanks.

Variation Colouring very variable. All-black specimens are known and so are females with green flanks. A number of subspecies have been described mainly on the basis of pattern and minor details of scaling, especially that of the snout and vent region. Most west European animals as far east as Denmark and the west Tyrol in Austria are referred to *L. a. agilis* although ones in the Pyrenees are sometimes distinguished as *L. a. garzoni*. Populations from further east (*L. a. argus*) are rather small and include red-backed individuals. Balkan animals (*L. a. bosnica*) often have a continuous narrow pale stripe along the centre of the back. Males from Romania to west Ukraine (*L. a. chersonensis*) may be almost entirely green. East of the Dneiper river including the Crimea (*L. a. exigua*) animals often have three clear pale stripes on body or are virtually unmarked, and the band of narrow scales along the mid-back is less pronounced. Several of these named forms intergrade.

Similar Species Viviparous Lizard (p. 144) and Green Lizard (p. 138).

Habits Occurs up to 2,000m in south, but a lowland species in north of range. Largely a ground lizard and, in many areas, lives in a wide variety of usually fairly dry habitats such as meadows, steppe, field-edges, road embankments, grassland with occasional low bushes, rough grazing, hedgerows, and even in crops and gardens. Can also occur in moister situations. In north, and in England, it is more restricted and is usually found on coastal sand-dunes with some plant cover and on sandy heaths (hence its English name). On heathland tends to live in dense old heather stands, in which it clambers and is very inconspicuous. In southern regions, it is partly montane and occurs in upland pastures and alpine situations. Like other Green Lizards usually found in or near at least some dense vegetation, but this is usually lower and sometimes less extensive than that required by the other species. May occur at densities of 10–300 animals per hectare in north of range. More intelligent than many lizards and capable of changing behaviour rapidly in response to circumstances.

Males threaten and fight each other vigorously in the breeding season and guard their mates after copulation, although both sexes often ultimately have several partners. Females lay 4–14 eggs (often 5–6, and more in large animals) which in north and central Europe are buried in sandy ground exposed to the sun, the mother sometimes looking after the nest for some time. Single clutches are common in cool areas but two per year may be laid elsewhere. Eggs usually hatch in 7–12 weeks and are 12–15mm x 7–10mm when first laid, swelling to about 20mm x 15mm. Babies are about 2–3.5cm from snout to vent; in northern Europe. Males often mature at 2 years (sometimes 1) and females at 3 at around 7–8cm from snout to vent. May live 12 years in the wild.

Small Lacertas
(*Lacerta* and *Podarcis*)

There are 32 species of lizard in Europe that are best called Small lacertas. They fall
into two distinct groups: Wall lizards (*Podarcis*) and the rest, namely Rock, Meadow
and Viviparous lizards (parts of *Lacerta* as presently understood; Rock lizards have
sometimes been called Archaeolacerta). In fact Wall lizards are generally very similar to each other and all appear to be descended from a single exclusive ancestral
species, whereas the other Small lacertas are often more different, not always closely related and share their common ancestor with a range of other lacertid lizards. The
two groups differ in many important internal features but are often almost impossible to tell apart when alive so it is easier to treat them as a single unit.

Small lacertas are one of the most difficult groups to identify with certainty. Many
of the species are closely related and very similar in appearance; at the same time
they show great variation both within populations and from area to area. This makes
it very difficult to produce a single key that will identify them all. Fortunately many
of them have restricted ranges and usually not more than a few occur at any one
locality. It is therefore easiest to deal with them by regions as follows.

Area

1 North, western and central Europe	p. 144
2 Iberian peninsula and Mediterranean France; Madeira	p. 147
3 Balearic Islands	p. 155
4 Corsica and Sardinia	p. 158
5 Italy, Sicily and Malta	p. 161
6 East Adriatic coastal area	p. 165
7 South-eastern Europe	p. 171

Their boundaries are shown in the figure opposite.

Identification within these areas often depends on examining details of scaling,
colour and pattern. The terms used for these features are shown in Figs. on pp. 111
and 112.

Note Most Small lacertas may have conspicuous blue or black spots on the outer
row of belly scales, or both. Because it is so widespread, this feature is usually not
very helpful in identification and has been omitted from most species texts. In the
descriptions, *streak* and *stripe* may include not only continuous marks but also rows
of spots or blotches. Small lacertas are a group where 'jizz' is very helpful in identifying difficult animals, once sufficient experience is gained, but do not place too much
reliance on it at first.

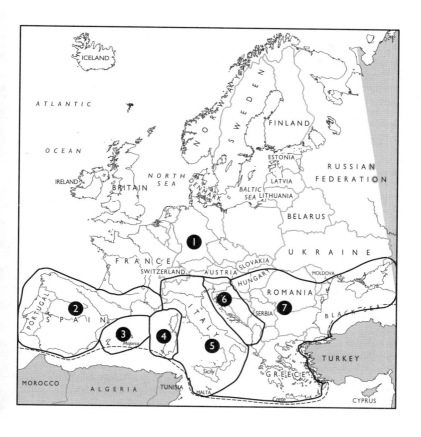

SMALL LACERTAS AREA 1:
North, western and central Europe

Although this area includes the greater part of Europe, it contains only two native small lacertas: the Viviparous Lizard and the Common Wall Lizard. These are the two most widespread members of the group and they are also likely to be encountered in other mainland areas (2, 5, 6 and 7). The only other native lacertids in area 1 are members of the Green Lizard group.

KEY

Scarcely flattened, short-legged and small-headed. Collar distinctly serrated; scaling coarse (25–37 dorsal scales across mid-body). Granules between supraocular and supraciliary scales absent or at most four on each side; 5–15 femoral pores present

Viviparous Lizard, *Lacerta vivipara* (below, Pl. 27)

Rather flattened, legs fairly long and head quite large. Collar not usually distinctly serrated; scaling fine (42–75 dorsal scales across mid-body). Granules between supraocular and supraciliary scales extensive, usually at least five on each side; 13–27 femoral pores

Common wall Lizard, *Podarcis muralis* (p. 145, Pl. 27)

VIVIPAROUS LIZARD *Lacerta vivipara*   **Pl. 27**

Range Most of Europe including Arctic Scandinavia, Britain and Ireland, but absent from the Mediterranean area: extends south to north Spain, north Italy, and Macedonia and south-west Bulgaria. Also through much of north Asia to Pacific coast and Sakhalin island. Map 104.

Identification Up to about 6.5 cm from snout to vent; tail 1.3–2 times as long. A long-bodied, almost unflattened, short-legged lizard with a small, rather rounded head and thick neck and tail. Collar distinctly serrated and back scales very coarse and usually keeled, only 25–37 across mid-back. (See key, above, for other distinctive features.) Pattern very variable: most animals basically brown but may be grey or olive. Females usually have dark sides and a vertebral stripe, often a number of light streaks (especially dorsolateral ones) and sometimes scattered light or dark spots or ocelli. Ocelli are frequently better developed in males, which often lack a continuous vertebral stripe. Underside with many dark spots in most males and in some females. Throat whitish or bluish; belly white, yellow, orange, or red. Young very dark, almost blackish-bronze.

Variation Great variation in pattern and details of scaling even within populations. Occasional individuals are yellow, greenish or black and, while some animals are fairly uniform, others are very heavily striped or ocellated. Populations from south-eastern Europe (north to east Austria and Hungary) are sometimes separated as *L. v. pannonica*.

Plate 1 SALAMANDERS 1
(x 2/3)

All these forms have very large paratoid glands

1. **Fire Salamander** *Salamandra salamandra*   **31**
   Usually brilliant black and yellow to black and reddish above; pattern very variable.

   1a. Spotted individual, eastern Europe (*S. s. salamandra*); spotted patterns also occur elsewhere.
   1b. Striped individual, Pyrenees (*S. s. fastuosa*); a degree of striping is common in France and neighbouring areas.
   1c. North-west Spain and Portugal (*S. s. gallaica*)

2. **Corsican Fire Salamander** *Salamandra corsica*   **33**
   Corsica only.

3. **Alpine Salamander** *Salamandra atra*   **33**
   Alps, mountains of east Adriatic coast to Albania.

   3a. Most populations; totally black (*S. atra atra*).
   3b. North-east Italy (Vicenza) only; varying amounts of pale colouring on back (*S. a. aurorae*).

4. **Lanza's Salamander** *Salamandra lanzai*   **34**
   Alps of South-east France and north-west Italy. Differs from Alpine Salamander in rounded tail-tip.

fastuosa   *almanzoris*   *terrestris*

Fire Salamanders, some pattern variations in Southwest Europe

1

1a

1b

1c

2

3a

3b

4

Plate 2   SALAMANDERS 2
(x ⅔)

1. **Golden-striped Salamander** *Chioglossa lusitanica*   **35**
   North-west Spain and Portugal only. Very slender.

2. **Spectacled Salamander** *Salamandrina terdigitata*   **35**
   Peninsular Italy only. Dark; underside of tail and legs bright red. Only four toes on hind feet.

   2a. Animal exposing bright 'warning' tail colour.
   2b. Underside.

3. **Ambrosi's Cave Salamander** *Speleomantes ambrosii*   **51**
   Toes stubby and partly webbed as in other cave salamanders. South-east France and north-west Italy (Liguria).

4. **Monte Albo Cave Salamander** *Speleomantes flavus*   **52**
   North-east Sardinia. Large.

5. **Gené's Cave Salamander** *Speleomantes genei*   **52**
   South-west Sardinia. Generally dark.

6. **Scented Cave Salamander** *Speleomantes imperialis*   **52**
   South-east Sardinia. Characteristic aromatic smell.

7. **Italian Cave Salamander** *Speleomantes italicus*   **51**
   Northern peninsular Italy.

8. **Supramontane Cave Salamander** *Speleomantes supramontis*   **52**
   East Sardinia.

2

Plate 3   NEWTS and SALAMANDERS
(1–5 x ⅔; 6 x ½)

1. **Sharp-ribbed Newt** *Pleurodeles waltl* **36**
Spain and Portugal only. Large (up to 30cm). Very flat head. A row of warts on each flank through which the ribs may project.

2. **Pyrenean Brook Newt** *Euproctus asper* **37**
Pyrenees only. Very rough skin. Dull ground colour, sometimes with lighter markings.

   2a. Female: pointed cloacal swelling.
   2b. Male: rounded cloacal swelling.

3. **Sardinian Brook Newt** *Euproctus platycephalus* **39**
Sardinia only. Toes not very stubby; compare with Cave salamanders (Pl. 2), the only other salamanders in the area.

   3a, 3b. Two males showing pattern variation. Males have blunt spurs on hind legs.

4. **Corsican Brook Newt** *Euproctus montanus* **38**
Corsica only. Never strongly contrasting black and yellow: compare with Corsican Fire Salamander (Pl. 1), the only other salamander in the area. Males have blunt spurs on hind legs.

5. **Luschan's Salamander** *Mertensiella luschani* **34**
South-east Aegean islands, the only salamander in the area. Large paratoid glands; brownish, often with pale spots. Males have a soft 'spike' on upper tail base.

6. **Olm** *Proteus anguinus* **53**
Subterranean waters of east Adriatic coastal area. Elongate with pink feathery gills. Only two toes on hind limbs.

3

1

2a

2b

3a   3b

5

4

6

Plate 4   **LARGE NEWTS**

(x ½)

1. Northern Crested Newt *Triturus cristatus*   **43**
Dark above, belly usually bright yellow or orange, with dark spots, flanks with white stippling, skin often rough. Widespread but not in Italy or south-west Europe

   1a. Breeding male: high spiky crest and pale tail flash.
   1b. Female.
   1c. Underside.

2. Italian Crested Newt *Triturus carnifex*   **44**
Little or no white stippling on flanks, belly often with large spots, skin fairly smooth; females may have yellow vertebral stripe. Italy and neighbouring areas, west Balkans.

   2a. Breeding male.
   2b. Underside.

2b

3

4

3. Balkan Crested Newt *Triturus karelinii*   **44**
Little or no white stippling on flanks, usually rather small spots on belly, throat pale with darker spots, skin fairly smooth and may have a bluish sheen. Balkans except west, also Crimea.

4. Danube Crested Newt *Triturus dobrogicus*   **44**
Little or no white stippling on flanks, brown or even reddish above, belly often reddish-orange and spots may join together to form one or two bands, skin very coarse and head small. Females may have yellow vertebral stripe. Tisza and lower Danube river valleys.

5. Marbled Newt *Triturus marmoratus*   **42**
Green and black above, belly sombre. SW Europe. *T. pygmaeus* similar.

   5a. Breeding male: fairly smooth-edged, barred crest and pale tail flash.
   5b. Female: orange vertebral stripe.
   5c. Underside.

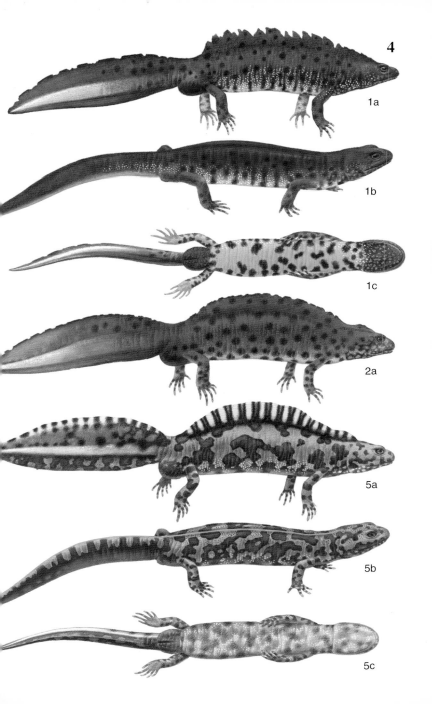

4

1a

1b

1c

2a

5a

5b

5c

Plate 5   SMALLER NEWTS I
(x ⅕)

1. Bosca's Newt *Triturus boscai*   **48**
Portugal and western Spain. Bright orange dark-spotted belly often with pale streaks at sides. May occur with Palmate and Alpine Newts but only in north of range.

  1a. Breeding male: crest only on tail, no toe fringes.
  1b. Breeding female: pointed cloacal swelling.
  1c. Underside.

2. Palmate Newt *Triturus helveticus*   **48**
West and south-west Europe, but not much of Portugal and Spain. Delicately coloured: belly with silvery-orange or yellow streak and weak spots, throat pale, unspotted often pinkish. Non-breeders easily confused with Common Newt.

  2a. Male: low crest; tail filament; dark, webbed feet.
  2b. Female.
  2c. Underside.

3. Common Newt *Triturus vulgaris*   **46**
Not south-west Europe. In west, can be confused with Palmate Newt but usually has more spots on belly and some on throat.

  3a. North and west Europe (*T. v. vulgaris*), breeding male: undulating crest, fringes on hind toes.
  3b. Female.
  3c. Underside of breeding male; spots are often smaller in females.
  3d. Italy and north-west Croatia (*T. v. meridionalis*), breeding male: crest low and smooth, tail ends in filament.
  3e. South Balkan peninsula (*T. v. graecus*), breeding male: small, crest low and smooth with an additional crest on each side of body, tail ends in filament; females heavily spotted.

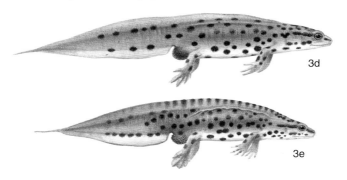

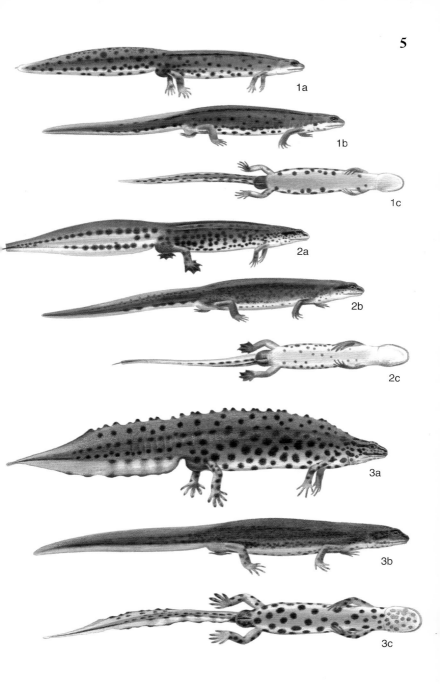

1a

1b

1c

2a

2b

2c

3a

3b

3c

Plate 6   SMALLER NEWTS 2
(x ⅔)

1. Italian Newt *Triturus italicus*   **49**

Southern Italy. Throat yellow or orange; belly paler. Head pattern usually characteristic (see Fig., below). Only likely to be confused with Common Newt (Pl. 5).

1a. Breeding male: crest low; toes unfringed.
1b. Female.
1c. Underside.

Italian Newt

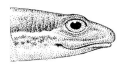

Common Newt

Small newts in Italy, head patterns

2. Alpine Newt *Triturus alpestris*   **44**

Underside uniform deep yellow to red; belly usually unspotted (rare exceptions). No grooves on head. Absent from most of south-west Europe.

2a. Breeding male: smooth crest with dark spots.
2b. Female.
2c. Neotenous individual (common in north-west Balkans).
2d. Underside: throat may be spotted.

3. Montandon's Newt *Triturus montandoni*   **45**

Underside yellow to reddish-orange, often with spots especially at sides. Grooves on head. Carpathian and Tatras mountains. Only likely to be confused with Alpine and Common Newts

3a. Breeding male: low crest, toes unfringed, body section very square.
3b. Female.
3c. Underside.

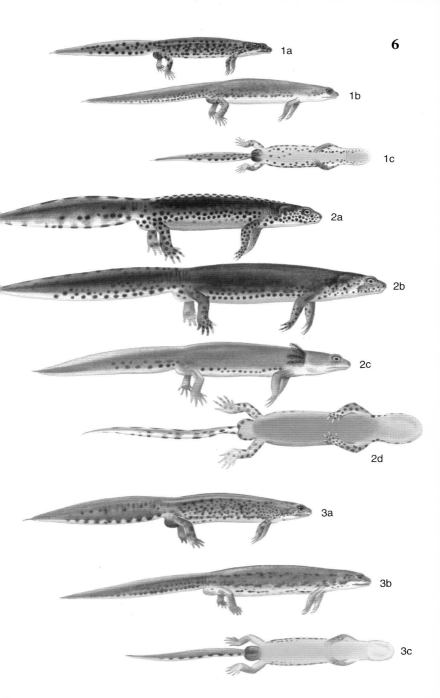

1a

1b

1c

2a

2b

2c

2d

3a

3b

3c

Plate 7 **PAINTED FROGS**

(x ¾; 4 x ½)

1. Painted Frog *Discoglossus pictus* **62**

Sicily, Malta, Mediterranean area of north-east Spain and adjoining France. Rather like Typical frogs (Rana, Pls.12, 13 and 14) but pupil not horizontal. West Iberian Painted Frog (*D. galganoi*, p. 62) and East Iberian Painted Frog (*D. jeanneae*, p. 63) are similar

1a, 1b. Colour variants: may be spotted or striped.

2. Tyrrhenian Painted Frog *Discoglossus sardus* **63**

Corsica, Sardinia, Iles d'Hyéres etc. Often a pale blotch on back; never striped but may be uniform.

3. Corsican Painted Frog *Discoglossus montalentii* **64**

Corsica only. Snout blunt with horizontal top (Fig., below); tip of fourth finger broader than its base.

Tyrrhenian and Corsican Painted Frogs, differences in snout profile

4. Clawed Toad *Xenopus laevis* **25**

Introduced in to Britain from South Africa. Smooth skin, eyes on top of head, very fully webbed feet.

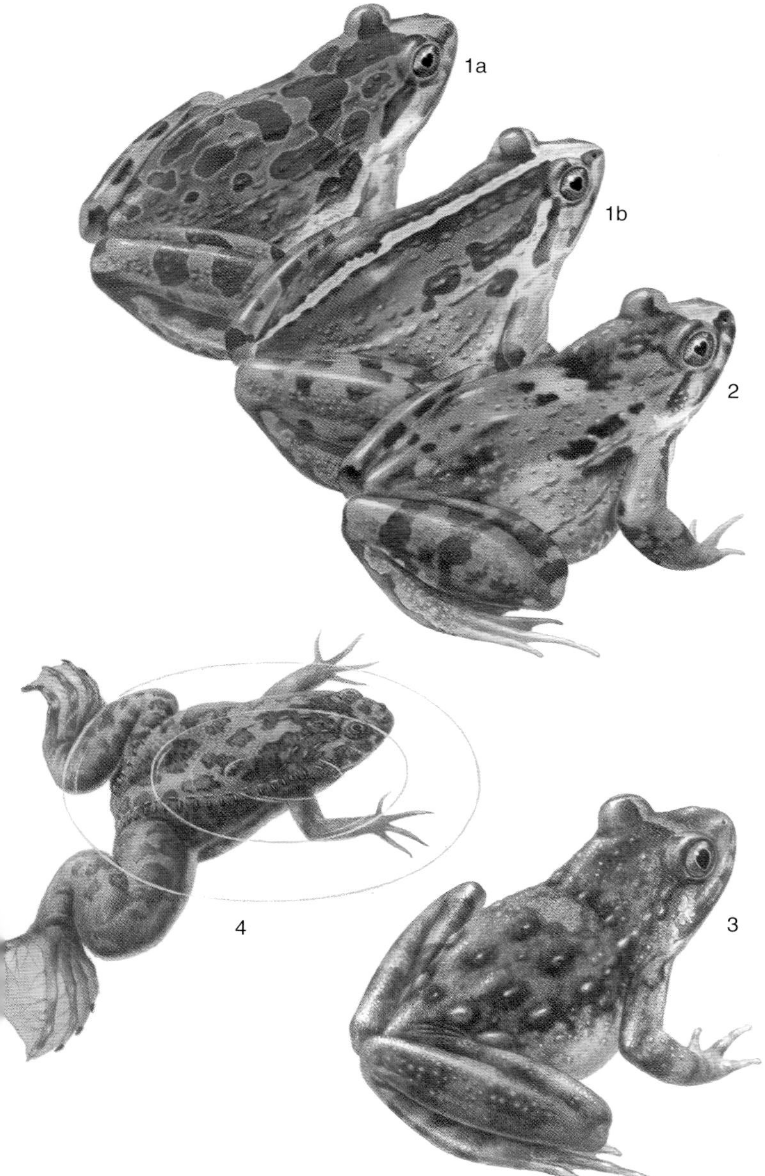
1a
1b
2
4
3

Plate 8   PARSLEY FROGS and MIDWIFE TOADS

1. **Parsley Frog** *Pelodytes punctatus*   **71**
   South-west Europe only. Small (up to 5cm), slender, long- legged; vertical pupil; green or olive spots on back. Iberian parsley Frog (*P. ibericus*, p. 72) is similar but tubercles under the base of the fingers are conical (Fig., below)

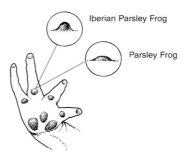

Hand of Parsley frog showing differences in tubercle shapes

2. **Common Midwife Toad** *Alytes obstetricans*   **64**
   Western Europe only. Small (up to 5.5cm), plump, short legged. Vertical pupil. Males may carry eggs. Three tubercles on hand (Fig., p. 66).

   2a. Male with eggs: typical northern animal (*A. o. obstetricans*).
   2b. Female: north-east Spain (*A. o. almogavarii*).
   2c. Male: north and west Spain, Portugal (*A. o. boscai*)

3. **Iberian Midwife Toad** *Alytes cisternasii*   **66**
   West and central Iberian peninsula only. Like Common Midwife Toad but only two tubercles on hand (Fig., p. 66).

4. **Southern Midwife Toad** *Alytes dickhelleni*   **66**
   Southern Spain only. Separate from Common Midwife Toad by range.

5. **Majorcan Midwife Toad** *Alytes muletensis*   **67**
   Majorca only. Slender.

8

Plate 9

SPADEFOOTS

(x ⅔)

Plump toads with big eyes and vertical pupils; a prominent 'spade' on the hind foot.

1. **Western Spadefoot** *Pelobates cultripes*   **68**
   South-west Europe only. Spade black. Two animals, showing variation in
   pattern.

2. **Common Spadefoot** *Pelobates fuscus*   **69**
   Not most of south Europe. Spade pale; a lump on top of head, behind
   eyes. Extensive webbing on hind feet. Two animals, showing variation in
   pattern.

3. **Eastern Spadefoot** *Pelobates syriacus*   **70**
   East and southern Balkans only. Spade pale. No lump on head. Webbing
   on hind feet indented (see Fig., below).

Common

Western

Eastern

Spadefoots, hind feet

9

Plate 10   TYPICAL TOADS
(x ⅔)

Often large toads with warty skins, large paratoid glands and horizontal pupils.

1. **Common Toad** *Bufo bufo*   73

   Large; up to 15 cm. Rather uniform. Paratoid glands oblique (Fig., below). Eye often copper coloured.

   1a. Southern female. (*B. b. spinosus*)
   1b. Northern male. (*B. b. bufo*)

2. **Natterjack** *Bufo calamita*   74

   Not south-eastern Europe or Italy. Usually a yellow stripe on back.

3. **Green Toad** *Bufo viridis*   76

   Not south-west Europe. Distinctive marbled pattern, especially in females.

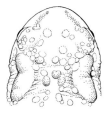

Common Toad
oblique

Natterjack and Green toads
parallel

Typical toads, paratoid glands

Natterjack male, calling

10

1a

1b

2

3

Plate 11 FIRE-BELLIED TOADS

(x ⅔)

Small, flattened toads with warty skins and brightly coloured bellies.

1. Yellow-bellied Toad *Bombina variegata*   **60**
   Belly usually yellow or orange with darker markings; light patch on palm
   often extends onto thumb (Fig., p. 61).

2. Fire-bellied Toad *Bombina bombina*   **61**
   Not western or much of southern Europe. Belly usually red or red-orange
   with dark grey or blackish markings enclosing many white dots; light
   patch on palm often does not extend onto thumb.

TREE FROGS

(x 2/3)

Small, often bright-green frogs with round pads on finger and toe tips. Colour can
change rapidly.

3. Stripeless Tree Frog *Hyla meridionalis*   **79**
   No stripe on flank.

4. Common Tree Frog *Hyla arborea*   **77**
   Distinct stripe along flank with upward branch. Italian Tree Frog (*Hyla
   intermedia*, p. 78) is similar.

   4a, 4b. Showing colour change.

5. Tyrrhenian Tree Frog *Hyla sarda*   **78**
   Only on Corsica, Sardinia, Elba and Capraia. Often spotted; stripe on
   flank without upward branch.

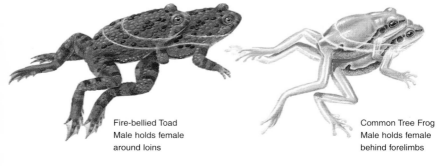

Fire-bellied Toad
Male holds female
around loins

Common Tree Frog
Male holds female
behind forelimbs

11

Plate 12   **BROWN FROGS 1**

(x ⅔; 3 x ¾)

Frogs with horizontal pupils, dorsolateral folds and a dark 'mask'. Often not very aquatic outside breeding season.

1. Common Frog *Rana temporaria*   **80**
Very variable in colour and pattern but not obviously striped. Dorsolateral folds close together (Fig., p. 58). Legs usually rather short.

    1a. Breeding male: bluish throat, thick arms, large nuptial pads on thumbs; extensive foot webs.
    1b. Male outside breeding season.
    1c. Breeding female: very plump; pearly granules on flanks.

2. Moor Frog *Rana arvalis*   **82**
Often very short-legged. Large hard metatarsal tubercle on hind foot (Fig., p. 82). Snout usually pointed. Sometimes striped (both sexes).

    2a. Male.
    2b. Female.
    2c. Partly blue breeding male. In south of range, more long-legged individuals occur and these might be confused with Agile Frog (Pl. 13).

3. Pyrenean Frog *Rana pyrenaica*   **81**
Found in Pyreneees with Common Frog but smaller (adults not more than 5cm) with a less distinct ear drum. Lives in and around mountain streams.

2c

3

12

Plate 13 BROWN FROGS 2
(x ⅔)

Frogs with horizontal pupils, dorsolateral folds and a dark 'mask'. Often not very aquatic outside breeding season.

1. Iberian Frog *Rana iberica*   **87**
West and central Iberian peninsula. Most likely to be confused with Common Frog (Pl. 12). Dorsolateral folds well separated. Two animals, showing variation in colouring.

2. Agile Frog *Rana dalmatina*   **83**
Legs often very long. Snout rather pointed. Ear-drum large and very close to eye. Yellow on groin; throat pale, or stippled at sides only.

3. Italian Agile Frog *Rana latastei*   **84**
North Italy and southern Switzerland etc. Like Agile Frog but ear drum smaller, sometimes reddish colouring on underside and throat dark. Distance between nostrils less than distance from nostril to eye (Fig., p. 85).

4. Balkan Stream Frog *Rana graeca*   **85**
Southern Balkans. Often yellow under hind legs; throat dark. Snout may be rounded; distance between nostrils greater than distance from nostril to eye. Usually in or near running water. Italian Stream Frog, *Rana italica*, (p. 86) is very similar.

| Iberian | Agile | Italian Agile | Stream |

Typical throat patterns: there is great variation in some forms and the Common Frog (Pl.12) may have all the patterns shown

13

Plate 14   **WATER FROGS**
(x ½; 4 x ¼)

Frogs with horizontal pupils, dorsolateral folds, no obvious dark 'mask', eyes close together, and vocal sacs at the corners of the mouth in males. Often aquatic and noisy and colouring very variable. This plate shows the three forms that are widespread in north, central and east Europe.

Another ten very similar forms occur further south (pp. 93–97). Distribution is very important in identifying them. As in the three forms illustrated here, useful features for identification are colour of the vocal sacs in males and of the backs of the thighs, length of the hind limbs, size and shape of the metatarsal tubercle, and song. American Bullfrog is also aquatic.

1. **Marsh Frog,** *Rana ridibunda*   **89**
   Naturally absent from parts of western Europe and from the Iberian peninsula and Italy. Often very large; up to 15cm. Back of thighs without yellow or orange colouring. Vocal sacs of male grey. Metatarsal tubercle small (Fig., p. 91).

   1a. On land.
   1b. In water, calling.

2. **Edible Frog,** *Rana* kl. *esculenta*   **92**
   Absent from southern Europe. Back of thighs often with yellow or orange colouring. Vocal sacs of male often white. Metatarsal tubercle large (Fig., p. 91). A hybridogenic form originating from crosses between the Marsh and Pool Frogs.

3. **Pool Frog,** *Rana lessonae*   **90**
   Similar distribution to Edible Frog and similar range of colouring including that of back of thighs, but often smaller and metatarsal tubercle very large (Fig., p. 91).

4. **American Bullfrog,** *Rana catesbeiana*   **97**
   Huge (up to 20cm), ear drum much larger than eye in males, no dorsolateral folds, vocal sac of male under chin.

   4a. Male
   4b. Female

4a

4b

14

1a

2

1b

3

Plate 15   TORTOISES
(x ½)

1. **Hermann's Tortoise** *Testudo hermanni*   **100**
   Large scale on tail-tip. Usually two supracaudal plates (Fig., below).

   1a. Adult
   1b. Hatchling (full size)

2. **Spur-thighed Tortoise** *Testudo graeca*   **101**
   Spurs on back of thighs. Usually one supracaudal plate (Fig., below).

3. **Marginated Tortoise** *Testudo marginata*   **102**
   Greece and Sardinia only. Rather like Spur-thighed tortoise but shell usually strongly flared in adults with characteristic pattern. See text for younger animals.

1b

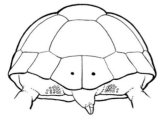

Hermann's
often two supracaudals,
large scale on tip of tail

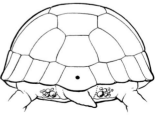

Spur-thighed
often one supracaudal,
spurs on thighs

Tortoises, back views

15

1a

2

3

Plate 16 TERRAPINS

(x ½)

Tortoise-like animals usually found in or near water.

1. **European Pond Terrapin** *Emys orbicularis* **103**
   Usually dark with yellowish spots and streaks.

2. **Spanish Terrapin** *Mauremys leprosa* **104**
   Spain and Portugal. Neck with obvious stripes; eye light; two dark spots
   on each bridge of shell.

   2a. Adult.
   2b. Plastron (highly variable).

3. **Balkan Terrapin** *Mauremys rivulata* **105**
   Southern Balkan peninsula and islands. Like Spanish Terrapin, but eye
   dark and plastron often almost entirely black or with dark plate edges.

4. **Red-eared Terrapin** *Trachemys scripta* **105**
   Introduced from USA. Red patch on each side of head.

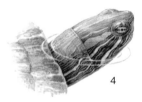

4

2   3

Plate 17 **SEA TURTLES 1**
(not to same scale)

1. Loggerhead Turtle *Caretta caretta*   107
See text and key (p. 98) for identification.

1a. Subadult. Shell about 80 cm.
1b. Young animal. Shell About 30cm.

2. Kemp's Ridley *Lepidochelys kempii*   107
Young animal. Shell about 30 cm. Adults do not occur in European
waters. 'Saw-backed' appearance also found in young Loggerhead Turtles.
See text and key for identification.

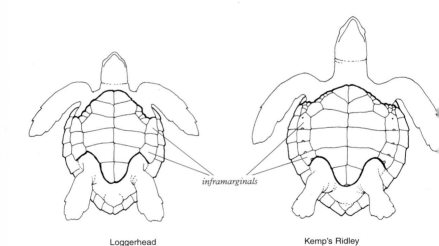

inframarginals

Loggerhead   Kemp's Ridley

Sea Turtles, undersides

1a

1b

2

Plate 18   SEA TURTLES 2
(not to same scale)

1. **Green Turtle** *Chelonia mydas*   **108**
   Adult. Shell about 150cm. See text and key (p. 98) for identification.

2. **Hawksbill Turtle** *Eretmochelys imbricata*   **109**
   Young. Shell grows to about 110cm. Plates on back usually overlap
   posteriorly. See text and key (p. 98) for identification.

3. **Leathery Turtle** *Dermochelys coriacea*   **110**
   Adult. Shell to 200cm or more, skin-covered with seven or five prominent
   ridges on upper surface.

18

Plate 19 AGAMA AND CHAMELEONS

(x ⅔)

1. **Starred Agama** *Laudakia stellio*   **118**
   Isolated areas in Greece and Aegean sea; also Malta. Spiny and rather flattened with small scales on head and belly. Often bobs head.

2. **Mediterranean Chameleon** *Chamaeleo chamaeleon*   **119**
   South Portugal and Spain, Sicily, Malta, Crete, Chios, Samos. Body flattened from side to side, bulging eyes, grasping feet, a flap-like lobe on each side of back of head.

3. **African Chameleon** *Chamaeleo africanus*   **120**
   Extreme southern Greece only. No flap-like lobes at back of head, males have spurs on the hind feet.

   3a. Hind foot of male showing spur.
   3b. Head.

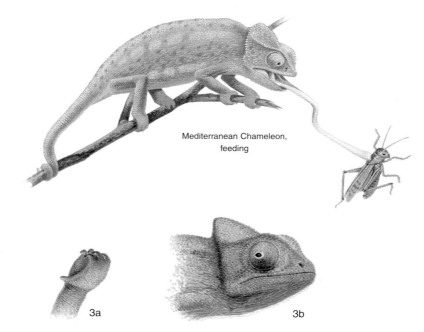

Mediterranean Chameleon,
feeding

3a   3b

Plate 20

19

GECKOS

(x ⅔)

Soft-skinned lizards with large eyes and vertical pupils. All eight European species are agile climbers confined to warm regions and are often nocturnal.

1. **Moorish Gecko** *Tarentola mauritanica*   **121**
   Robust. Adhesive pads extend along whole length of toes and are undivided beneath (Fig. below); large claws only on third and fourth toes of each foot. Four similar species occur in the Canary islands (p. 122–123)

2. **Turkish Gecko** *Hemidactylus turcicus*   **124**
   Often rather translucent. Adhesive pads do not extend to toe-tips and are divided beneath.

3. **European Leaf-toed Gecko** *Euleptes europaea*   **124**
   Corsica, Sardinia and nearby islands only. Small; no tubercles on back; adhesive pads limited to toe-tips.

   3a. Animal with intact tail.
   3b. Animal with regenerated tail.

4. **Kotschy's Gecko** *Cyrtopodion kotschyi*   **125**
   South-east Europe. No adhesive pads, but toes characteristically kinked.

| Moorish | Turkish | European Leaf-toed | Kotchy's |
| (underside) | (underside, side) | (underside) | (side) |

Gecko toes

Plate 21   MISCELLANEOUS LACERTIDS

1. **Spiny-footed Lizard** *Acanthodactylus erythrurus*   129
   Iberian peninsula only.

   1a. Adult: usually with some striping; no occipital scale (Fig., p. 112);   129
   subocular scale narrowed (Fig., p.115).
   lb. Juvenile: black with pale stripes and red tail.

2. **Large Psammodromus** *Psammodromus algirus*   127
   South-west Europe. Usually characteristic pattern and long tail; scales
   very large, pointed and keeled; no collar.

3. **Spanish Psammodromus** *Psammodromus hispanicus*   128
   South-west Europe. Small (up to 5.5cm from snout to vent); scales
   pointed and keeled; weak collar. Two animals, showing variation in
   pattern.

4. **Steppe Runner** *Eremias arguta*   130
   Romania to southern Ukraine. Large, robust; 14–20 large scales across
   belly.

5. **Snake-eyed Lacertid** *Ophisops elegans*   130
   North and east Aegean area only. No collar; eye without obvious eye-lids.

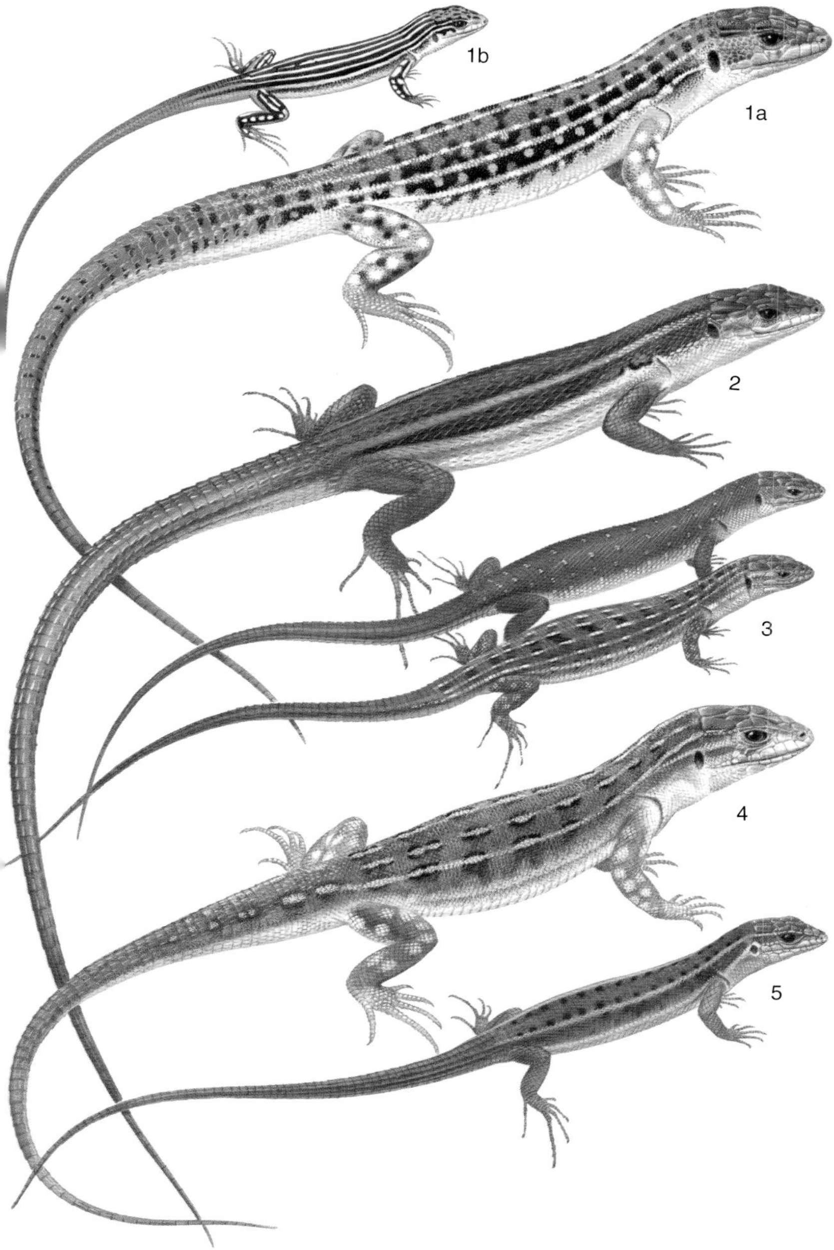
1b
1a
2
3
4
5

21

Plate 22　　　ALGYROIDES

Small lizards with large, keeled scales and a distinct collar. Often found in or near semi-shaded places.

1. **Dalmatian Algyroides** *Algyroides nigropunctatus*　　　**131**
   Adriatic coast to Ionian islands. Back scales large, keeled and blunt; flank scales granular. Male with yellow to red belly and brilliant blue or green throat.

2. **Greek Algyroides** *Algyroides moreoticus*　　　**131**
   Southern Greece and Ionian islands. Back and flank scales all large, strongly keeled and pointed.

   2a. Female: largely uniform.
   2b. Male: dark with light-spotted sides.

3. **Spanish Algyroides** *Algyroides marchi*　　　**133**
   South-east Spain. Scales on back large, blunt and weakly keeled; flank scales granular. Belly yellowish at least in males.

4. **Pygmy Algyroides** *Algyroides fitzingeri*　　　**133**
   Corsica and Sardinia. Very small (about 4cm from snout to vent or less). Scales on back and flanks all strongly keeled and pointed.

22

1

2a

2b

3

4

Plate 23   GREEN LIZARDS 1
Ocellated and Schreiber's Green Lizard
(x ⅔)

Large, robust lizards with serrated collars. Confined to south-western Europe. See Pl. 26 for juveniles. Other Green Lizards in north of this area: Sand Lizard (Pl. 25) and Green Lizard (Pl. 24).

1. Ocellated Lizard *Lacerta lepida*   **136**

Iberian peninsula, south France etc. Adults very large (up to 20cm or more from snout to vent). Often blue spots on flanks. Occipital scale nearly always wider than hind edge of frontal scale (Fig., p. 135). Male (illustrated) has more massive head than female.

2. Schreiber's Green Lizard *Lacerta schreiberi*   **137**

Hilly areas of north-west, west, central and south Iberian peninsula. Up to about 13.5 cm snout to vent. Occipital scale narrower than in Ocellated Lizard. In north of range often distinguishable from Green Lizard by having eight (not usually six) rows of large, often dark-spotted scales across belly.

2a. Male: usually green with black spots.
2b. Female: often at least partly brownish with dark blotches.

23

1

2a

2b

Plate 24   GREEN LIZARDS 2
Green and Balkan Green Lizards
(x ⅔)

Large, robust lizards with serrated collars. Absent from most of Iberian peninsula. See Pl. 26 for juveniles.

1. Balkan Green Lizard *Lacerta trilineata*   **139**

Balkans only. Up to 16cm from snout to vent or more. Often difficult to distinguish from Green Lizard(see text) but frequently bigger. Young and subadult animals may have characteristic pattern of three or five narrow light stripes (Pl. 26).

1a. Male: large head.
1b. Nearly mature female.

2. Green Lizard *Lacerta viridis*   **138**

Usually under 13cm snout to vent. Mostly green or brown. Some animals, especially young, have two or four narrow light stripes.

2a. Male: large head.
2b. Female.

24

1a

1b

2a

2b

Plate 25 GREEN LIZARDS 3

Sand Lizard *Lacerta agilis* **140**

Not Italy, or most of Greece and Iberian peninsula. Medium-sized, robust lizard with deep, rounded head, serrated collar and overlapping belly scales. Scales along mid-back usually narrowed and rarely green. Colouring very variable. See Pl. 26 for juvenile.

a. Typical male: green flanks, brighter in breeding season.

b. Typical female.

c. Largely green male: frequent pattern in Romanian area to west Ukraine (*L. a. chersonensis*).

d. Female with continuous, pale, vertebral streak: frequent pattern in Balkans (*L. a. bosnica*).

e. 'Red-backed' phase: not found in England but common in *L. a. argus*.

Sand Lizards, about to mate

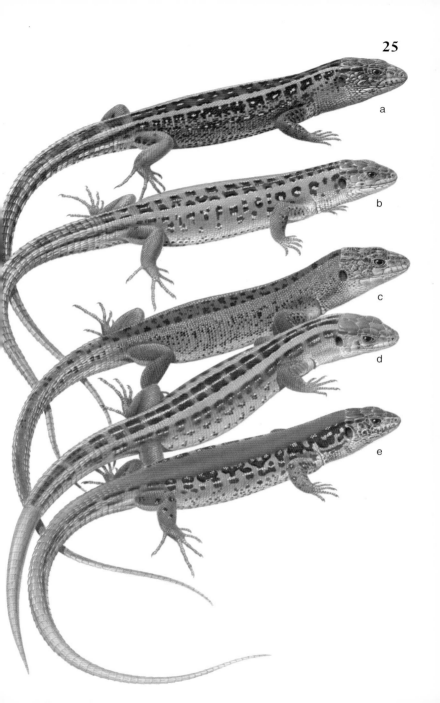

a

b

c

d

e

Plate 26   GREEN LIZARDS
Juveniles

Young Green Lizards often have distinctive patterns. They can be distinguished from adult Small lacertas by their typical juvenile shape (large rounded head, large eyes). As in adults, the collar is serrated and the belly scales overlap strongly.

1. **Ocellated Lizard** *Lacerta lepida*   **136**
   Iberian peninsula, south France etc. Light, often dark-edged ocelli all over, or uniform. Occipital scale as wide as hind margin of frontal scale (Fig., p. 135).

2. **Schreiber's Green Lizard** *Lacerta schreiberi*   **137**
   Iberian peninsula. Light, dark-edged ocelli or bars especially on sides.

3. **Green Lizard** *Lacerta viridis*   **138**
   Not most of Iberian peninsula. Either fairly uniform, or with two or four narrow pale stripes.

4. **Balkan Green Lizard** *Lacerta trilineata*   **139**
   Often rather dark: fairly uniform, or with three or five narrow pale stripes, or with dark spots on midback, or light spots on sides.

5. **Sand Lizard** *Lacerta agilis*   **140**
   Often weak version of adult pattern. Ocelli frequently present.

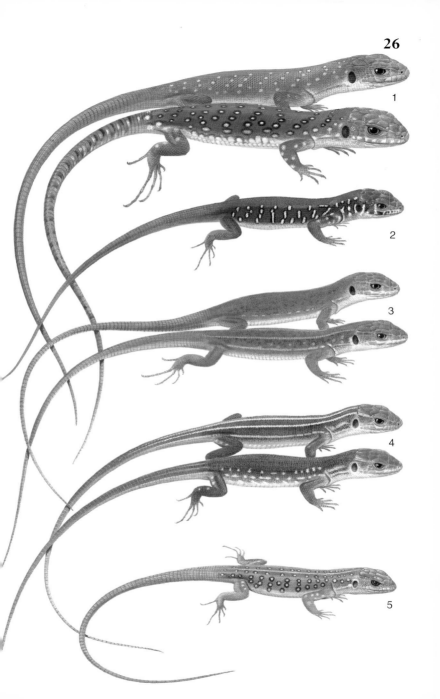

26

1

2

3

4

5

Plate 27   SMALL LACERTAS 1
Viviparous Lizard and Common Wall Lizard

The only two small lacertas native to north, west and central Europe; they also occur in many other continental areas.

1. Viviparous Lizard *Lacerta vivipara*   **144**

Small head and short legs. Serrated collar. Few supraciliary granules (see Fig., p. 112). Pattern very variable: often striped; no blue on flanks or belly scales.

1a. Baby: very dark colouring.
1b. Male.
1c, 1d. Females: head and legs smaller than male.
1e. Underside of breeding male: colour varies.

2. Common Wall Lizard *Podarcis muralis*   **145**

Head and limbs larger than in Viviparous Lizards of the same sex, collar more or less un-notched, and more supraciliary granules. Pattern very variable.

2a. Male: widespread pattern.
2b. Female: widespread pattern.
2c. Male: more heavily marked animal.
2d. Male: Rome area (*P. m. nigriventris*). Heavy markings and greenish ground colour. North-west Italian animals and ones from nearby islands are rather similar but often have weaker dark markings and ground colour is not always green.
2e. Breeding male: underside of weakly marked animal.
2f. Breeding male: underside of west Italian animal.

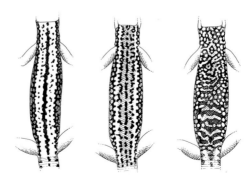

Common Wall Lizard, some variations in back pattern

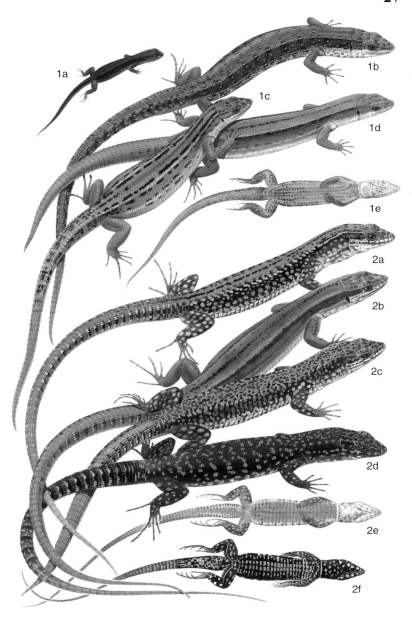

27

1a

1b

1c

1d

1e

2a

2b

2c

2d

2e

2f

Plate 28   SMALL LACERTAS 2a
Iberian peninsula and Mediterranean France

See also Pl. 29 and Viviparous Lizard (Pl. 27, Pyrenees and Cantabrians etc.); Common Wall Lizard (Pl. 27, France, north and central Spain); Italian Wall Lizard (Pl. 32, only Almeria, Santander and French coast). Other lacertid lizards present are Sand Lizard (Pl. 25), Ocellated Lizard and Schreiber's Green Lizard (Pl.23); Green Lizard (Pl. 24); Spiny-footed Lizard, Large and Spanish Psammodromus (Pl. 21) and Spanish Algyroides (P1. 22).

The north Iberian peninsula, especially the north-west, is one of the most confusing areas for small lacertas because most species are highly variable.

1. **Iberian Rock Lizard** *Lacerta monticola*   **148**
   Mainly high mountains of north-west, west and central Iberian peninsula but down to sea level in Galicia. Very variable but underside often green or yellow-green.

   1a. Young: blue tail (also found in some Iberian Wall Lizards).
   1b. Male: back often bright green (*L. m. monticola*, Serra da Estrela)
   1c. Female: a common pattern in several areas including Cantabrian mountains (*L. m. cantabrica*).
   1d. Underside of breeding male.

2. **Iberian Wall Lizard** *Podarcis hispanica*   **151**
   Iberian peninsula and Mediterranean France. Often flattened and delicately built. Pattern highly variable. Unlike many Common Wall Lizards, dark dorsolateral stripe often stronger than vertebral one (if any). Throat pale, frequently with small dark spots.

   2a. Male.
   2b. Female.
   2c. Underside of breeding male.

3. **Bocage's Wall Lizard** *Podarcis bocagei*   **152**
   North-west Spain and north Portugal south to Douro valley. Like Iberian Wall Lizard but more robust and less flattened. Back often green in males, and belly often yellow, salmon or orange, typically with heavy dark spotting and with no blue spots on the outer belly scales.

   3a. Male.
   3b. Female.
   3c. Underside of breeding male.

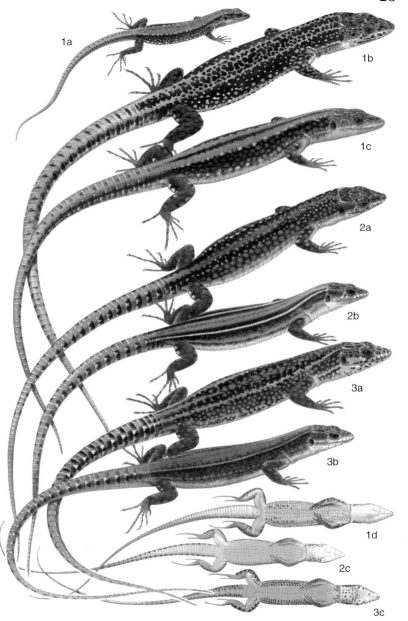

28

1a

1b

1c

2a

2b

3a

3b

1d

2c

3c

Plate 29   SMALL LACERTAS 2b
Iberian peninsula and Madeira

1. **Pyrenean Rock Lizard** *Lacerta bonnalli*   **150**
   Central Pyrenees above 1,700m Underside often yellowish or greenish
   and unspotted. For other differences from Common Wall Lizard, which
   occurs at lower altitudes, see p. 150.

2. **Aurelio's Rock Lizard** *Lacerta aurelioi*   **151**
   North-west Andorra and neighbouring France and Spain above 2200m.
   Similar to Pyrenean Rock Lizard but back more often dark spotted and
   pale stripes more frequently present.
   Aran Rock Lizard (*Lacerta aranica*, p. 151) is similar.
   1a. Adult.
   1b. belly.

3. **Carbonell's Wall Lizard** *Podarcis carbonelli*   **153**
   Portugal south from Douro valley and adjoining areas of Spain. Similar to
   Bocage's Wall Lizard but largely separable by distribution. Males usually
   with less heavily marked backs which are often not green, whereas the
   flanks often are; belly often whitish.

4. **Columbretes Wall Lizard** *Podarcis atra*   **154**
   Columbretes islands off eastern Spain. The only lacertid lizard in this
   archipelago.

5. **Madeira Lizard** *Lacerta dugesii*   **154**
   Madeira, Porto Santo and Selvages islands. The only native lacertid lizard
   in this archipelago.

5

29

Plate 30   SMALL LACERTAS 3
Balearic Islands

Only other lacertid lizard present is the Italian Wall Lizard (Pl. 32).

1. Liford's Wall Lizard *Podarcis lilfordi*   **157**
Majorca, Minorca and nearby small islands. Highly variable in colouring. Scales very fine and smooth. Often dark spots on throat. Italian Wall Lizard on Minorca has clearly keeled back scales between hind legs and no dark spots on throat.

  1a. Melanistic animal (*P. l. lilfordi* from Isla del Ayre, Minorca).
  1b. Green-backed animal (*P. l. sargantanae* from Isla de Sargantana, Minorca).

2. Ibiza Wall Lizard *Podarcis pityusensis*   **158**
Ibiza, Formentera and nearby islands, introduced into small areas of Majorca. Highly variable in pattern. Most easily separated from Lilford's Wall Lizard by usually coarser back scales which may be lightly keeled.

  2a, b. Adults from different islands.
  2c. Small-island form basking with body flattened so increasing rate of heat intake.

3. Moroccan Rock Lizard *Lacerta perspicillata*   **156**
Minorca only. Ten large scales across mid-belly; 'window' in lower eye-lid. Pattern usually reticulated (3a), but uniform animals (3b) have been reported.

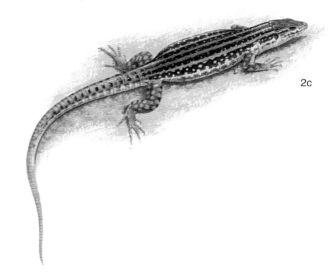

2c

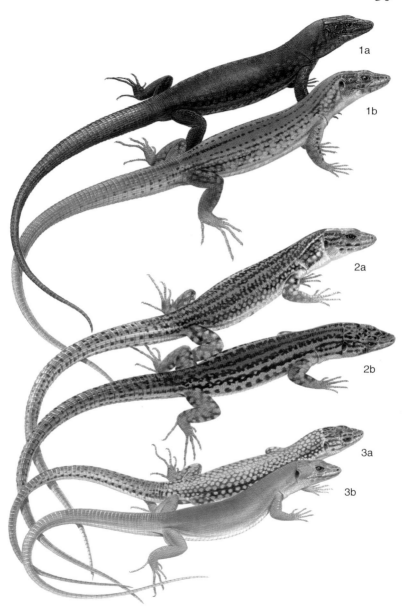

30

Plate 31   SMALL LACERTAS 4
Corsica and Sardinia

Italian Wall Lizard (Pl. 32) also present; only other lacertid lizard found is the Pygmy Algyroides (Pl. 22). Italian Wall Lizard differs from most individuals of the other small lacertas in usually having distinct keeling on the back scales between the hind legs and no dark spots on throat (see key, p. 159, and texts for other differences).

1. **Bedriaga's Rock Lizard** *Lacerta bedriagae*   **159**
   Body flattened. Often reticulated.

2. **Tyrrhenian Wall Lizard** *Podarcis tiliguerta*   **160**
   Body not obviously flattened. Often striped, but colouring highly variable.

   2a. Male: reticulated
   2b. Male: striped
   2c. Female

SMALL LACERTAS 5:
Malta

3. **Maltese Wall Lizard** *Podarcis filfolensis*   **165**
   Malta and nearby islands. Only lacertid lizard in the area; highly variable.

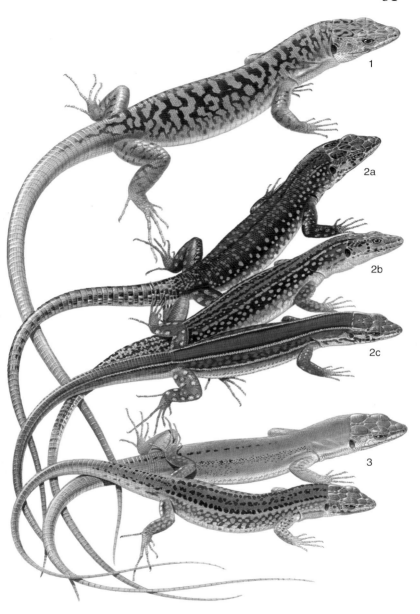

31

Plate 32   SMALL LACERTAS 5:
Italy and Sicily

Other small lacertas present are Viviparous Lizard (extreme north Italy only, Pl. 27) and Common Wall Lizard (not Sicily, Pl. 27). Only other lacertid lizard is Green Lizard (Pl. 24).

1. Italian Wall Lizard *Podarcis sicula*   **162**
(Also present in many areas outside Italy.) Highly variable but often green or olive-backed; underside usually whitish, greyish or greenish, unspotted. Common Wall Lizard typically has dark markings on throat and often on belly, which may be red.

   1a. Female: typical pattern in north Italy.
   1b. Male: typical pattern in north Italy.
   1c. Male: reticulate pattern (common in south Italy, Sicily, Sardinia, Minorca etc.).
   1d. Male: uniform phase (often occurs with reticulated individuals).

2. Sicilian Wall Lizard *Podrcis wagleriana*   **163**
Sicily only. Differs from Sicilian examples of the Italian Wall Lizard in better defined, pale, dorsolateral streaks, often simple row(s) of dark spots on back and red, orange or pink belly in breeding males.

   2a. Male.
   2b. Female.

3. Aeolian Wall Lizard *Podarcis raffonei*   **164**
Aeolian islands only. Back usually brown and throat dark spotted.

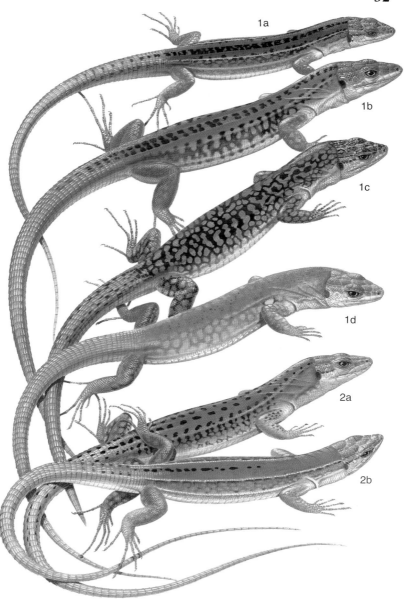

32

1a

1b

1c

1d

2a

2b

Plate 33   SMALL LACERTAS 6
East Adriatic coastal area

For this area see also Viviparous and Common Wall Lizard (Pl. 27), Balkan Wall Lizard (Pl. 34) and Italian Wall Lizard (Pl. 32). Other lacertid lizards present are Sand Lizard (Pl. 25), Green and Balkan Green lizards (Pl. 24), and Dalmatian Algyroides (Pl. 22).

1. Horvath's Rock Lizard *Lacerta horvathi*   **167**
Mountains of north-west Croatia, Slovenia and immediately neighbouring areas. Flattened, head blunt. Throat white and unspotted; belly often yellow (lb). See text for separation from Common Wall Lizard.

2. Mosor Rock Lizard *Lacerta mosorensis*   **168**
Mountains of southern Croatia and Bosnia, and Montenegro. Strongly flattened. Typically brownish above and yellow or orange below (2b).

3. Sharp-snouted Rock Lizard *Lacerta oxycephala*   **168**
Southern Croatia and Bosnia, and Montenegro. Strongly flattened. Either dappled above (3b) or black (3a), especially at high altitudes and on some small islands; underside bluish.

4. Dalmatian Wall Lizard *Podarcis melisellensis*   **169**
Not flattened. Often with light stripes or rather uniform; underside frequently orange-red in breeding males. For separation from Common, Balkan and Italian Wall Lizards see key (p. 165) and text (p. 169).

4a. Male: striped phase.
4b. Female: uniform phase.
4c. Female: lightly striped individual.
4d. Breeding male: underside.

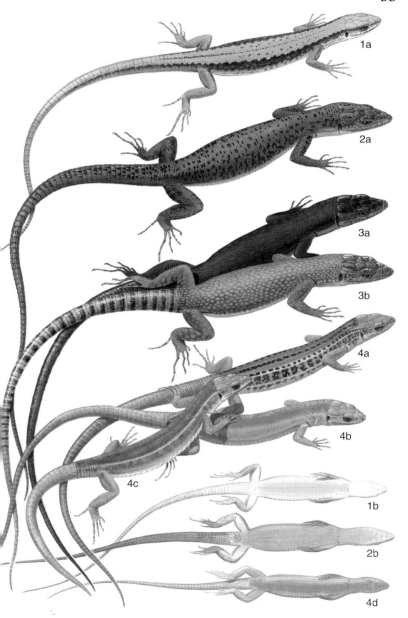

33

Plate 34 SMALL LACERTAS 7a
South-eastern Europe, mainland.

For this area see also Viviparous Lizard and Common Wall Lizard (Pl. 27), Erhard's Wall Lizard (south only, Pl. 36), Peloponnese Wall Lizard, and Greek Rock Lizard (South Greece only, Pl. 35), Italian Wall Lizard (Turkey only, Pl. 32). Other lacertids present are Sand Lizard (Pl. 25), Green and Balkan Green lizards (Pl. 24), Snake-eyed Lacertid and Steppe Runner (Pl. 21), and Greek and Dalmatian Algyroides (Pl. 22).

1. **Meadow Lizard** *Lacerta praticola* **173**
   Limited range. Small; coarse scales. Distinctive pattern. Belly often green or yellow (1b). Most likely to be confused with Viviparous Lizard.

2. **Balkan Wall Lizard** *Podarcis taurica* **175**
   Collar serrated. Green or olive above, often with pale stripes; belly usually orange or red in males. Most likely to be confused with Italian Wall Lizard (Pl. 32) in east European Turkey and Dalmatian Wall Lizard (Pl. 33) in northern Albania; see text.

   2a. Male: typical pattern for north of range.
   2b. Male: frequent pattern in south-west of range.
   2c. Male: uniform phase commoner in south of range.
   2d. Underside of breeding male.
   2e. Female.

   Skyros Wall Lizard (*Podarcis gaigae* p.178), confined to Skyros and Piperi islands in the northern Sporades is often similar to 2b.

3. **Caucasian Rock Lizard** *Lacerta saxicola* **179**
   Crimea only. Rather flattened, no obvious stripes.

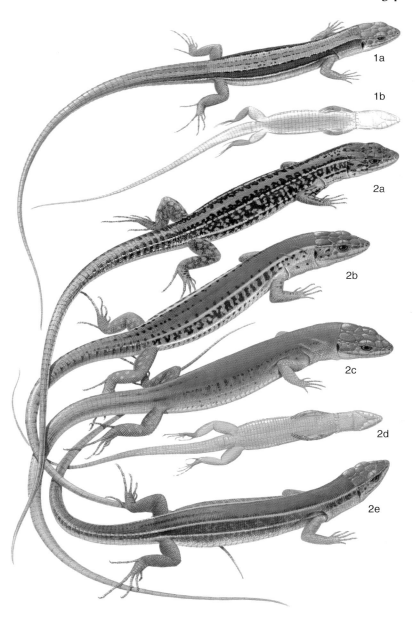

34

1a

1b

2a

2b

2c

2d

2e

Plate 35   SMALL LACERTAS 7b:
Southern Greece (Peloponnese)

For this area see also Common Wall Lizard (mountains, Pl. 27), Balkan Wall Lizard (Pl. 34), Erhard's Wall Lizard (Pl. 36). Other lacertid lizards present are Balkan Green Lizard (Pl. 24) and Greek Algyroides (Pl. 22).

1. Greek Rock Lizard *Lacerta graeca*   **174**

Rather flattened. Not striped. Belly often yellow or orange with dark spots. No large scales on cheek.

1a. Female.
1b. Male.
1c, 1d. Underside.

2. Peloponnese Wall Lizard *Podarcis peloponnesiaca*   **177**

Often strongly striped, but variable in pattern. No dark spots on throat. Young may have blue tail. Confusable with Erhard's Wall Lizard (Pl. 36) but fewer supraciliary granules (0–7 instead of 10–17; see Fig., p. 177). Balkan Wall Lizard has a serrated collar.

2a, 2b. Males: massive head; often blue on sides.
2c. Female: strongly striped (young Balkan Green Lizard (Pl. 26) is stockier).
2d. Breeding male: underside sometimes whitish.

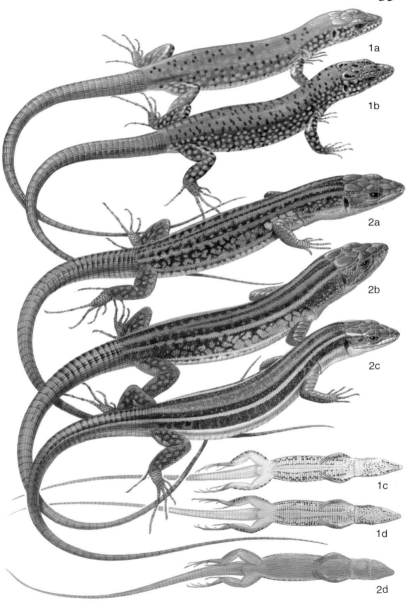

35

1a

1b

2a

2b

2c

1c

1d

2d

Plate 36   SMALL LACERTAS 7c
South Balkans and Aegean Islands.

See also Pls. 34 and 35 for other lacertid lizards in area.

1. Milos Wall Lizard *Podarcis milensis*   **178**
Milos and nearby islands: the only small lacerta in the region. Throat and sides of head largely black in males.

    1a. Male.
    1b. Female.
    1c. Underside of male.

2. Erhard's Wall Lizard *Podarcis erhardii*   **176**
South Balkans and Aegean islands (not Skyros and Milos Groups or islands near Asiatic Turkey): on islands where it occurs, nearly always the only small lacerta present. On mainland, dark dorsolateral streaks if present nearly always better developed than vertebral one (in contrast to Common Wall Lizard). In south Greece, Peloponnese Wall Lizard can be confused but has fewer supraciliary granules (0–7, not 10–17, see Fig. p. 177).

    2a, 2c. Males.
    2b, 2d. Females.
    2e. Underside of breeding male (colour quite variable).

3. Anatolian Rock Lizard *Lacerta anatolica*   **179**
Islands off coast of Asiatic Turkey, including Rhodes. The only small lacerta present.

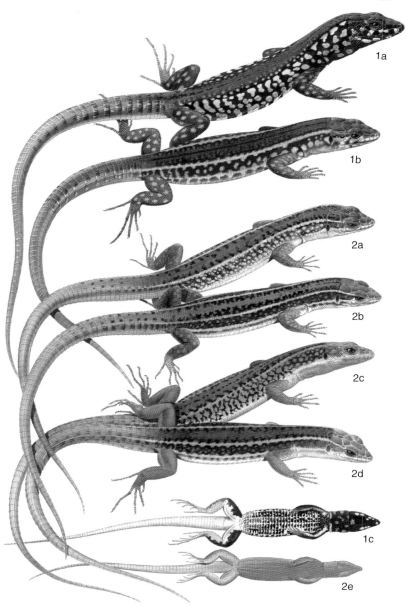

36

1a

1b

2a

2b

2c

2d

1c

2e

Plate 37 CANARY ISLAND LIZARDS 1

(x ½)

Small to large species of lacertid lizards that differ between islands.

1. Tenerife Lizard *Gallotia galloti*   **182**
   Tenerife and La Palma only. Males very dark, females lighter.

   1a. Male from northern Tenerife (*G. g. eisentrauti*); southern animals (*G. g. galloti*) are more sombre but with prominent blue spots on flanks; those from La Palma (*G. g. palmae*) are small and dark with a blue throat, jowls and chest.
   1b. Female.

2. Boettger's Lizard *Gallotia caesaris*   **182**
   El Hierro and La Gomera only. Smaller than the Tenerife Lizard and males often largely blackish.

   2a. Male.
   2b. Female.

3. Atlantic Lizard *Gallotia atlantica*   **181**
   Eastern Canary islands, where it is the only native lacertid lizard. Small (less than 11cm from snout to vent); scales on back very coarse.

   3a. Male from Lanzarote (*G. a. atlantica*).
   3b. Female.

37

1a

1b

2a

2b

3a

3b

Plate 38 CANARY ISLAND LIZARDS 2
(1, 2, 4 x ⅓; 3 x ½)

Large to very large species of lacertid lizards that differ between islands.

1. Gran Canaria Giant Lizard *Gallotia stehlini* **184**
The only native lacertid lizard on Gran Canaria. Very large; throat reddish in males.

2. El Hierro Giant Lizard *Gallotia simonyi* **184**
Males very large and dark with yellow spots on flanks.

3. Tenerife Speckled Lizard *Gallotia intermedia* **183**
Extreme north-west Tenerife only. Often spotted pattern and usually greater number of large belly scales across mid-body distinguish it from Tenerife Lizard.

4. La Gomera Giant Lizard *Gallotia gomerana* **185**
Very large and dark with whitish underparts; jaw and often sides of the neck also this colour in adults.

38

1

2

3

4

Plate 39   SKINKS 1
(x ½)

Small or medium-sized lizards with cylindrical bodies, thick neck, short legs and very smooth, shiny scales.

1. **Ocellated Skink** *Chalcides ocellatus*   **187**
   Up to 30cm. Pattern of dark-edged ocelli.

   1a. Greece, Aegean (*C. o. ocellatus*)
   lb. Sardinia, Sicily, Malta, Naples area (*C. o. tiligugu*)

2. **Bedriaga's Skink** *Chalcides bedriagai*   **188**
   Spain and Portugal only. Like small a small Ocellated Skink (see also Fig., p. 187).

3. **Snake-eyed Skink** *Ablepharus kitaibellii*   **186**
   Balkan area. Very small (usually under 13cm total length). Eye lacks normal eye-lids and cannot close.

4. **Western Three-toed Skink** *Chalcides striatus*   **188**
   Portugal, Spain, south France and adjoining Italy. Snake-like with tiny limbs; 9–13 dark lines.

5. **Italian Three-toed Skink** *Chalcides chalcides*   **189**
   Italy, Sicily Sardinia. Snake-like with tiny limbs; 5-6 dark lines sometimes with broad pale dorsolateral stripe, or without any markings at all.

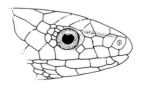

Snake-eyed Skink

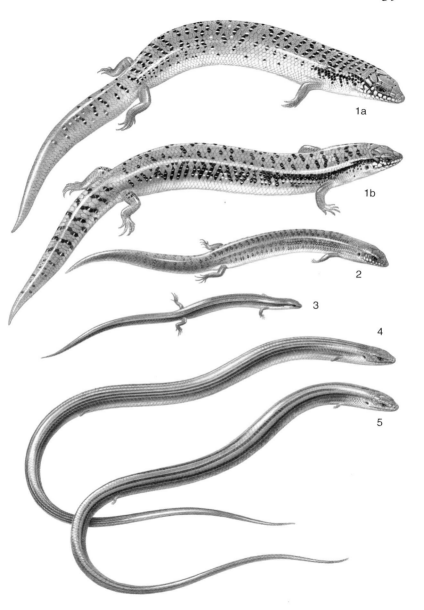

39

1a

1b

2

3

4

5

Plate 40   SKINKS 2
(x ½)

Small or medium-sized lizards with cylindrical bodies, thick necks, short legs and very smooth, shiny scales.

1. **Eastern Canary Skink** *Chalcides simonyi*   **189**
   Fuertaventura and Lanzarote in the Canary islands.

2. **Gran Canaria Skink** *Chalcides sexlineatus*   **190**
   Gran Canaria in the Canary islands

   2a. Southern animal (*C. s. sexlineatus*)
   2b. Northern animal (*C. s. bistriatus*)

3. **West Canary Skink** *Chalcides viridianus*   **190**
   Tenerife, La Gomera and El Hierro in the Canary islands.

4. **Levant Skink** *Mabuya aurata*   **191**
   Rhodes and perhaps neighbouring islands. Distinctive pattern.

40

Plate 41 LIMBLESS LIZARDS
(1 x ½; 2–4 x ⅔)

European limbless lizards have closable eye-lids, breakable tails and unenlarged belly scales.

1. European Glass Lizard *Ophisaurus apodus* **193**
South-east Europe. Very large (up to 140cm). Prominent groove on flank.

1a. Adult: colour uniform.
1b. Young: pale with dark bars.

2. Slow Worm *Anguis fragilis* **192**
Very smooth scales. Blunt head. Peloponnese Slow Worm (*A. cephallonica*, p. 193) is similar.

2a. Male: often rather uniform; may have blue spots.
2b. Female: often with dark flanks.
2c. Young: gold or silvery usually with very dark vertebral stripe, flanks and belly.

3. Limbless Skink *Ophiomorus punctatissimus* **191**
Greece only. Small (up to about 18cm). Pattern strongest on tail. Head viewed from above is pointed.

4. Iberian Worm Lizard *Blanus cinereus* **194**
Iberian peninsula. Blunt, short tail. Ring-like grooves on body.

5. Anatolian Worm Lizard *Blanus strauchi* **194**
Islands close to Asiatic Turkey. Similar to Iberian Worm Lizard but upper jaw overhangs lower one.

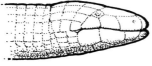
Spanish

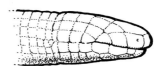
Anatolian

Worm Lizards, heads

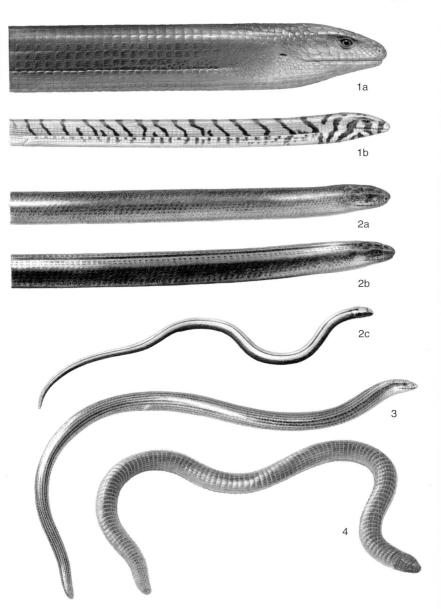

41

1a

1b

2a

2b

2c

3

4

Plate 42 WORM SNAKE, SANDBOA, MONTPELLIER SNAKE and HORSESHOE WHIP SNAKE

(3 and 4 x ⅔)

1. **Worm Snake** *Typhlops vermicularis* **200**
South Balkans. Like dry, shiny worm. Eyes minute; tail thicker than small head.

2. **Sand Boa** *Eryx jaculus* **201**
South Balkans. Tail blunt, eyes small with vertical pupil, belly scales narrow (Fig., p. 197).

   2a. Head and neck.
   2b. Tail.

3. **Montpellier Snake** *Malpolon monspessulanus* **202**
Brow-ridges give penetrating expression. Frontal scale narrow (Fig., p. 203).

   3a. Adult: often fairly uniform.
   3b. Young: may be uniform or spotted.

4. **Horseshoe Whip Snake** *Coluber hippocrepis* **204**
Iberian peninsula, south-west Sardinia, Pantellaria. Bold pattern of dark-edged blotches; a row of small scales under eye. Adults often generally dark; juveniles brighter.

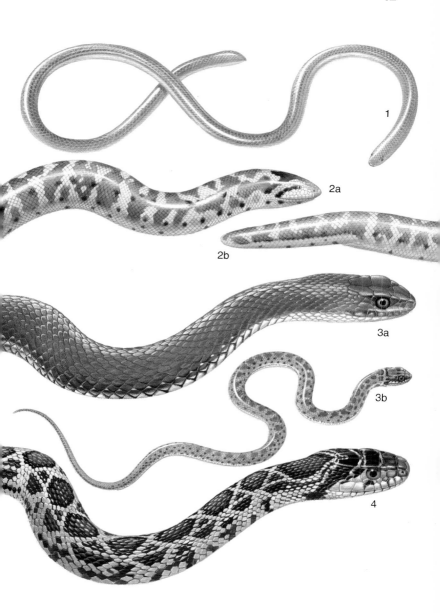

42

1

2a

2b

3a

3b

4

Plate 43   WHIP SNAKES

(⅔; 4 x ½)

Active, slender diurnal snakes. The first three species on this plate have fine striping on hind parts (1d) and are easily confused, so check range and texts carefully.

1. **Western Whip Snake** *Coluber viridiflavus*   **207**
   Absent from most of east Europe (overlaps with Balkan Whip Snake in north-west Croatia). Up to 150 cm.
   1a. Adult with heavily marked foreparts.
   1b. Uniform, black adult.
   1c. Young, bold head pattern.
   1d. Tail showing characteristic striping.

2. **Balkan Whip Snake** *Coluber gemonensis*   **208**
   West and south Balkans. Usually under l00cm. Foreparts of adults with dark spots and usually small white flecks on some scales. Belly pale yellowish or whitish, frequently with some dark spots. Pattern of young similar to that of Western Whip Snake.

3. **Large Whip Snake** *Coluber caspius*   **209**
   Balkans. Up to 200 cm or more.
   3a. Adult: large; fine stripes all over body; underside unspotted, yellow to orange-red.
   3b. Young: usually well spaced bars on back; no bold light marks on head.

4. **Coin-marked Snake** *Coluber nummifer*   **210**
   Islands close to Asiatic Turkey. Pattern of bold spots (4a), tail striped (4b).

5. **Algerian Whip Snake** *Coluber algirus*   **205**
   Malta only. Widely separated bars on back; ground colour may be grey brown or tinged yellow or pink.

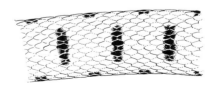

Algerian Whip Snake, alternative pattern

1a

1c

1d

1b

2

3b

3a

4a

4b

5

Plate 44

WHIP, DWARF
AND RAT SNAKES

$(\frac{2}{3}; 3 \times \frac{1}{2})$

1. **Dahl's Whip Snake** *Coluber najadum*   **205**

West and south Balkans, east Aegean islands. Very slender; a row of ocelli on each side of neck and often other scattered spots on foreparts; skin in front of and behind eye light.

2. **Reddish Whip Snake** *Coluber collaris*   **206**

South-east Bulgaria and adjacent Turkey. Very slender; a dark, light-edged bar across neck and a variable pattern of scattered spots on foreparts; skin in front of and behind eye not light.

3. **Dwarf Snake** *Eirenis modestus*   **210**

European Turkey and east Aegean islands. Less than 60cm long and very slender; characteristic head pattern; only 17 back scales across mid-body.

4. **Steppe Snake** *Elaphe dione*   **213**

South-east Ukraine. Pattern of dark cross-bars and pale stripes along body; head often with dark U-shaped mark on crown.

5. **Leopard Snake** *Elaphe situla*   **211**

Balkans, Crimea, southern Italy, Sicily and Malta. Variable pattern of brown to red, dark-edged markings, consisting of spots or stripes. Juveniles have similar markings to adults.

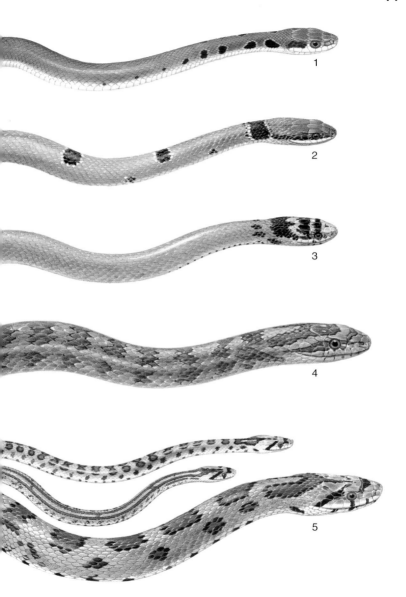

Plate 45   RAT SNAKES
(x ½)

Medium-sized to large snakes. Young often have distinctive patterns. Adults eat many small mammals and often climb.

1. Ladder Snake *Elaphe scalaris*   **215**
Iberian peninsula and Mediterranean France only. Young usually with bold 'ladder' pattern which is reduced to two simple stripes in adults. Rostral scale pointed behind.

2. Aesculapian Snake *Elaphe longissima*   **214**
Not most of Iberian peninsula. Slender. Juveniles spotted, with yellow blotch on side of head (rather like Grass Snake, Pl. 46, but scales not keeled). Adults uniform or with faint stripes, often with small white flecks.

3. Italian Aesculapian Snake *Elaphe lineata*   **215**
South Italy only. Few if any pale flecks on body, if stripes present they are narrower than spaces in between; belly often greyish.

4. Four-lined Snake *Elaphe quatuorlineata*   **211**
Balkans and Italy. Back scales lightly keeled in adults, giving rather rough appearance.

   4a. Western Four-lined Snake (*Elaphe q. quatuorlineata*). Italy, west and south Balkans to south-west Bulgaria; adults with four dark stripes. Animals from the central and south Cyclades (*E. q. muenteri*) are small with narrow stripes.

   4b. Blotched snake (*E. q. sauromates*) North-east Greece to Ukraine. Adults with patterns of bold blotches. Best treated as a separate species.

   4c. Young Four-lined Snake. Usually boldly spotted, but pattern variable.

45

1

2

3

4c

4a

4b

Plate 46 WATER SNAKES
(x ⅔)

Medium to large snakes with clearly keeled body scales, large scales on head and round pupils. Belly pattern often chequered. Typically found in fairly moist places, or in water.

1. Dice Snake *Natrix tessellata*   **220**
Not western Europe. Pattern variable but often dark spotted. Nostrils directed upwards, snout rather pointed. Usually 19 rows of scales at mid-body.

2. Viperine Snake *Natrix maura*   **218**
West and south-west Europe. Pattern variable but often ocelli on sides. Nostrils directed upwards. Usually 21 rows of scales at mid-body. Two animals, showing variation in pattern.

3. Grass Snake *Natrix natrix*   **216**
Usually a yellow, orange or white collar with dark border behind. Often no keeling on tail scales. Nostrils not directed upwards. Colouring very variable.

3a. Typical west European pattern (*N. n. helvetica*)
3b. Widespread pattern in north and north-east Europe (*N. n. natrix*).
3c. Striped morph found in south-east Europe (*N. n. persa*).
3d. Uniform colouring common in Spanish and Portuguese Snakes (*N. n. astreptophora*).

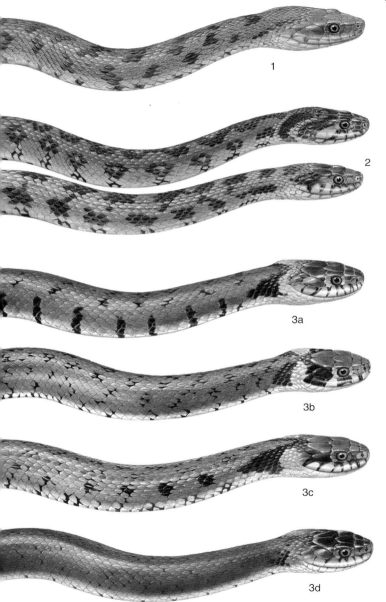

46

1

2

3a

3b

3c

3d

Plate 47 SMOOTH SNAKES, FALSE SMOOTH SNAKE and CAT SNAKE

1. **Smooth Snake** *Coronella austriaca* 221
   Stripe from side of neck to eye and usually to nostril. Third and fourth labial (upper lip) scales often border eye. Belly rather uniform (Fig., below). Two animals, showing pattern variants.

2. **Southern Smooth Snake** *Coronella girondica* 222
   Stripe from side of neck to eye, rarely to nostril. 'Bridle' on snout. Fourth and fifth labial (upper lip) scales border eye. Belly chequered or striped (Fig., below).

3. **False Smooth Snake** *Macroprotodon cucullatus* 223
   Spain and Portugal except much of north, Majorca and Minorca. Often a dark collar or hood. Rostral scale low; largest upper labial scale usually touches or comes close to parietal scale (see Figs., p. 224).

   3a. Spain and Portugal (*M. c. ibericus*), common pattern.
   3b. Balearic islands (*M. c. mauritanicus*), pale individual.

4. **Cat Snake** *Telescopus fallax* 225
   South-east Europe and Malta. Pupil vertical; body scales smooth (in contrast to vipers).

   4a. Typical pattern.
   4b. Pale individual from south Aegean islands.

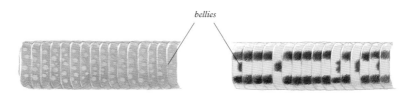

bellies

Smooth Snake   Southern Smooth Snake

47

1

2

3a

3b

4a

4b

Plate 48   VIPERS 1

Usually thick-set snakes, with very well-defined heads, vertical pupils and keeled body scales. Often a zig-zag stripe on back. All are venomous.

1. **Orsini's Viper** *Vipera ursinii*   **228**
   Limited range (see text). Rather narrow head, rough skin texture, nostril low down in nasal scale, and vertebral stripe often black-edged.

2. **Adder** *Vipera berus*   **230**
   Not much of southern Europe including Portugal, Spain and most of Italy. Rather variable. Nostril in centre of nasal scale, and vertebral stripe usually not dark-edged. Males tend to be greyer and brighter than females.

   2a. Male.
   2b. Female.

3. **Seoane's Viper** *Vipera seoanei*   **231**
   North-west Iberian peninsula. Colouring very variable (sometimes like that in 4a). Illustrated animal has the *bilineata* pattern.

4. **Asp Viper** *Vipera aspis* **(see also Pl. 49)**   **232**
   North-east Spain, France, south-west Germany, Switzerland, and Italy. Highly variable in pattern. Usually has turned-up snout and often two rows of small scales below the eye.

   4a. South-west France and central Pyrenees (*V. a. zinnikeri*): typical pattern.
   4b. North and east France, northern Italy, Switzerland: barred pattern frequent; bars on one side may be staggered relative to those on the other and are often narrower and completely separated.

48

1

2a

2b

3

4a

4b

Plate 49   VIPERS 2

1. Asp Viper *Vipera aspis* (see also Pl. 48)   **232**

    1a. Black phase, especially common in Switzerland and north-west Italy. Black individuals of other vipers species also occur, particularly of the Adder.

    1b. South Italy and Sicily (*V. a. hugyi*). Typical pattern: broad, wavy, brownish stripe or series of blotches. Animals from Montecristo island (*V. a. montecristi*) are similar.

2. Lataste's Viper *Vipera latasti*   **234**
Iberian peninsula except north. Usually a distinct horn on nose.

3. Nose-horned Viper *Vipera ammodytes*   **235**
Balkan peninsula and neighbouring regions. Horn on nose.

    3a. Male.
    3b. Female.

4. Ottoman Viper *Vipera xanthina*   **236**
North-east Greece, European Turkey, east Aegean islands. Bold head pattern. Scales on top of head small, except for a large one over each eye.

5. Milos Viper *Vipera schweizeri*   **237**
West Cyclades islands in the Aegean Sea. All scales on top of head small, even those over eyes.

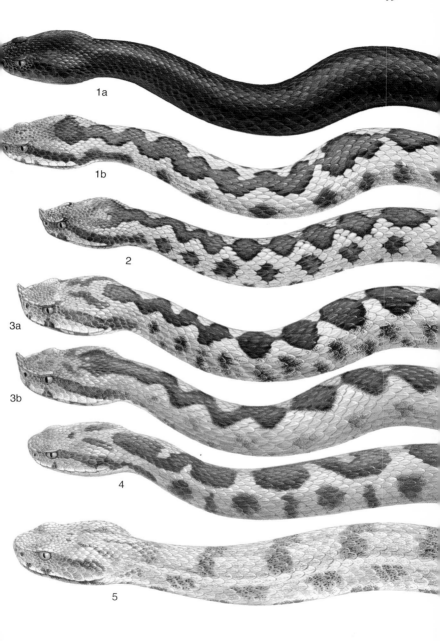

49

1a

1b

2

3a

3b

4

5

Similar Species Meadow Lizard (p. 173). Common Wall Lizard (below) is flatter with (in each sex) a larger head and longer legs (see key, opposite for further differences). The Sand Lizard (p. 140), even when young, is more robust with a larger, deeper head, usually a band of narrow scales along the mid-back, greater overlap of belly scales and a different range of patterns.

Habits Essentially a ground-dwelling lizard, although it may climb occasionally, especially in vegetation. Requires a fairly humid environment and is typically found among grass or other dense herbaceous plants. In the south of its range, where it is often montane occurring up to 2,500m in some places, it is largely confined to moist situations: alpine meadows, wet ditches, marshes, edges of damp woods, rice fields etc. In the north, it is more widespread, being found in open woods, field-edges, heaths, bogs, grassland and sand-dunes, on sea-cliffs, hedge-banks and railway embankments and even in gardens. Occurs further north than any other reptile reaching 70°N in Norway, over 350km beyond the Arctic circle. May sometimes occur at densities of 100–1,100 animals per hectare, at least in northern Europe. Swims well and often alternates basking with active hunting forays.

In most places, the Viviparous Lizard gives birth to 3–11 (often 7–8) fully formed young which are born after 6–13 weeks wrapped in a transparent membrane which they soon break. Pregnant females bask extensively, speeding development of young, and are less active at this time. In Spain and adjoining south-west France this lizard usually lays a single clutch of 1–13 (average 5–7) eggs, about 10–12mm x 8–10mm, which may sometimes be deposited communally and develop quickly, in about 4–5 weeks. Egg laying also occurs in Slovenia. Babies are very small, being 1.5–2.5cm from snout to vent. In north of range, males become sexually mature after their second hibernation and females after their third, but in parts of France 50 per cent of animals are mature the year after birth. After the first year, life expectancy is good, especially in some parts of northern range, and animals have been known to live 12 years in the wild.

Other names. Common Lizard, *Zootoca vivipara*.

COMMON WALL LIZARD *Podarcis muralis*   **Pl. 27**

Range Mainland Europe north to France, south Belgium and extreme southern Netherlands, Rhine Valley, south and east Austria, Slovakia and Romania. South to central Spain, southern Italy and south Balkan peninsula. Occurs on islands off Atlantic coast of Spain and France (including Channel islands) and islands off north-west Italy. Absent from the Aegean except for Samothraki and perhaps Thasos. Also north-west Asiatic Turkey. Map 105.

Identification Up to about 7.5cm from snout to vent, but usually smaller, tail l.7–2.3 times as long. A small, often rather flattened wall lizard, usually with a smooth-edged collar and lightly keeled scales. Very variable in pattern and, for identification outside area 1, see **Variation** and regional keys. Most individuals are brownish or grey (occasionally tinged green) often with black and white bars on sides of tail. Females usually have dark flanks, sometimes pale dorsolateral streaks, which are best developed on the neck, and frequently a dark vertebral stripe or series of spots; this is nearly always better developed than the dark dorsolateral stripes, if any. The vertebral stripe may be replaced or accompanied by dark spots or the back may be unmarked. Males are sometimes similar but pattern is typically more complex: sides are often light spotted and back is more boldly marked. Reticulated animals may occur. Ground colour of belly may be whitish or pale buff, but often with at least some red, pink or orange, especially in males. Throat is usually whitish or cream with rusty markings and typically has a variable amount of black pigment which often extends on to belly as well and is best developed in males. Juveniles are more or less like females but tail is sometimes light grey.

Variation Considerable variation in pattern, even within populations; blackish or nearly uniform specimens occur occasionally and, rarely, the belly can be yellowish. Tendency in several areas for increase in size and in amount of dark pigmentation, and for development of green on back: these colour trends may be largely confined to males or affect both sexes. Principal regional variations are as follows:

Atlantic coast and islands off Spain and France (*P. m. brogniardi*). Lizards tend to be large with bold markings that sometimes form a reticulation; ground colour may be olive. Populations from western France to extreme north-west Germany that show these tendencies to a lesser extent, may also be assigned to this subspecies.

French Riviera (*P. muralis merremia*). Some but not all populations have small, almost uniform individuals with few dark markings. Animals from central Spain, eastern France and the upper- and mid-Rhine area may also be referred to this form.

North Italy to north-west Croatia (*P. m. maculiventris*). Animals tend to be large and heavily marked, especially below, and are usually brown backed.

West Italy, south to Naples (*P. m. nigriventris*). Variable in the north of this area but often very dark and males frequently have bright green backs: Further south both sexes and young are extremely dark, with ground colour of back frequently reduced to yellow-green spots. More normal animals occur in mountains. Populations on islands off north-west Italy (Elba etc.) are generally like west Italian ones but back colour, degree of darkening and sexual differences vary from island to island and dark back markings often form cross bands.

South Italy (*P. m. breviceps*). Animals in this isolated population are a little like the Viviparous Lizard with a rounded head and a tendency for the collar to be notched.

Other Areas Lizards are referred to *P. m. muralis*.

Similar Species Females can look superficially like the Viviparous Lizard (p. 144). More, similar small lacertas exist in areas 2, 5, 6, and 7: see relevant keys and descriptions.

Habits Widespread over most of its range, but restricted to sheltered sunny localities in the north, and often to mountainous areas in the southern part of its distribution, where it may occur up to 2,500m. Frequently quite common and may sometimes reach densities of 1,400 animals per hectare. Typically found in much drier, less grassy habitats than the Viviparous Lizard, but in south often encountered in rather humid, semi-shaded places. Usually a climbing species seen in rocky situations, on boulders, outcrops, field and garden walls, parapets, and even on tree trunks. Southern populations often exist alongside more specialised climbing species (Iberian Wall Lizard, p. 151, and various rock lizards) and then the Common Wall Lizard climbs less high and on less precipitous surfaces, often occurring on overgrown screes, path sides, road banks, cliff bases, and sunny slopes in open broad-leaved and coniferous woods. In general, this species is very active, alert, and usually more adventurous and opportunistic than its relatives. More than any other small lacerta, it occurs near human habitations.

Males are territorial and dominant ones may defend a territory of about 25 square metres. Females often lay 2–3 clutches a year but often only one in mountain areas and possibly as many as six in warmer parts of range. 2–10 (often around 6) eggs are laid at a time that are about 10–12mm x 5–8mm swelling to 14–15mm x

11–12mm. These hatch in 6–11 weeks producing young 2.5–3cm from snout to vent that may mature in a year and occasionally live up to 7 years in the wild.

Note It is likely that The Common Wall Lizard originated in Italy and, on an evolutionary time scale, has spread over the rest of its large range relatively recently.

Other names. *P. m. brueggemanni* was used for green backed populations from north-west Italy.

SMALL LACERTAS AREA 2:
Iberia and Mediterranean France; Madeira

In this area there are thirteen small lacertas of which nine are confined to it: the Iberian, Pyrenean, Aurelio's's and Aran rock Lizards; the Iberian, Bocage's, Carbonell's and Columbretes wall Lizards; and the Madeira Lizard. Two others, the Viviparous and Common Wall Lizards are very widespread elsewhere in Europe. Two further species have been introduced in very small areas: the Italian Wall Lizard (Almeria, Santander, Toulon, Ile d'If off Marseilles, coastal Provence) and the Ibiza Wall Lizard (Barcelona, Basque Coast Lisbon); only the former which is longer established is included in the key below. The north-west of the area is one of the most difficult regions for positively identifying small lacertas, because it contains at least four very variable species that can often look alike. Other lacertids found in the Iberian peninsula are three species of Green lizards (p. 134), the Spiny-footed Lizard (p. 129), the Spanish Algyroides (p. 133) and two Psammodromus species (p. 127, 128).

KEY

1. Northern mountains and adjoining France etc. Collar serrated, granules between supraocular and supraciliary scales few or absent (0–4), dorsal scales coarse (25–37 across body) and usually keeled

    Viviparous Lizard, *Lacerta vivipara* (p. 144, Pl. 27)

    Collar not distinctly serrated, granules between supraocular and supraciliary scales 5 or more, dorsal scales finer, usually more than 40 across body   **2**

2. Almeria, Santander and French Coast areas only. Fairly large; adults often 6cm or more from snout to vent; frequently green-backed; underside whitish or with greenish tinge, unspotted. Scales on back between hind legs usually distinctly keeled

    Italian Wall Lizard, *Podarcis sicula* (p. 162, Pl. 32)

    All areas. In places where Italian Wall Lizard occurs, usually less than 6cm from snout to vent, not green-backed, belly may be white, buff, pink or red; throat often with small black spots. Scales on back between hind legs not clearly keeled   **3**

3. Note. Because of the great variation within species, the rest of the key will not identify all animals with certainty, so species texts should be checked carefully.

    High mountains of north-west, west and central Iberian peninsula (usually above 1,100m but lower in Galicia). Underside frequently green or yellow-green

    Iberian Rock Lizard, *Lacerta monticola* (p. 148, Pl. 28)

Mountains of north and central Iberian peninsula, Atlantic coast and France. Belly whitish, buff, pink or red; throat whitish with at least some dark blotches and rusty markings (Fig., p. 152). If present, vertebral stripe is usually better developed than dark dorsolateral stripes or series of marks

Common Wall Lizard, *Podarcis muralis* (p. 145, Pl. 27)

High Pyrenees (usually above 1700m). Underside generally unspotted and often yellow or greenish; rostral scale often meets frontonasal (Fig., p. 150)

Pyrenean Rock Lizard, *Lacerta bonnali* (p. 150, Pl. 29)

Aurelio's Rock Lizard, *Lacerta aurelioi*, (p. 151, Pl. 29)

Aran Rock Lizard, *Lacerta aranica* (p. 151)

Widespread. Belly whitish, buff, pink, red, yellow or orange; throat often with well-defined, small spots, especially at sides Fig., p. 152. If present, vertebral stripe is usually weaker than dorsolateral stripes or series of marks (exceptions frequent in north-east and east) 4

4. Widespread. Belly whitish, buff, pink or red (occasionally yellow). In north-west Spain and Portugal, typically quite small and flattened usually with a brown or grey back

Iberian Wall Lizard, *Podarcis hispanica* (p. 151, Pl. 28)

Portugal and north-west and extreme west Spain. Belly usually whitish, yellow, salmon or orange; back or flanks or both may be green in males. More robust and less flattened than Iberian Wall Lizards within range and throat and belly often more heavily spotted 5

5. North-west Spain and Portugal south to Douro valley. Males often with green back and dorsolateral stripes and strong dark markings, flanks brownish, belly often yellow, salmon or orange without blue spots at sides

Bocage's Wall Lizard, *Podarcis bocagei* (p. 152, Pl. 28)

Portugal from Douro valley southwards and adjoining Spain. Males often with brown or grey back and weak markings, flanks often dappled green, belly frequently whitish with blue spots at sides

Carbonell's Wall Lizard, *Podarcis carbonelli* (p. 153, Pl. 29)

IBERIAN ROCK LIZARD *Lacerta monticola*   **Pl. 28**

Range Mountains of the north-western, west and central Iberian penininsula (Cantabrian, Estrela, Peña de Francia, Bejar, Gredos and Guadarrama ranges); also lower areas in Galicia. Map 106.

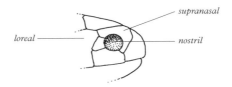

Identification Up to 8cm snout to vent, tail about twice as long. A small to medium sized, rather robust small lacerta, often with a distinctive green underside. Pattern varies: sometimes a vertebral streak or double series of bold spots on midback. Males sometimes green above. Young have a blue tail. A very variable species that is most likely to be confused with the Common Wall Lizard (p. 145) but, if present, green underside is diagnostic of the Iberian Rock Lizard which usually also has flatter, shinier back scales. Some animals, especially immature ones, have whitish bellies and are more difficult to identify with certainty.

Variation Considerable regional variation exists and the problems of identification also vary from place to place.

  Gredos Bejar and Guadarrama ranges in central Spain (*L. monticola cyreni*). Differs from Common Wall Lizard, which occurs in the Guadarrama range, in its usually unspotted underside; males are sometimes green above; rostral scale usually contacts frontonasal (Fig., p. 150).

  Peña de Francia range in west Spain (*L. monticola martinezricae*). At least some animals look like Iberian Wall Lizards (p. 151), males sometimes having reticulations on the sides that may extend on to the back; they presumably lack the typical throat pattern of the wall lizard consisting of small dark spots (Fig., p. 152).

  Estrela range in central Portugal (*L. m. monticola*). No Common Wall Lizards present. Rather like central Spanish populations but belly may have a slight pink flush and is often heavily spotted; rostral scale often does not contact frontonasal scale.

  Galicia, and León and west Cantabrian ranges in north-west Spain (*L. m. cantabrica*). The eastern part of the range can be one of the most difficult areas for identification as animals may look very like the Common Wall Lizard which occurs here. However there is little or no dark spotting on the underside, some males are green above; rows of granules between supraocular and supraciliary scales are frequently complete (often incomplete in the Common Wall Lizard). Although often very different in appearance from Estrela animals, the north-west Spanish populations are closely related to them.

Similar Species Common Wall Lizard (see above). Bocage's, Carbonell's and Iberian Wall Lizards lack green bellies and usually have their dark dorsolateral stripes or series of spots better developed than the vertebral stripe if this is present.

Habits Mainly limited to a few mountainous areas where not usually seen below 1,100 m (higher in south where can reach nearly 2,600m). Often found near or above the tree-line in a variety of sometimes rather damp habitats: screes, large boulders, rocky hillsides with pine, heather and juniper scrub, in and around boulder-filled stream beds etc. A very cold-resistant lizard, largely restricted to areas where winters are long and summers often interrupted by rain, fog and even snow. Usually occurs at generally higher altitudes than Wall lizard species. Where it is found with the Common Wall Lizard, it tends to climb rather more than this species. May occur at quite high densities ranging from 50 per hectare in the Cantabrian mountains to 1,500 per hectare in the Serra da Estrela. In Galicia, populations are exceptional in occurring almost down to sea level.

  Breeding males are often territorial. Females lay from 3–10 eggs, 10–16mm x 6–10mm, the average number varying from area to area and ranging from about 5 to 7. These are often laid under stones, sometimes, communally, and take 6–8 weeks to hatch. Two clutches may be produced in northern Spain but only one at high altitudes in the centre of the country. Babies mature after 1–2 years at a snout vent length of about 5cm.

Note Central Spanish populations have sometimes been treated as a separate species, *Lacerta cyreni*.

PYRENEAN ROCK LIZARD *Lacerta bonnali*   **Pl. 29**

Range Central Pyrenees mountains: in Spain from El Portalé pass (Huesca) to Bonaigua (Lerida) and in France around Lac Bleu de Bigorre. Map 107.

Description Up to about 6cm from snout to vent; tail about twice as long. Grey-brown above without strong stripes but dark flecks sometimes present; flanks totally dark or with small light flecks. Underside whitish, yellowish or greenish.

Similar Species Aurelio's and Arán Rock Lizards (below) have different ranges. Common Wall Lizard (p. 145) lives in same areas although generally at lower altitudes with little or no overlap. Pyrenean Rock Lizard can be separated from this wall lizard by its wholly or largely unspotted, often yellowish or greenish underside, consistent lack of well defined light dorsolateral stripes, rostral scale meeting frontonasal (Fig., below), and supranasal scale contacting loreal (Fig., p. 148).

Habits Only found in alpine habitats at very high altitudes, 1,700–3,000m, where it has a very short period of activity in the summer. Lives mainly on and around rock formations and screes with many crevices, such as ones made up of limestone, slate and schist where it may be abundant; it is less common on granitic rocks. Often occurs on rocky areas at the edges of alpine meadows where snow is absent only in the summer and is frequently encountered close to glacial lakes and torrents. In spite of its cool habitat, this lizard often reduces activity in the middle of the day, perhaps avoiding the strong ultraviolet radiation present at high altitudes.

Females lay eggs about 14mm x 9mm which may hatch in 4–5 weeks; babies are 2–2.5cm from snout to vent.

Other names. Originally described as a subspecies of Iberian Rock Lizard, *Lacerta monticola bonnali*.

Note This and the next two species are very similar in appearance but differences in their chromosomes and DNA indicate that they have had long separate histories.

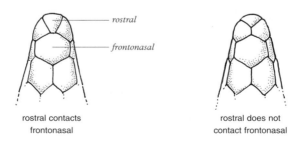

rostral contacts
frontonasal

rostral does not
contact frontonasal

Small Lacertas, tops of snouts

AURELIO'S ROCK LIZARD *Lacerta aurelioi*   **Pl. 29**

Range Pyrenees around the area where France, Spain and north-west Andorra meet: massifs of Pica d'Estats, Montroig, Coma Pedrosa and Tristaina. Map 108.

Identification Up to about 6cm from snout to vent. Very similar to Pyrenean Rock Lizard but more commonly has dark spots along sides of back which are often separated from the dark flanks by pale dorsolateral stripes; underside often yellow.

Similar Species Best separated from Pyrenean and Aran Rock Lizards by range. For distinction from Common Wall Lizard see Pyrenean Rock Lizard.

Habits Similar to Pyrenean Rock Lizard. Found from 2,200–3,000m in mainly south-facing glaciated valleys

ARAN ROCK LIZARD *Lacerta aranica*

Range Central Pyrenean mountains: the Mauberme massif, between the Arán and Ariége valleys. Map 109.

Identification Up to about 6cm from snout to vent. Grey-brown or greyish above often with two dorsolateral rows of dark spots on foreparts which are separated from the dark flanks by a lighter area. On the cheek, a large scale contacts both the masseteric and tympanic scales.

Similar Species Best separated from Pyrenean and Aurelio's Rock Lizard by range. For distinction from Common Wall Lizard see Pyrenean Rock Lizard.

Habits Similar to Pyrenean Rock Lizard. Known only from an area of 26 square kilometers where it occurs from 1,900–2,500m.

IBERIAN WALL LIZARD *Podarcis hispanica*   **Pl. 28**

Range Iberia and west Mediterranean coast of France inland to Toulouse and Lyon. Also north-west Africa, from Morocco as far east as Tunisia. Map 110.

Identification Up to about 6.5cm snout to vent but usually less, tail about twice as long. A rather small, delicately-built, often flat Wall Lizard; typically with a fairly pointed snout, brown or greyish ground colour (occasionally green) and very variable pattern. Belly usually whitish, buff, pinkish or red (occasionally yellow) and throat pale often with small clearly-defined spots, especially at sides (Fig., p. 152). Pattern often basically striped: in most areas dark vertebral streak absent or weaker than the dark dorsolateral streaks or series of marks (exceptions in north-east and east). Typically, females have strong regular stripes while males are more spotted and blotched. Males may be reticulated especially in north-west. Young may have blue tail.

The only small lacerta over most of the southern half of Iberia. Here, Carbonell's Wall Lizard occurs only in the Atlantic region and the Italian Wall Lizard only near Almeria.

Variation Great variation in colouring and some in adult size and proportions. For instance, animals in parts of southern Spain have green backs and deep heads, while those from parts of the the east have only restricted dark markings and may be quite uniform in colour. Animals on islands off the north coast of Spain tend to be large and dark. This lizard is also sometimes bigger in towns than in the surrounding country, presumably because food is more abundant.

Similar Species Common Wall Lizard (p. 145) sometimes has strong vertebral stripe (in most cases stronger than dark dorsolateral stripes or series of spots), usually blotched or heavily marked throat and often orange red belly; it is less delicately built than Iberian Wall Lizard. See also Bocage's (p. 152), and Carbonell's (p. 153), Wall lizards, and the Iberian Rock Lizard (p. 148).

Habits Found up to 2,500m in south of range. Usually a climbing species encountered on rocks, road-cuttings, walls, parapets, outcrops and even tree-boles in some places. Often replaced in less precipitous habitats by other lacertids (Spiny-footed Lizard, Psammodromus species, other wall lizards etc.). Where they are found together, Common Wall Lizard may climb less than Iberian Wall Lizard and can be found on overgrown screes, road banks etc. Ecological separation of these two forms is often complex and other factors are involved: the Common Wall Lizard tends to predominate at higher altitudes, in moister places and near human habitations.

Males are not very territorial. Females lay 1–5 eggs, 9–15mm x 5–8mm which hatch in about 8 weeks producing babies 2–2.5cm from snout to vent. These lizards only rarely survive 4 years in natural conditions.

Other names. *Podarcis hispanicus*.

Note Recent research shows the Iberian Wall Lizard is not a single species and there may be at least five in the Iberian peninsula. From here the group has invaded north-west Africa at least twice, presumably being carried across the sea on natural rafts.

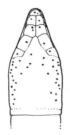

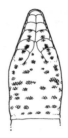

Iberian Wall Lizard
often at least some well
defined spots

Common Wall Lizard
usually blotches or heavier
markings

Wall Lizards, throat patterns

BOCAGE'S WALL LIZARD *Podarcis bocagei*   Pl. 28

Range Extreme north-west Spain and north Portugal as far south as the Douro valley. Map 111.

Identification Up to 7cm snout to vent, tail about twice as long; males larger than females. Closely related to the Iberian and Carbonell's Wall lizards. Differences from the Iberian Wall Lizard vary somewhat from locality to locality. Generally Bocage's Wall Lizard is more robust and less flattened and the back of males is often green as is their light dorsolateral stripes. The flanks may be brownish or yellowish and the belly is frequently yellow, salmon or orange (otherwise whitish) with often quite strong dark spotting but without blue spots on the outer scales. Females may have yellowish light dorsolateral stripes but in general are not as strongly striped above as those of the Iberian Wall Lizard.

Variation Very variable in colour and pattern. Some populations on islands off the coast of Spain have very heavy spotting on underside. Dark dorsolateral streaks are sometimes close together.

Similar Species Iberian Wall Lizard (p. 151); Carbonell's Wall Lizard (below).
Habits Occurs largely outside areas with a Mediterranean climate in places with a cooler, less dry summer. Often in open deciduous woods or their degraded remains and where they have been replaced by scrub; but also in gorse thickets near the sea; babies tend to be in places with more herbacious vegetation than adults. Extends up to 1,850m and may reach densities of 50–300 animals per hectare.

Habits like Iberian Wall Lizard but does not climb as much and is usually found in less precipitous places where it often hunts on the ground among vegetation.

Females may lay 3 or even 4 clutches of 2–9 eggs, 9–14mm x 6–8mm, which hatch in 2–3 months. Babies are 2–2.5cm from snout to vent and take 1–2 years to mature at about 4.5–5cm from snout to vent; they may live up to 5 years in the wild.

CARBONELL'S WALL LIZARD *Podarcis carbonelli*   **Pl. 29**
Range Portugal, from Douro valley through centre and south of country, in mountains and along coast; also neighbouring Spain (Peña de Francia and Gata ranges; Salamanca area; southwest coast). Additionally present on Berlingas islands off Portuguese coast. Map 112.
Identification Up to about 6.5cm from snout to vent, intact tail about twice as long; females may be rather larger than males at some localities. Differs from Bocage's Wall Lizard in usually having weaker dark markings on the back which is often brown or grey, although it may sometimes be green, especially in males where the flanks are also frequently this colour with a dark reticulation. The belly is usually whitish rather than yellow or orange and there are frequently blue spots on the outer belly scales.
Variation Animals from the Berlingas islands (*P. c. berlingensis*) are large, males being bigger than females, with a heavily spotted whitish underside.
Similar Species Bocage's Wall Lizard (opposite) which only occurs with Carbonell's Wall Lizard in the Douro valley. The Iberian Wall Lizard (p. 151) is less robust and more flattened than Carbonell's Wall Lizard, and usually lacks green flanks and heavy markings beneath.
Habits Like Bocage's Wall Lizard, climbs less than Iberian Wall Lizards within its range and is often largely ground dwelling. In mountain areas, at least in the north of its range, occurs in open oak woods and scrub areas produced by their degradation. Found up to at least 1,100m. Often seen on the banks of forestry roads where it may be present in high densities. Takes refuge in cracks and holes between stones and among the roots of trees and shrubs. Also occurs near the sea on sand dunes. Sometimes found at densities of 330–1,600 animals per hectare and 4,000–9,000 per hectare on the Berlingas Islands. Feeds mainly on arthropods but Berlingas animals eat many small snails, especially in autumn. Here some males each live with perhaps four females in a small area of less than a square metre, while other, unattached males move over large distances sometimes poaching the females of their sedentary colleagues.

In north-east of range, females may have a single clutch per year which averages about 2 eggs, whereas in the Berlingas islands there may be 3 clutches of 1–4 large eggs which are laid 2–6 weeks after mating and take 10–15 weeks to hatch.
Other names. Originally described as a subspecies of Bocage's Wall Lizard, *Podarcis bocagei carbonelli*.
Note Although Carbonell's Wall Lizard is most similar to Bocage's Wall Lizard, they both seem to be independently derived from the Iberian Wall Lizard.

COLUMBRETES WALL LIZARD *Podarcis atra*   Pl. 29
Range Columbretes islands, off eastern Spain. Map 113.
Identification Up to 7.5cm from snout to vent; tail about 1.5–2 times as long and often thick; males bigger than females. Grey, brown or greenish above, often with a pattern of dark speckles or spots that may be irregular or arranged in lines along the body. Females especially may have dark flanks and light dorsolateral stripes. Uniformly coloured animals also exist. Underside white, orange or red with dark spots on the throat that may sometimes extend on to belly; bright belly colour may be present even in babies.
Variation Differences exist between populations on the three islands that make up the Columbretes. Frequent features are as follows. Columbreta Gran: dark vertebral stripe, reticulated pattern, white or orange belly. Montcolibre: absence of vertebral stripe and reticulation, often greenish colouring, white belly. La Foradada: any vertebral stripe weak, white or yellow spots on back, orange belly.
Similar Species None within presently accepted range.
Habits Found in a wide range of habitats, the adults occurring especially in rocky areas with low dense bushy vegetation. Young animals are found among more herbaceous plants and grass. Slower moving than related populations on mainland of Spain. Abundance is very variable but reaches 1,000 animals per hectare in some places. Eats a wide range of food including some vegetable matter and scraps left by people; adults often cannibalise eggs and babies.

About 1–3 weeks after mating, females lay 1–5 eggs, approximately 13mm x 7mm, which may be deposited in sandy soil, sometimes communally. Several clutches may be produced in a season and the eggs take 6–10 weeks to hatch. Babies are 2.5–3cm from snout to vent and mature in under a year at lengths of 5–6cm.
Other names. *Podarcis ater*. Previously a subspecies of the Iberian Wall Lizard, *P. hispanica atra*.
Note Although Carbonell's Wall Lizard is most similar to Bocage's Wall Lizard, they both seem to be independently derived from the Iberian Wall Lizard.

MADEIRA LIZARD *Lacerta dugesii*   Pl. 29
Range Atlantic islands of Madeira, Porto Santo, Deserta Grande, Bugio and Great and Little Selvages. Introduced to the Azores and the harbour area of Lisbon in Portugal. Map 114.
Identification Up to 8cm from snout to vent, tail up to 1.7 times as long. Like many small lacertas, has fine scaling on the back and six large scales across the mid-belly. Distinctive features include: two postnasal scales, fine scaling on the temple without an enlarged masseteric scale, large scales in centre of lower eyelid. Usually brown or grey above but sometimes green. In young and females, the centre of the back is dark speckled, bordered by paler areas and the flanks are darker with light flecks. Adult males are also speckled and flecked but more uniform overall. General colouring above varies, tending to match the specific habitat in which the lizards are found. Most animals are cream or yellowish below sometimes with darker spots, but some males have a blue throat and bright orange or red bellies. However, this bright colouring is said to sometimes fade within seconds if the lizard is disturbed.
Variation Animals from most of the range are robust (*L. d. dugesii*). Those from Porto Santo (*L. d. jogeri*) are a little smaller with rather coarser scaling, and these trends are continued on the Selvages (*L. d. selvagensis*). Black animals occur sporadically and are especially common on Deserta Grande and Bugio. On Madeira, animals from lowland pop-

ulations, especially near the sea, tend to be bigger and darker with longer limbs and tail.
Similar Species No other native lacertid lizard exists in the natural range of the
Madeira Lizard, but the Tenerife Lizard (p. 182) is introduced to the Botanic
Gardens at Funchal on Madeira. The introduced population of Madeira Lizard in
Lisbon can be distinguished from local wall lizards by having two postnasal scales,
fine scaling on the temple and a range of different patterns.
Habits Often quite tame and extremely common, sometimes with up to 4 individuals per square metre. Found in a very wide range of habitats, from high mountains
up to 1,850m to desert islets and sea coasts wetted by spray. Occurs commonly in
rocky, scrubby places and woods where it may climb up to 4m on trees; also in cultivated areas including vineyards and may commonly live close to houses and tourist
facilities such as picnic areas. Babies often occur in poorer habitats than adults. Diet
is very varied and differs from place to place. Overall a relatively high proportion of
ants and vegetation is taken, but many other insects are often eaten and on the coast
food may include sea woodlice (*Ligia*). Also forages opportunistically, eating scraps
from tourists and even the tails of its own species. The tail is easily shed but regenerates relatively slowly in animals that have a largely vegetarian diet. Tameness, wide
habitat range, abundance and a broad diet are all common features of lizards on
oceanic islands where there were virtually no predators until they were colonised by
people.

Females have 2–3 clutches a year. The eggs hatch in 6–8 weeks and babies are 3–3.5
cm from snout to vent; they probably first breed after two years.
Other names. *Podarcis dugesii, Teira dugesii*
Note Related to the Moroccan Rock Lizard (p. 156) and a natural colonist of
Madeira from North Africa.

SMALL LACERTAS AREA 3: Balearic islands

Small lacertas of the Balearic islands are potentially confusing: the two endemic
species, Lilford's Wall Lizard and the Ibiza Wall Lizard, have many often well differentiated island populations and some 47 subspecies are presently recognised. There
are also two introduced species, the Italian Wall Lizard and the Moroccan Rock
Lizard. However, distribution is very helpful in identification as most islands have
only one or two species as follows:

Ibiza, Formentera and nearby small islands: Ibiza Wall Lizard only.
Majorca: Lilford's Wall Lizard and Ibiza Wall Lizard.
Minorca: Lilford's Wall Lizard, Italian Wall Lizard and Moroccan Rock Lizard.
Small islands near Majorca and Minorca: Lilford's Wall Lizard (Ibiza Wall Lizard on
Les Illetes, near Majorca).

KEY
Colour descriptions refer mainly to the larger islands and do not cover all variations.

1. Minorca only. Ten rows of scales across belly; lower eyelid with large
'window'
Moroccan Rock Lizard *Lacerta perspicillata*, (p. 156, Pl. 30)
Six (rarely eight) rows of scales across belly; lower eyelid without large
window   **2**

2. Minorca only. Back scales between hind limbs (and on upper surface of lower hind leg) clearly keeled; scaling fairly fine (60–75 scales across mid-back, rarely more). Underside whitish or greenish without dark marks. Dorsal pattern variable; often boldly reticulated and back sometimes with an irregular vertebral streak; some individuals may be fairly uniform. Light dorsolateral streaks often not clear and may be broken up.

Italian Wall Lizard, *Podarcis sicula* (p. 162, Pl. 32)

Minorca, Majorca and nearby islands. Back scales between hind limbs (and on upper surface of lower hind leg) smooth; scaling often very fine (70–90 across mid-back, rarely fewer). Underside whitish, reddish, yellow, blue or black; throat often with dark markings. Dorsal pattern variable, not boldly reticulated; a pair of lightish dorsolateral stripes frequently distinct, often enclosing three dark streaks or rows of spots; some individuals may have very weak pattern. Some populations very dark

Lilford's Wall Lizard, *Podarcis lilfordi*, (p. opposite, Pl. 30)

Ibiza, Formentera, nearby islands and parts of Majorca. Back scales between hind limbs (and on upper surface of lower hind leg) often feebly keeled; scaling rather coarse (usually fewer than 70 scales across mid back, rarely more). Underside whitish or brightly coloured; throat sometimes with dark marks. Dorsal pattern variable; a pair of light dorsolateral streaks frequently distinct, often enclosing three dark streaks or rows of spots. Some populations dark

Ibiza Wall Lizard, *Podarcis pityusensis*, (p. 158, Pl. 30)

MOROCCAN ROCK LIZARD *Lacerta perspicillata*   Pl. 30

Range In Europe only Minorca where it was first reported in 1928 but may have been introduced considerably earlier. Occurs naturally in north-west Africa, mainly in the mountains of Morocco but also on coast of Algeria. Map 115.

Identification Adults up to 6cm from snout to vent, tail up to about 1.7 times body length. A small, rather flattened rock lizard. Minorcan animals are usually buff, greyish or greenish above with a dark reticulation that reduces the ground colour to isolated pale spots. Tail may be bluish and is usually so in young. In north-west Africa other colour forms exist including animals with two broad pale stripes along back. Underside bluish or whitish. Easily distinguished from other European small lacertas by the combination of a completely transparent 'window' in lower eye-lid (Fig., below) and 10 or 12 large belly scales across mid-body (instead of 6 or 8).

Variation Uniformly coloured animals occur in North-west Africa and have been reported from Minorca in the past

Similar Species No really similar species within range.

eyelid window

Moroccan Rock Lizard

Habits An agile, climbing lizard, usually seen on and around rock-faces, cliffs, large boulders, rocks in scrub areas, quarries, dry-stone walls, the surfaces of cement or rock water cisterns and human habitations. May also be encountered on the boles of trees. An excellent climber on vertical surfaces but also hunts on the ground usually close to such structures. Tail rather fragile and often regenerated. Females lay 1–3 eggs, about 12–17mm x 6–8mm.

Other names. *Podarcis perspicillata, Teira perspicillata.*

LILFORD'S WALL LIZARD *Podarcis lilfordi* **Pl. 30**

Range Balearic islands. Restricted to small offshore islets off Majorca and Minorca and to the Cabrera archipelago south of Majorca. Occasional animals found on the main islands are likely to be introductions. Map 116.

Identification Up to 8 cm from snout to vent but usually smaller; tail may be 1.8 times as long. Typically a rather robust small lacerta with a pointed snout and very fine, smooth scaling (70–90 scales across mid-back, rarely fewer). In many populations, most animals are brownish or greenish above, often with lightish dorsolateral streaks between which are three dark, often broken, stripes. Sides are sometimes reticulated and some individuals have only a very faint pattern. Underside whitish, yellow or reddish, often with some dark marks, especially on throat. Tail of babies may be blue-green.

Variation Some island populations vary greatly in size, shape and colouring. Maximum snout to vent length ranges from under 6cm to 8cm. Members of some populations are fairly slender but most are stout with a rather pointed head, a thick (often 'turnip-shaped') tail and shortish legs. Melanistic populations are common, being black, bluish, or dark brown above and often largely blue or more rarely black beneath. Other variants in dorsal colouring occur, for instance the back can be light brown or bright green and the tail of adults may be greenish. Some 23 subspecies are presently recognised. One of the more distinctive of these was found on Les Rates islet in Maó harbour, Minorca, but this was completely destroyed by dynamite in the last century to improve navigation.

Similar Species Most likely to be confused with the Ibiza Wall Lizard (p. 158) and perhaps the Italian Wall Lizard (p. 162 and Key, p. 155).

Habits Mainly confined to small, rocky islets, many of which are very close to the shores of the main islands. May occur on quite bare surfaces, in low scrub, and in woods on Cabrera. A very tough lizard able to tolerate extremely hostile environments. Usually ground dwelling, although does climb to some extent. Often more than 300 animals per hectare and in limited areas may reach extraordinary densities of 44,000 per hectare. On some islets this lizard competes with a large species of spider and is predated by gulls and kestrels. It is often tame. Food is made up of arthropods, snails and a varying proportion of vegetable material including fruit and flowers. Pollen and nectar are also taken and some endemic Balearic plants depend on the lizard for pollination. Populations on some islands exploit nesting birds, eating fragments of food disgorged by gulls for their young, and also feeding on discarded prey of Eleonora's falcon in a similar way to Skyros (p. 178) and Erhard's (p. 176) Wall Lizards. This lizard may also be cannibalistic, eating its own babies and tails of other individuals.

Females may lay three clutches in a season containing 1–4 eggs, about 14–19mm x 8–13mm, which hatch in about 8 weeks, the babies being about 3–3.5cm from snout to vent.

Note Until less than 2,000 years ago, this lizard was abundant on the main islands of Majorca and Minorca. Its disappearance was probably due to the accidental human introduction of predators which have not yet reached the offshore islets. Among them is possibly the False Smooth Snake (p. 223) and the Weasel.

IBIZA WALL LIZARD *Podarcis pityusensis*   **Pl. 30**
Range Balearic islands of Ibiza and Formentera and numerous small islets nearby. Also introduced to Majorca (Palma, and Les Illetes islets in Palma bay), Barcelona, the coast of Spain, Basque and Lisbon. Map 117.
Identification Ibiza (*P. p. pityusensis*). Up to 7 cm from snout to vent but usually smaller; tail may be about twice length of body. A robust, rather short-headed, unflattened small lacerta with relatively coarse, usually very slightly keeled scales (typically under 70 across mid-body, rarely up to 75). Back is green in most animals although it can also be brown or grey. Lightish dorsolateral stripes are usually well defined, and there are often three dark streaks or rows of spots between these. Underside is whitish, grey, yellow, orange or pink and throat and sometimes belly may be dark spotted.

Formentera (*P. p. formenterae*). Generally similar to Ibiza form but up to 8 cm snout to vent, green on back often brighter and more extensive and belly is greenish-grey, often with some yellow or red towards tail.
Variation Small-island populations are extremely varied. Adult size ranges from slightly over 6cm snout to vent to 9cm. Members of many populations are extremely robust but on some islands they are rather slender. Some are coloured more or less like large-island forms, others are light brown and a number tend to be melanistic, being black, dark brown or bluish above with an often bluish belly. Dorsal pattern is frequently broken up so that dark stripes and light dorsolateral stripes are no longer clear. In some cases the flanks are brightly coloured, for instance. blue or orange. A total of about 24 subspecies are presently recognised.
Similar Species Most likely to be confused with Lilford's Wall Lizard (p. 157).
Habits A widespread, versatile wall lizard found on the main islands mainly in well-vegetated sites, including cultivated fields, gardens, coastal situations, and on and around drystone walls, ruins and human habitation. Occurs in pine woods but usually at low densities. Small-island populations may exist on almost bare rock or salt-flat with very little vegetation. Often very common, frequently reaching densities of 200 animals per hectare on Ibiza and as many as 30,000 per hectare on some small islets. Food consists of invertebrates, sometimes including a high proportion of ants, and a substantial amount of vegetation especially in summer.

Females lay 1–4 eggs, about 15mm x 9mm; hatchlings are 3–3.5cm from snout to vent and probably become sexually mature at around 1.5–2 years.

SMALL LACERTAS AREA 4:
Corsica and Sardinia

The large Mediterranean islands of Corsica and Sardinia have only two endemic Small lacertas: Bedriaga's Rock Lizard and the Tyrrhenian Wall Lizard. A third, more widespread species, the Italian Wall Lizard, also occurs. The only other lacertid is the Pygmy Algyroides (p. 132) easily distinguished by its small size, coarse scales and usually sombre colouring.

KEY

Distinctly flattened; supratemporal scales not turned down onto side of head. Back scales between hind legs unkeeled, more or less flat. Underside greyish, yellowish, green or red, often with dark markings especially on throat or completely dark. Upperside rarely striped, usually boldly reticulated, sometimes spotted

Bedriaga's Rock Lizard, *Lacerta bedriagae*,(p. below, Pl. 31)

Not distinctly flattened; supratemporal scales clearly turned down onto side of head in adults. Back scales between hind legs usually not or only lightly keeled, unflattened. Underside whitish, yellow, orange, salmon or red, typically with dark spots, especially on throat. Upperside often striped: light dorsolateral streak or row of spots usually visible. Dark dorsolateral stripe or series of spots is better developed than dark vertebral streak (if present)

Tyrrhenian Wall Lizard, *Podarcis tiliguerta*, (p. 160, Pl. 31)

Not distinctly flattened, supratemporal scales turned down on to side of head in adults. Back scales between hind legs usually distinctly keeled, unflattened. Underside pale without clear spots, although throat may be grey. If striped, dark dorsolateral stripe or row of spots is less well developed than vertebral streak

Italian Wall Lizard, *Podarcis sicula*, (p. 162, Pl. 32)

BEDRIAGA'S ROCK LIZARD *Lacerta bedriagae* Pl. 31

Range Corsica and small isolated areas in Sardinia. Sardinian localities include the coast round Punta Falcone, Monte Limbara, Monte Albo, Sopramonte di Oliena, Catena del Marghine, the Gennargentu massif and the Sette Fratelli forest in the south-east of the island. Map 118.

Identification Up to about 8 cm from snout to vent; tail about 1.5–2 times as long. A medium-sized, distinctly flattened lizard with pointed snout and often bulging cheeks. Back scales flattish and unkeeled and supratemporal scales not clearly turned down onto side of head. Above usually greenish, yellow-green, brownish or greyish, with black or dark brown markings that often form a reticulation. This may be so extensive that the lizard is very dark with small light spots of ground colour, or so reduced that the lizard is almost unmarked. Neck, forepart of back and sides sometimes overlaid with brownish tinge. Very rarely, an animal may have some tendency to striping. Underparts whitish, greyish, yellowish, greenish, reddish or red; often with some dark marks especially on throat; in dark-backed individuals belly is also often dark. Young are like adults but tail frequently a bright blue-green.

Variation Over most of Corsica (*L. b. bedriagae*), animals tend to be large and robust often with a coarse pattern involving interconnected transverse dark bars, but in the extreme south of Corsica and in Sardinia they are more delicately built with a finer reticulation enclosing small spots. Three subspecies have been named from Sardinia but they may all be referable to one, *L. b. sardoa*.

Similar Species In Sardinia, most likely to be confused with the large often reticulated Italian Wall Lizards found there. However, this species is less flattened, usually has the back scales between the hind legs quite strongly keeled, nearly always a pale whitish, greyish or greenish underside without clear spots, and the supratemporal scales are clearly turned down on to the side of the head in adults.

Habits Largely a montane form. Over much of Corsica usually found over 550m and extending up to at least to 2,550m although it is commonest between 1,000m and 1,800m, increasing in abundance as the vegetation starts to disappear. In Sardinia occurs up to 1,800m in the Gennargentu massif. Essentially a climbing species most often seen on rocks and stony surfaces, especially eroded crystalline ones such as granite, although it may occur on limestone in Sardinia. Found on cliffs, boulders, outcrops, natural pavements, rocks along mountain streams, on dry-stone walls, bridges and even buildings. These habitats may be situated in light woods including pine, scrub, and heathland above tree line. In mountain areas avoids very dry places and extremely sunny situations, often occurring in semi-shade and frequently found relatively close to water. On Corsica can sometimes be encountered at high densities and up to 30 animals have been observed taking refuge in a single crevice. In extreme southern Corsica, and northern Sardinia including several of the La Maddalena islands, this lizard lives at sea level in rocky situations including granite boulders.

Females lay 3–6 elongate eggs, about 15–17mm x 6–12mm, which hatch in about 8–9 weeks; the babies are 3–3.5cm from snout to vent and become sexually mature at around 6–7cm.

TYRRHENIAN WALL LIZARD *Podarcis tiliguerta*   Pl. 31
Range Corsica, Sardinia and neighbouring small islands. Map 119.
Identification Up to about 6.5 cm from snout to vent (rarely larger); tail up to about twice as long. A small unflattened Wall lizard usually with convex, smooth or feebly keeled back scales, the supratemporal scales turned downwards onto the side of head (at least in adults), and usually a basically striped pattern which is very variable in form. Females tend to be some shade of brown with pale dorsolateral streaks; these are each often bordered above by a dark dorsolateral stripe or series of spots that is usually better developed than the dark vertebral streak (if present). Males may also be brown but are often green on the back or sides or on both. Sides may be reticulated, sometimes with blue spotting. Dark marks on back may be more broken up and often more extensive than in females. In some cases, they form a general reticulation that covers the whole of the body, but light dorsolateral streaks often still show through as a series of spots. Occasionally, animals are almost unmarked and may have only a very faint light dorsolateral streak or series of spots. Underside may be whitish, yellow, orange, salmon or even red, the bright colour sometimes being restricted to the throat. Most animals have some dark spotting beneath, especially on throat and lips.
Variation Populations on some small islands are very variable and 10 subspecies are recognised; these may vary in body size and some are very dark.
Similar Species Italian Wall Lizard (p. 162); Bedriaga's Rock Lizard (p. 159). The Italian Wall Lizard is often larger, usually has more strongly keeled scales, and is typically whitish or greenish beneath without spots on belly and throat. Corsican animals often have a broad brown or dark vertebral stripe that is better developed than the dark dorsolateral stripe, if any. Sardinian animals are often boldly reticulated, or almost unmarked. Bedriaga's Rock Lizard is flatter with flatter body scales, and the supratemporal scales are not clearly turned down on to side of head.
Habits Found in dry and sometimes moist situations, including stony places in scrub, open woods, field-edges, dry-stone walls, and path- and road-sides. Also encountered in sandy places with vegetation near the sea and on screes and among dwarf juniper bushes above the tree-line; frequently occurs close to people but tends to avoid dense woods and scrub. Widely distributed from sea-level to over 1,800m

but commonest at middle altitudes. Climbs to some extent but not so extensively as Bedriaga's Rock Lizard. Often very common: 700–900 animals per hectare may be found in rocky degraded forest on main islands and densities of 2,000 per hectare occur on some offshore islets where there may locally be 10 lizards per sq. m. In some lower areas, especially under 400m, is partly replaced by the Italian Wall Lizard which may predominate in richer grassy areas, in moist valleys in fields, and often in and near human settlements. This species has replaced the Tyrrhenian Wall Lizard on the eastern coastal plain of Corsica where it was probably accidentally introduced.

Females lay clutches of 6–12 eggs, 10–16mm, long that hatch in 8–12 weeks to produce young 5–6cm in total length

SMALL LACERTAS AREA 5:
Italy, Sicily and Malta

There are six small lacertas in this area of which three are confined to the region. These are the Sicilian, Aeolian and Maltese Wall Lizards. The Italian Wall Lizard now also occurs in areas 1, 2, 3, 4, 6 and 7. Two much more widespread species are also present: the Viviparous Lizard and the Common Wall Lizard. The only other lacertid lizard in the area is the Green Lizard (p. 138).

KEY
ITALY
1. Alps etc. only. Short-legged and small-headed. Collar distinctly serrated. Scaling coarse (25–37 dorsal scales across mid-body). Granules between supraocular and supraciliary scales absent or at most 4 on each side. 5–15 femoral pores present
Viviparous Lizard, Lacerta vivipara (p. 144, Pl. 27)
All regions. Legs relatively long and head quite large. Collar not usually distinctly serrated, scaling fine (42–75 dorsal scales across mid-body). Granules between supraocular and supraciliary scales usually extensive, at least 3 on each side;13–27 femoral pores   **2**

2. Underside usually with some dark spotting, at least on throat. Belly may be red and throat may have some rusty markings   **3**
Underside nearly always without dark markings. Belly usually whitish or with greenish tinge (rarely red, especially on mainland; can be blue or blackish in some island populations)
Italian Wall Lizard, Podarcis sicula (p. 162, Pl. 32)

3. Aeolian islands only. Under parts whitish; brown above.
Aeolian Wall Lizard, Podarcis raffonei (p. 164; Pl. 32)
Italy. Belly often red in males and throat may have some rusty markings.
Common Wall Lizard, Podarcis muralis (p. 145, Pl. 27)

SICILY
Usually well defined, light dorsolateral streaks. Markings on back may consist of a simple vertebral streak (especially over loins), similar simple dark dorsolateral streaks or rows of spots, or both. Sometimes uniform. Throat

may be spotted; throat and belly often red, orange or pink in males; (for Egedi islands see text)

Sicilian Wall Lizard, *Podarcis wagleriana* (p. 163, Pl. 32)

Light dorsolateral streaks usually absent or broken up. Sides and back often at least partly reticulated, or pattern weak and body more or less uniform. Throat not clearly spotted; belly whitish or greenish, occasionally grey

Italian Wall Lizard, *Podarcis sicula* (p. below, Pl. 32)

MALTA, GOZO, FILFOLA etc.

Only one highly variable species present.

Maltese Wall Lizard, *Podarcis filfolensis*, (p. 165, Pl. 31)

ITALIAN WALL LIZARD *Podarcis sicula* Pl. 32

Range Italy, Sicily, Corsica, Sardinia, Minorca, south France (Ile d'If near Marseilles, Toulon, perhaps Provence), Spain (Almeria, Santander), east Adriatic coast (south to Dubrovnik), Turkey (Sea of Marmara area). Also Elba and many small islands in the Tyrrhenian and Adriatic seas. Other isolated colonies have been reported: on coast of Libya and perhaps Tunis and in the USA (Long Island, New York; Philadephia, where it is now extinct). Map 120.

Identification Up to 9cm but usually smaller. A highly variable Wall Lizard, usually with a rather deep, often fairly long head and robust body. Underside usually whitish, greyish, or with greenish tinge, nearly always without dark spots. Typically green, yellowish, olive or light brown above. Females tend to be smaller, and smaller-headed, than males with a more obviously striped pattern. Individuals in some populations slowly change colour, becoming browner in summer. Main regional variations are given below.

Most of Italy, Elba, Corsica (*P. s. campestris*). Often under 7cm in North. Usually basically streaked. Vertebral stripe may consist of a brownish area with darker markings; light dorsolateral streaks or series of spots often clear. Intergrades with more southern populations.

East Adriatic coast. As north Italian animals but rather large, dorsolateral streaks often not very clear; unpatterned animals occur occasionally.

Southern Italy and Sicily (*P. s. sicula*), **European Turkey** (*P. s. hieroglyphica*). Up to 8.5cm or even bigger. Generally like north Italian animals but often larger, more robust and bigger-headed. If present, vertebral streak is usually black and light dorsolateral streaks are not clear. Sides frequently reticulated or chequered, markings may extend over back. Some animals have a very faint pattern or are nearly uniform.

Sardinia also south Corsica (*P. s. cetti*) **and Minorca**. Like south Italian animals; often reticulated, or with very faint pattern, or uniform.

Small island populations. Highly variable. In Tyrrhenian area, some populations are dark, even blackish, or blue, or with a blue belly. Other belly colours occur, such as red or yellow, especially on islands off north Croatia, and there are often large differences in back pattern between populations.

Variation Great variation in pattern, even within populations. Animals with red, yellow or grey undersides may uncommonly occur on mainlands. Around 48 subspecies have been accepted in recent years.

Similar Species Italy: only species likely to be confused is Common Wall Lizard (p. 145) but this usually has dark markings on underside (at least on throat), and a different range of back patterns. **Sicily**: Sicilian Wall Lizard (opposite).

Corsica, Sardinia and other Tyrrhenian islands: other species present have distinct black spotting on throat and often belly; this is visible even in most animals with dark undersides but is lacking in the Italian Wall Lizard. For other areas, see regional keys and descriptions.

Habits Very variable. A vigorous, opportunistic lizard. In north Italy, Corsica etc., typically found in grassy places, roadside verges, wood edges etc. at low altitudes. In other areas, often also occurs in very open fields, flat derelict areas, sandy places, near sea, vineyards etc. and may exceptionally extend up to 2,000m, for instance on Sicily, although usually below 1,000m. Tolerates close proximity to people better than many small lacertas and often found in parks and gardens of towns.

Tends to climb less than some other small lacertas in range, but more than the Balkan Wall Lizard (p. 175), Sicilian Wall Lizard (below) or Dalmatian Wall Lizard (p. 169) etc. Hunts on ground and is capable of running long distances to shelter, but may return to dry-stone walls, bushes etc. for refuge. In fact can climb quite efficiently and, in the absence of better adapted species, will occupy rock-faces with some vegetation, ruins etc. Often abundant, sometimes occurring at densities of more than 10,000 and even 16,000 individuals per hectare. Quite aggressive and males in populations that occur outside the natural range of the species will sometimes attack other kinds of lizards with green coloration. Sometimes eats a fairly high proportion of vegetable food and on small islands may occasionally eat its own young.

After mating, females may lay up to 4 or even 5 clutches but only 1 or 2 in lizards breeding for the first time; the clutches are laid every 12 days or so and contain 2–12 eggs (often about 5–6) which are about 10–12mm x 5–6mm when first laid but swell to 14–15mm x 11–12mm. Populations on small islands off north Croatia lay fewer eggs (2–4) than those on the neighbouring mainland (4–7). The eggs are also larger and hatch into bigger babies with relatively short legs. This is probably advantageous in islet conditions where there are few predators but restricted food. Eggs generally hatch in about 5–7 weeks producing babies 3–3.5cm from snout to vent. Males mature in 1 year, and females in 1 or 2 when about 5cm from snout to vent. Has lived 13 years in captivity.

Note The natural range of this species appears to be Italy and adjoining Croatia. Its presence elsewhere is almost certainly due to accidental human introduction. Ability to survive in and around towns means the Italian Wall Lizard is likely to be inadvertently transported, to survive in the port areas to which it is taken and expand from them. For instance, on Corsica there seem to have been at least three separate colonisations by this species. Its introduction may have resulted in the loss of most populations of the Aeolian Wall Lizard (p. 164) and possibly the total loss of a distinctive form on San Stefano island near Naples that was originally regarded as just a subspecies of the Italian Wall Lizard, *P. s. sanctistephani*.

Other names. *Podarcis siculus*.

SICILIAN WALL LIZARD *Podarcis wagleriana* Pl. 32

Range Sicily except the north-east; Egedi Islands west of Sicily (Marettimo, Levanzo, Favignana), also Isola Grande dello Stagnone and Maraone. Map 121.

Identification Up to about 7.5cm, females smaller. A deep headed species that is similar to the Balkan and Dalmatian Wall Lizards. Frequently green above but females especially sometimes olive or brown. Light white, yellow or greenish dorsolateral streaks are typically well-defined and there is frequently a simple dark vertebral streak or row of spots (often just over loins), or similar dark dorsolateral streaks or spots, or both. The sides are dark-spotted or dappled in males, more uniformly dark in females. Animals without any markings above are quite common. Underside is white, but offten red, orange, or pink in males; throat occasionally spotted.

Variation The population on Marettimo, the most western of the Egedi islands, is heavily marked and often reticulated with a dark spotted throat and is distinguished as *P. w marettimensis*. Some hybridisation with the Italian Wall Lizard apparently occurs here.

Similar Species Only likely to be confused with the Italian Wall Lizard (p. 162). In many animals the pattern on the upper parts is different, Italian Wall Lizards often being reticulated. Both species have individuals with virtually no pattern and in such cases the Sicilian Wall Lizard can often be distinguished by bright belly colour, deeper head and more graceful build.

Habits On Sicily, mainly a ground-dwelling lizard often found in places with quite rich vegetation of grass and small shrubs in which it takes refuge. Habitats include dry bushy slopes, edges of broad-leaved woods, pastures and cultivated land and gardens, especially irrigated ones. Occurs in some relatively drier environments in the south of Sicily. Found up to 1,200m. Does not climb much on walls or on rocky slopes, in contrast to the Italian Wall Lizard in this area which also occupies barer and more open places. Tends to be the predominant species inland, while the Italian Wall Lizard is commoner near the coast. Shyer and less aggressive than this species. The population on Marettimo occurs mainly on the scrub-covered higher slopes of the 680m high island, where frequent cloud cover provides some moisture. On Isola Grande dello Stagnone occurs among halophytic plants.

Females lay clutches of 4–6 eggs, 11–13mm x 7–9mm, which are often buried at the base of small shrubs and hatch in about 8 weeks; babies are 5.5–6cm in total length.

Other names. *Podarcis waglerianus*.

AEOLIAN WALL LIZARD *Podarcis raffonei* **Pl. 32**

Range Aeolian islands, north of eastern Sicily. Limited to Vulcano island and three tiny islets: Strombollichio north-east of Stromboli, La Canna west of Filcudi, and Scoglio Faraglione west of Salina. Map 122.

Identification Adults up to 7cm, tail about 2–2.3 times as long, males bigger than females. Brown above without obvious pale stripes but with prominent dark spots or a reticulation. Underside whitish with dark markings on the throat, breast and sides of belly. Back scales only lightly keeled.

Variation Different populations are regarded as separate subspecies. The ones on the small islets are relatively dark. On Vulcano, this lizard interbreeds with the Italian Wall Lizard.

Similar Species Italian Wall Lizard (p. 162).

Habits On Vulcano found on just one peninsula on dry volcanic rocks with little vegetation. This lizard was probably once more widespread in the Aeolian archipelago but has been displaced from most inhabited larger islands by the Italian Wall Lizard which was probably accidentally introduced by people. This species is now widespread on Vulcano.

Other names. Was previously regarded as subspecies of Sicilian and Italian Wall Lizards, under the names *Podarcis wagleriana antoninoi*, *P. sicula alvearioi*, *P. s. cucchiarai* and *P. sicula raffonei*.

MALTESE WALL LIZARD *Podarcis filfolensis*   **Pl. 31**
Range Malta, Gozo and nearby islets; Filfola Rock. Also Linosa island and
Lampione close to Lampedusa. Map 123.
Identification Malta and Gozo. Adults up to about 6.5cm but often smaller, tail
about twice as long. The only lacertid within its range. A small wall lizard with rather
fine dorsal scaling (60–85 scales across mid-back). Colouring highly variable:
ground colour grey, brown or green. Males may have dappled sides, light dorsolateral streaks and two or three dark stripes or rows of spots on back; sometimes heavily reticulated but may be only lightly marked or plain, especially females. Underside
white, yellow, orange or red, sometimes with spots, especially on throat.
Variation Animals from populations on small islands tend to be bigger and more
heavily marked than animals from the larger ones. On Filfola rock, south of Malta,
lizards grow to 8.5 cm and can be predominantly black with small greenish spots, or
brown with heavy dark markings. This population is named *P.f. filfolensis*, while those
on Malta and Gozo are called *P.f. maltensis*. Additional subspecies have been named
for other islands.
Similar Species None within range.
Habits A typical wall lizard found in a wide range of rather dry habitats including
those associated with people: gardens, dry-stone walls, stone-piles, road banks, rocky
slopes. Abundant, especially on small islands.

SMALL LACERTAS AREA 6:
East Adriatic coastal area

The eastern side of the Adriatic sea from the Italian border region to north Albania
includes a total of eight Small lacerta species. Of these, four are confined to the area,
being found mainly near the coast and not extending inland more than 130km (usually less). They are Horvath's Rock Lizard, the Mosor Rock Lizard, the Sharp-snouted Rock Lizard and the Dalmatian Wall Lizard. Of the more wide-ranging species,
the Italian Wall Lizard is almost entirely restricted to the coast and islands in this
area, while the Common Wall Lizard is limited to rather cooler habitats and avoids
the coast except in the north; the Balkan Wall Lizard is found only in north Albania.
The final species, the Viviparous Lizard, is confined to humid mountain regions. The
other lacertid lizards in the area are the Dalmatian Algyroides (p. 131), distinguished
by its large scales and sombre colouring above, and some Green lizards (p. 134).

KEY

1. Mountain areas. Collar distinctly serrated, granules between supraocular and
   supraciliary scales few (0–4) or absent. Dorsal scales coarse (25–37 across
   body).
                  Viviparous Lizard, *Lacerta vivipara* (p. 144, Pl. 27)
   Collar usually not very distinctly serrated, granules between supraocular and
   supraciliary scales often five or more. Dorsal scaling not very coarse (more
   than 40 scales across body)   **2**

2. Central pairs of scales on lower side of tail wide (largest at least twice as wide
   as adjoining scales, Fig., p. 160). Underside bluish. Dorsal scales smooth and
   flat. Head and body very flattened. Back buff-grey or greenish-grey with
   dappled pattern and tail banded blue-green and black, or entire animal

blackish above

      Sharp-snouted Rock Lizard, *Lacerta oxycephala* (p. 168, Pl. 33)
Central pairs of scales on lower side of tail not or only slightly wider than
adjoining scales. Underside not usually bluish   **3**

3. Mountain areas only. First supratemporal scale large, often emarginating
parietal scale (Fig., below). Rostral scale usually contacts frontonasal (Fig.,
p. 150). Scales on back flattish, unkeeled. Belly usually white, yellow or orange.   **4**

First supratemporal scale shallow, does not emarginate parietal scale. Rostral
usually does not usually contact frontonasal (Fig., p. 150). Scales on back
often convex and at least lightly keeled. Belly rarely yellow   **5**

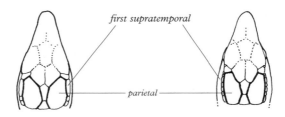

first supratemporal emarginates   first supratemporal does not
(cuts into) parietal   emarginate parietal

4. North of about Sibenik. Usually one postnasal scale and five pairs of chin
shields. Dorsal scales on lower hind leg smaller than back scales. Back rather
pale, contrasting strongly with dark sides. Belly often yellow and throat
white, in adults. Head blunt and short

      Horvath's Rock Lizard, *Lacerta horvathi* (opposite, Pl. 33)
South of about Sibenik. Often two postnasal scales and six pairs of chin
shields. Dorsal scales on lower hind leg about as big as those on back. Back
often mottled and may not contrast strongly with sides. Belly and throat
usually both yellow or orange in adults. Head usually rather long

      Mosor Rock Lizard, *Lacerta mosorensis*, (p. 168, Pl. 33)

5. Not on islands, or coast except in north. Underside usually with some dark
marks, at least on throat; belly typically whitish, pink or red, throat whitish
often with rusty pigment. Upper-side brownish (almost never green) with
darker markings. Eye often coppery. Tail may have conspicuous black and
white spots on sides. Head sometimes rather flattened

      Common Wall Lizard, *Podarcis muralis* (p. 145, Pl. 27)
Underside usually unspotted (or rarely more than a few spots at sides of
neck), commonly greenish, white, orange or red. Back usually green or
brownish. Some animals almost without other markings. Vertebral stripe if

present may contain areas of lighter colour. Eye not coppery. Head usually rather deep

Dalmatian Wall Lizard, *Podarcis melisellensis* (p. 169, Pl. 33),
Italian Wall Lizard, *Podarcis sicula* (p.162, Pl. 32)
Balkan Wall Lizard, *Podarcis taurica* (p. 175, Pl. 34)

HORVATH'S ROCK LIZARD *Lacerta horvathi*   **Pl. 33**

Range North-west Croatia (north-east Istria; Kapela and Velebit ranges south to Sibenik), Slovenia, and small adjoining areas of north-east Italy and Austria (border of Kärnten). Also recently reported from the border region of southern Germany. Map 124.

Identification Up to about 6.5cm from snout to vent, tail about 1.5–2 times as long. A distinctly flattened Small lacerta with a short blunt head. Back usually pale grey-brown with sharply contrasting dark sides, often with a wavy upper edge; sometimes also a dark vertebral streak or irregular spotting on back. Underside unspotted, whitish or with yellow on belly. Iris pale beige. Young like adults, but tail often green or greenish-grey. See sections 3 and 4 of key (opposite) for details of scaling.

Variation Some minor differences in pattern and shape between populations.

Similar Species Similar to some Mosor Rock Lizards (below) but ranges almost separate. Some Common Wall Lizards (p. 145) are very like Horvath's Rock Lizard but usually have dark markings on throat, often a reddish belly in males and, in the range of Horvath's Rock Lizard, the eye is often very coppery red. Also, supratemporal scales are shallow and do not clearly emarginate the parietal scale (see Fig., opposite); rostral scale usually separated from frontonasal (Fig., p. 150), supranasal scale does not contact loreal as it does in most Horvath's Rock Lizard (Fig., below).

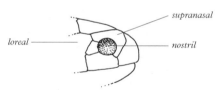

Habits Confined to rather moist mountainous areas, usually above 500m and up to 2,000m, although commonest between 800m and 1,200m. Occurs in humid places and often found in open beech or conifer forests or above the tree-line in desolate karst often with low alpine scrub; occurs in deep wooded canyons at lower altitudes. Sometimes active in cloudy conditions. Strictly a rock lizard and occurs on cliffs, karst pavements, outcrops, boulders, road cuttings, bridges etc. Like other east Adriatic rock lizards, very agile, often leaping into air to catch prey, which it may ambush from the crevices in which it also hunts and customarily takes refuge. Frequently occurs with Common Wall Lizard at lower altitudes where this species tends to occupy less precipitous habitats. Is more static than the Common Wall Lizard and unaggressive.

Females lay clutches of 3–5 (usually 3–4) eggs, about 18mm x 8mm, that hatch in approximately 5–6 weeks. Hatchlings are about 2.5cm from snout to vent and become sexually mature in two years.

Note The closest relatives of this species are the Rock lizards of the Pyrenees and the Iberian peninsula.

MOSOR ROCK LIZARD *Lacerta mosorensis* **Pl. 33**

Range. Southern Croatia (north to Split), southern Bosnia (Hercegovina), south-west Serbia (Montenegro). Map 125.

Identification Adults up to about 7cm, from snout to vent; tail about 1.7–2.3 times as long. A distinctly flattened small lacerta with a long head and long slender tail. Brown, grey-brown or olive with an oily gloss and dark spotting or mottling. Flanks often rather darker than back. In some populations, unmarked animals occur and others are known with very dark sides and spots limited to centre of the back. Underside unspotted, sometimes white or grey but usually deep yellow or even orange in adults. Young similar to adults, but always with a pale belly. For details of scaling, see sections 3 and 4 of key (p. 166). Babies sometimes have a bluish tail.

Variation Considerable variation in markings and substantial differences between some populations.

Similar Species Sharp-snouted Rock Lizard is only other species in the area with a strongly flattened body. Common Wall Lizard (p. 145) is less flattened, has a short-er head, and is rarely yellow below; see key for further details.

Habits Largely confined to areas of high rainfall at some altitude, usually from 450m to 1900m. Primarily a rock lizard, and often found on limestone outcrops in desolate bare karst areas above the tree-line with little vegetation. Also occurs in open woods and areas with juniper bushes. Does not usually climb as high as the Sharp-snouted Rock Lizard and, especially where the two species occur together, is found in less sunny and more humid habitats than this species. For instance in open decid-uous woods, north-facing rocky slopes, open shady holes in karst pavements, around springs, and on fissured outcrops in humid places with quite rich vegetation. Like other Rock lizards, regularly takes refuge in narrow crevices.

Females retain the eggs in their bodies for around 5 weeks after mating. They lay a single clutch a year of pale pink and elongate eggs, often up to 17mm x 8mm, which frequently number about 4, although there may be as many as 8 At laying the embryos are quite well developed and hatching usually takes place in just 17–20 days. The babies are rather less than 3cm from snout to vent with a very long tail. Similar retention of the eggs within the female, followed by a short incubation period after laying, occurs in egg-producing populations of the Viviparous Lizard.

SHARP-SNOUTED ROCK LIZARD *Lacerta oxycephala* **Pl. 33**

Range Southern Croatia (as far north as Sibenik) including islands, southern Bosnia, Serbia (Montenegro); possibly also extreme northern Albania. Map 126.

Identification Adults usually up to about 6.5cm, from snout to vent; tail 1.5–2 times as long. A distinctly flattened, delicately built small lacerta with a pointed snout, noticeably raised eyes, a rather short body, and a slender tail. Hind toes rela-tively short and distinctively kinked. The only species within its range to have the central pairs of scales on the underside of tail very wide (see Fig., below).

Colour varies. Most animals at low altitudes on mainland (main-ly under 600m but sometimes to 900m) are light buffish grey above (greenish in some lights) with a reticulated pattern, and the tail is conspicuously banded black and turquoise green (this pattern is lost on regrown sections of the tail). Animals from highland areas and some islands are much darker and may be entirely black above. Mainland populations at middle altitudes have animals of varied colouring. Some individuals may also change through the year, becoming darker in cooler conditions. Underparts are blue, more intense in males and in dark animals. Young are coloured like lowland adults, but tail is more brightly marked.

Variation Some regional variation in colour, see above.

Similar Species None very similar within range. Easily distinguished by colour, shape and broad scales under tail.

Habits Occurs from sea level to about 1,500m, and is very cold-tolerant, sometimes being seen in the mountains while snow is still on the ground. Its hibernation in such areas is shorter than most other lizards present. Found in a wide variety of habitats, but most often encountered on sunny cliffs, boulders, rock-pavements, stone piles, screes, walls, and buildings; occasionally also seen on tree trunks. Frequently enters towns and villages. Climbs higher and better than any other European Small lacerta sometimes being seen 20m or 30m up on walls and roofs, but normally much lower. May sometimes occur at high densities in favourable habitats. Usual pale colouring camouflages this species well on the pale grey karst limestone on which it often lives, but black animals are frequently surprisingly inconspicuous too, as there are many dark rock cavities and patches of very dark moss in their habitat. Retreats into rock crevices or among stones when disturbed, and resists extraction from these by clinging with its claws, pushing its back against the crevice ceiling and deepening the head so that it holds the lizard in place. Other lacertid rock lizards often behave similarly. Brightly coloured tail is frequently waved nervously and may be thrashed from side to side if the lizard is cornered by a predator. Presumably this and the bright colouring of the tail increases the chance the predator will grasp this expendable organ rather than the head or body of the lizard. The tail is very fragile in the present species and many other Rock lizards. This lizard eats a wide range of invertebrates including a high proportion of flying forms that land on rock surfaces

Males are very aggressive towards others in the breeding season. Females lay 2–4 (usually 3–4) eggs in a clutch which are about 16mm x 7mm and quite sausage-shaped, like those of some other Rock lizards. Possibly this is because species, that are flattened in connection with using narrow crevices as refuges, have a very low pelvis and their eggs must have a small diameter to pass through it. The eggs hatch in about 6–7 weeks and the babies are about 2cm from snout to vent, apparently becoming sexually mature after their second spring.

DALMATIAN WALL LIZARD *Podarcis melisellensis* Pl. 33

Range East Adriatic coastal region and islands from Italian border area through Croatia and western Bosnia and Montenegro to northern Albania. Map 127.

Identification Mainland. Up to 6.5 cm snout to vent, but often smaller, tail about twice as long as body. A small, fairly stocky lizard with a rather deep, shortish head. Underside white or orange, or red, usually unspotted (if present, spots are few and, in most cases, confined to sides of neck region). Usually light-brownish above or with a green back. Pattern varies: some animals, especially on coast, almost without markings; others with prominent narrow, light dorsolateral streaks, and often a dark vertebral stripe, at least on hind-back, which may contain small areas of lighter pigment. Rest of pattern rather variable but females tend to be more regularly striped than males. Babies are never green.

Variation Considerable differences in pattern and size and around 20 mainly island subspecies have been recognised in recent times. Populations on islands in shallow water relatively close to the shore are generally similar and closely related to mainland animals which are named as *P. m. fiumana*. Populations on more distant deep-water islands are more distinct. Lizards on Brusnik near Sveta Andrija Island (*P. m. melisellensis*) grow to 7.5cm snout to vent and are black with large blue blotches on lower flanks. Those from Vis and nearby islets (*P. m. lissana*) tend to have the sides heavily marbled.

Similar Species Common Wall Lizard (p. 145 and key p. 165). Balkan Wall Lizard in northern Albania (p. 175). Italian Wall Lizard is restricted to coast and islands; on mainland and large islands can usually be distinguished from Dalmatian Wall Lizard as follows: often larger, adults frequently over 6.5cm (rarely over 6cm in Dalmatian Wall Lizard); underside whitish (or tinged green) almost never brightly coloured; light dorsolateral streaks often broken up or faint; animals without any dark markings are uncommon; masseteric shield often not very big (large in most Dalmatian Wall Lizards and may contact supratemporal scales (Fig., below); body often more flattened, and head longer. Habitat can also be a useful clue. On small islands both species are much more variable and can sometimes be difficult to identify. Here 'jizz' is helpful, as is previous experience on the mainland and large islands. On small islands normally one or other species occurs, rarely both.

Habits Found in a variety of dry habitats: path and road banks, dry ditches, open woodland, scrub, dry stony pastures and their overgrown edges. Occurs up to about 1,400m. Largely ground-dwelling but may climb on low walls and stone-piles. Certainly climbs far less than the Common Wall Lizard and rock lizards. Tends to be replaced by Common Wall Lizard inland in moister, shadier places and at higher altitudes. On coast, occurs with Italian Wall Lizard but this species tends to be confined to richer areas (often near human habitation), fields, vineyards, open road verges and even gardens in cities and villages; here the Dalmatian Wall Lizard occupies more broken, often uncultivated country. Rarely, the two species may occur together on small islands, in which case the Dalmatian Wall Lizard is sometimes predominant in coastal areas.

Mainland females produce 2–8 (usually 3–5) eggs at a time and may lay several clutches. Babies are about 2.5cm from snout to vent (around 6cm in total length).

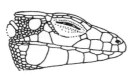

Dalmatian Wall Lizard

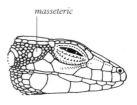

masseteric

Italian Wall Lizard

East Adriatic Wall Lizards, common arrangements of scales on cheek

SMALL LACERTAS AREA 7: South-eastern Europe

South-eastern Europe, as defined here (see Fig., p. 143), contains a wide variety of Small lacertas, there being twelve species in all. Two of these, the Viviparous and the Common Wall Lizard, are widespread in the rest of Europe as well. One species, the Italian Wall Lizard, is probably an ancient introduction from the Italian region, while the remaining nine are largely endemic to the area. Many of the species have restricted ranges and this is often helpful in identifying them. Other small lacertid lizards in south-east Europe are the Greek Algyroides (p. 132), Steppe Runner (p. 130), and the Snake-eyed Lacertid (p. 130), all of which are quite easily distinguished from the Small lacertas. Three kinds of Green lizard (p. 138-141) are also present.

KEY

1. Collar usually distinctly serrated, throat scales fairly coarse, body scales usually well keeled   2
   Collar more or less smooth-edged, body scales often not very distinctly keeled, or smooth   4

2. Usually fairly dry places. Back often green; body scales small. 14–25 femoral pores on each thigh; typically 45–62 dorsal scales across back
                         Balkan Wall Lizard, *Podarcis taurica* (p. 175, Pl. 34)
   (Note. the Skyros Wall Lizard (*Podarcis gaigeae*, p. 178), may key out here; locality, Skyros and Piperi islands in the north Aegean sea, confirms its identity)
   Usually rather lush places. Brownish (rarely greenish), with coarse body scales. 5–15 femoral pores on each thigh; less than 44 dorsal scales across mid-back   3

3. Restricted range. Pattern fairly characteristic and constant. Only one row of scales bordering pre-anal plate (Fig., p. 173). Scales on flanks rather smaller than those on back. A fairly long row of supraciliary granules (3–11)
                         Meadow Lizard, *Lacerta praticola* (p. 173, Pl. 34)
   Much of northern Balkans, particularly mountains. Pattern variable. Scales bordering preanal plate usually irregular. Little variation in size of dorsal scales. Only a few supraciliary granules (0–4)
                         Viviparous Lizard, *Lacerta vivipara* (p. 144, Pl. 27)

4. South Greece and islands close to coast of Asiatic Turkey only. Two scales (postnasals) behind nostril, usually no really large scales on cheek, first supratemporal scale usually larger than others and cuts into edge of parietal scale (Fig., p. 174)   5
   Nearly all areas. Not as above. Sometimes striped. Usually only one scale behind nostril and often one or more large cheek scales   6

5. South Greece only. Brownish, unstriped, flattened rock lizard; often with blue flank spots, at least over shoulder. Underside of adults usually orange or yellow, typically with black spots, at least on throat.

> Greek Rock Lizard, *Lacerta graeca* (p. 174, Pl. 35)

Islands close to south-west coast of Asiatic Turkey only. Often striped or with rows of light spots; underside pale but throat may be red and belly may be spotted at sides.

> Anatolian Rock Lizard, *Lacerta anatolica* (p. 179, Pl. 35)

6. Crimea only. Distinctly flattened; brown, or green with no clear light stripes in body pattern.

> Caucasian Rock Lizard, *Lacerta saxicola* (p. 179, Pl. 34)

All other areas   7

7. Milos and nearby islands in south-west Cyclades only. Males with very characteristic pattern: flanks, sides of head, and throat black usually with large light spots

> Milos Wall Lizard, *Podarcis milensis* (p. 178, Pl. 36)

All other areas   8

8. Coastal areas and islands in Sea of Marmara only. Robust, often large wall lizard with a rather flattened body. Often greenish above, with dark stripe on midline of back or reticulated; light narrow stripes or series of spots not usually very obvious. Belly whitish or tinged greenish.

> Italian Wall Lizard *Podarcis sicula* (p. 162, Pl. 32)

All other areas   9

9. Mainland and Samothraki island only (but perhaps also Thasos island); mainly montane in south of range. At least slightly flattened. Usually brownish, often without clear light narrow dorsolateral stripes. Dark dorsolateral stripes, if present, are generally weaker than vertebral streak, if any. In males, underside usually dark spotted, belly often red and throat white with rusty marks. Females have at least some throat spots

> Common Wall Lizard, *Podarcis muralis* (p. 145, Pl. 27)

Southern mainland and islands, except Milos and Skyros groups. Not very obviously flattened, often with light dorsolateral stripes. Dark dorsolateral stripes usually stronger than vertebral streak if present. Throat may have spots, but belly unspotted and often red, orange or yellow in males   10

10. Peloponnese only. Granules along edge of supraocular scales few in number (0–7); often a row of large scales under forearm

> Peloponnese Wall Lizard *Podarcis peloponnesiaca* (p. 177, Pl. 35)

Mainland south to Peloponnese, Aegean islands but not those close to Asiatic Turkey. In area of overlap with Peloponnese Wall Lizard this species often has more supraocular granules (10–17) and usually few large scales under forearm

> Erhard's Wall Lizard, *Podarcis erhardii* (p. 176, Pl. 36)

MEADOW LIZARD *Lacerta praticola*   **Pl. 34**

Range Confined to a few small, isolated areas in the Balkans: parts of eastern Serbia, south-west and south-east Roma   nia, Bulgaria, north and west European Turkey and extreme north-east Greece. Also Caucasus. Map 128.

Identification Adults up to about 6.5cm from snout to vent; tail 1.7–2 times as long; females bigger than males. A small, fairly deep-headed lizard generally reminiscent of a Viviparous Lizard, with short legs, usually a serrated collar and rather coarse keeled scales. Upperparts greyish, brown, or olive-brown, with darker flanks; a reddish brown band, often edged with dark spots, along centre of back. Underside unspotted; throat whitish; belly greenish in males, yellowish in females; males also often have gold spots along the upper part of the dark flank band.

Variation Little obvious variation in Europe where the populations are referred to the subspecies *L. praticola pontica* which is distinguished from Caucasian *L. p. praticola* by having six pairs of chin shields instead of five.

Similar Species Easily distinguished from Viviparous Lizard by characteristic pattern; only one semi-circle of small scales around preanal plate (Fig., below); more supraciliary granules (3–11, compared with 0–4 in Viviparous Lizard); scales on flanks narrower than those on back. Other Small lacertas in range (Common and Balkan Wall Lizards p. 145, 175) lack distinct band along mid-back (although a dark streak may be present especially in the Common Wall Lizard), have more femoral pores on each thigh (13–27, instead of 9–13), and finer scaling (over 41 scales across body, instead of 35–41, rarely 43). If belly is brightly coloured in these species, it is usually reddish or orange.

Habits A mainly ground-dwelling lizard although it is capable of climbing. Usually found in often hilly country in and around open broad-leaved woods, especially of oak although other tree species such as beech may be involved. Here it generally likes somewhat moist places with lush vegetation and may occur in clearings, glades and damp meadows in the woods, wood edges and on well vegetated stony slopes. Also sometimes occurs on stream-banks, marsh-edges, etc. Extends up to altitudes of 800m. Spends time around bushes and in leaf litter where it is well camouflaged. Often seen basking on fallen tree trunks, isolated stones etc. Sometimes digs its own burrow which often has two exits.

Females lay 4–6 eggs at a time which hatch in about 6–7 weeks, the babies having a total length of around 3cm.; they mature at a snout-vent length of about 4.5–5cm

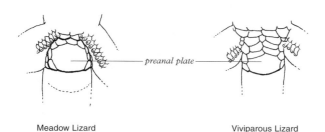

Meadow Lizard   Viviparous Lizard

Small Lacertas, vent regions

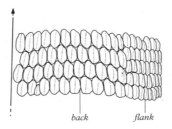

Meadow Lizard, variation in size of dorsal scales

GREEK ROCK LIZARD *Lacerta graeca*   **Pl. 35**

Range South Greece (Peloponnese) only. Map 129.

Identification Adults up to about 8cm from snout to vent; tail usually 2–2.5 times as long. A moderate-sized, flattened lizard with a rather 'rangy' appearance, the head, legs, and tail being long. Usually appears glossy grey-brown (sometimes rather yellowish), but bronzy-brown in some lights. Males have dark flanks with well marked light spots and irregular dark spots or blotches on back. Females are smaller-headed with only poorly-defined light spots on sides and the darker markings are more scattered. Often one or two blue spots present above the shoulder (these may extend along flanks in males). Underside usually black-spotted (at least on throat) and deep orange or yellow, but may be paler. Shape and colouring are quite characteristic. Head scaling is also distinctive: two scales immediately behind nostril, no large scales on cheek, first supratemporal scale usually larger than others and emarginating the parietal scale (see Fig., below).

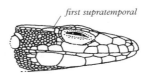

Greek Rock Lizard

Variation Some variation in patterning.

Similar Species Other Small lacertas within range are less flattened, often striped, and lack characteristic head scaling of Greek Rock Lizard. These features together with shorter head and often reddish or pinkish belly will distinguish the Common Wall Lizard, (p. 145), superficially the most similar form.

Habits Largely confined to mountain areas and most common from 300–700m, although it ranges from near sea level to 1,600m. Often occurs fairly near water, or at least not in very dry situations. Climbs extensively and well on rocky outcrops, screes, walls, road-parapets and cuttings, tree-boles etc. Avoids very hot sun, and can sometimes be found in light woods etc. Although mainly a climbing species, may hunt on ground near rocky outcrops etc.

Females lay 1–6 (usually 3–4) eggs that take around 6 weeks to hatch; a single clutch is laid each year. Animals mature the year after hatching at 6cm or rather less from snout to vent.

BALKAN WALL LIZARD *Podarcis taurica* Pl. 34

Range Balkans but not east Adriatic area north of Albania. Extends north to Hungary, south and east Romania and the Black Sea coast to the Crimea; also Corfu, the Ionian and Strofades islands and Thasopoula near Thasos island in the north Aegean. Found in one or two localities on coast of NW Asiatic Turkey. Map 130.

Identification Adults up to 8cm from snout to vent (smaller in north of range); tail sometimes twice body length. A rather robust, deep-headed, small lacerta often reminiscent of a miniature Green Lizard. Usually easily separated from other wall lizards and from Rock lizards in its range by its distinctly serrated collar. Back of adults often bright grassy green in spring (more olive or brown in summer). Rest of dorsal pattern varies geographically. In most populations (*P. t. taurica*), animals have a pair of narrow light dorsolateral stripes (or occasionally rows of spots) bordered towards centre of back by a brown area overlaid by black blotches. Populations in Albania, western Greece and the Ionian islands (*P. t. ionica*) tend to lack brown areas, and the light stripes may be faint; a vertebral streak or row of spots may be present especially over the hind quarters. Some animals in this area are almost uniform with a green back and some stippling on the flanks. Underside in all areas often unspotted whitish, but bright orange, red or yellow with greenish throat in breeding males. Young lack green on back and are frequently more strongly striped than adults.

Variation Considerable variation in size and colouring (see above). Collar serration occasionally rather weak (for instance on Corfu). Males very large on Strofades islands, sometimes being over 9cm from snout to vent. Animals on Thasopula (*P. t. thasapoulae*) are rather long-limbed and regularly climb on rocks.

Similar Species Skyros Wall Lizard (p. 178). Dalmatian Wall Lizard (p. 169) is similar to Balkan Wall Lizard and occurs close to it in the north Albanian region. However it is smaller, usually does not have such a distinctly notched collar and vertebral stripe, if present, often contains light areas. Italian Wall Lizard (p. 162) is confined to Sea of Marmara area; it is larger, has a more or less smooth collar, somewhat flattened body, typically no clear narrow light stripes in pattern and nearly always a pale, often greenish tinged belly. Adult Green Lizards (pp. 138–140) are bigger with deeper supratemporal scales; their juveniles have much bigger heads and are stockier than equivalent sized Balkan Wall Lizards. For Crimea, see also Caucasian Rock Lizard (p. 179).

Habits Nearly always almost entirely ground-dwelling and usually climbs far less than other Wall lizards in range. Typically found on flattish dry ground with at least some covering of grass or other low vegetation; habitats include meadows, steppes, field borders, crops, sparsely vegetated dunes by the sea and road and path verges. In south, can occur in more broken habitats, in brambles and sometimes by water. Mainly a lowland species found below 800m but reaches 2,350m in southern Greece. Very conspicuous when basking on stones, open ground etc., but colouring hides it well when hunting in vegetation and adapts through the year, changing as the summer advances. Takes refuge in holes in ground, in bushy plants, and sometimes under stones etc.

Northern animals lay clutches of 2–6 (usually 2–3) eggs while in south Greece the number is 2–10 (usually 3–5); clutch size is smaller on islands. Usually there are two clutches a year, the eggs hatching in 8–9 weeks. Babies are about 3cm from snout to vent and females mature in 18–20 months at a length of around 5.5cm from snout to vent.

Other names. *Podarcis tauricus*.

ERHARD'S WALL LIZARD *Podarcis erhardii*   Pl. 36

Range South Balkans north to Albania, Macedonia and south Bulgaria. Also many Aegean islands but not Milos and Skyros groups, Samothraki, Thasos and Thasopoula, and some other inshore islands, including those close to Asiatic Turkey. Mainland range is rather broken up. Map 131.

Identification Mainland. Adults up to about 7cm snout to vent; tail about twice as long. A medium sized Wall lizard often with a fairly deep head (which is also rather short in south of mainland range), smooth collar and more or less smooth body scales. Pattern variable but usually basically striped, especially in females. Light dorsolateral stripes are often present, and dark dorsolateral stripes or series of markings are nearly always better developed and broader than vertebral stripe, if any. Some males largely reticulated. Ground colour greyish or brown; occasionally greenish. Belly and often throat white, yellow, orange or red, brightest in breeding males; dark spotting, if present, is typically confined to chin and sides of throat. Sometimes large blue spots present on hind legs.

Variation May really be more than one species. Three subspecies are recognised from the Balkan mainland: *P. e. riveti* as far south as N Greece, *P. e. thessalica* in Thessalia province and *P. e. lividiaca* in central Greece, Euboa and the Peloponnese.

Island populations. About 25 subspecies have recently been accepted on the Aegean islands. Here there is considerable variation in size (up to 8cm from snout to vent), shape (robustness, length of limbs, tail-thickness) and colouring. Ground colour may be brown, grey, green, or very dark black-brown. Pattern varies: in extreme cases can be reticulated, or dark markings reduced, confined to front of body, or completely absent. Underside may be white, grey, yellow, red, greenish, bluish and sometimes hind parts are more strongly coloured than rest. Locality is very useful in recognising island Erhard's Wall Lizards: it is the only small lacerta species on most north Sporades and Cyclades islands but it is replaced by other species in the Skyros and Milos archipelagos and on islands in the north of the Aegean sea and those lying close to the mainland of Asiatic Turkey.

Similar Species Common Wall Lizard (p. 145), present on mainland only, is usually flatter and often has a different pattern: if dark streaks or rows of spots are present on back, the vertebral stripe or row is usually better developed than the dorsolateral ones; clear light dorsolateral stripes are uncommon. Belly may be pink or red but throat is usually whitish with rusty blotches and dark markings which are often heavy and may extend to belly. Peloponnese Wall Lizard (opposite) in Peloponnese only. Milos Wall Lizard (p. 178) has completely separate range.

Habits On mainland is found in often stony or rocky places, usually with low, dense, bushy vegetation sometimes including brambles. Climbs to some extent, but not as much as most Wall lizards. Tends to be in places with at least some humidity but in north of range is replaced by Common Wall Lizard in moister situations, at higher altitudes and near human habitation. Mainly a lowland species but can reach 1,000m. Island populations are often associated with dense growth of *Pistacea* bushes in which they hide, but can sometimes occur in more open places such as overgrown sand dunes. On some islands in the northern Sporades and one off north-east Crete, this lizard occurs with Eleonora's Falcon (*Falco eleonorae*). On the Cretan island at least, most animals in the large lizard population move into the nesting area of the falcon once its breeding begins in late summer. Here they feed on remains of its avian prey and associated flies and perhaps the external parasites of the young falcons.

Continental females lay 1–5 (usually 2–4) eggs in a clutch but numbers may be smaller for at least some island populations. Eggs are about 13–17mm x 8–10mm and babies measure approximately 3cm from snout to vent probably maturing in about 1.5–2 years.

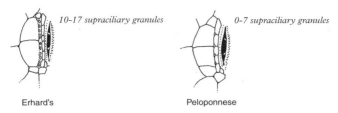

Erhard's   Peloponnese

10-17 supraciliary granules   *0-7 supraciliary granules*

Wall Lizards in Southern Greece, area above eye

PELOPONNESE WALL LIZARD *Podarcis peloponnesiaca*   Pl. 35
Range Southern Greece (Peloponnese) only. Map 132.

Identification Up to 8.5cm snout to vent; tail about twice as long; females smaller than males. A rather large, robust Wall lizard with a massive head in adult males, only a few supraciliary granules present or none at all, a smooth collar and often a row of distinctly enlarged scales under the forearm. Pattern very variable but usually distinctly striped, females more strongly than males. Typically, light dorsolateral streaks are present, bordered above by dark stripes or rows of marks that are often broad and nearly always better developed than the vertebral stripe (if any). Back of males frequently greenish. Often one or more blue spots above forelimb and blue may extend along flanks. Underside without dark spots, frequently white, orange or red. Juveniles more or less like females, but tail may be blue.

Variation Animals from the southern and central Peloponnese (*P. p. peloponnesiaca*) often have a complete dark vertebral stripe, males are green on neck and forebody and the underside is red or orange. Unmarked individuals are also found in this area. Those from the north-west (*P. p. lais*) are similar but the continuous vertebral stripe reaches only the shoulder and the red on the underside is weaker or reduced to small spots often best developed on the throat. In the north-east (*P. p. thais*), the vertebral stripe does not reach the shoulder, there is little green on males and the underside is white.

Similar Species Most similar to Erhard's Wall Lizard (opposite) but in the Peloponnese this species tends to be smaller and less boldly marked; it also has a more complete row of supraciliary granules (typically 10–17 compared with usually 0–7 in the Peloponnese Wall Lizard) and few obviously enlarged scales under the forearm. Striped young of Balkan Green Lizard (p. 139) could possibly be mistaken for female Peloponnese Wall Lizards, but they have a notched collar.

Habits The Common Wall Lizard of the Peloponnese. Found in a wide variety of dry, often broken habitats such as rocky, scrub-covered hillsides, olive groves, vineyards, road-banks, ruins, stone-piles etc. Occurs from sea-level to 1,600m, sometimes being found above tree line. Frequently seen perched on raised objects; low walls, stones, tree trunks etc, but hunts extensively on the ground. Often climbs even on sheer cliffs, although rather clumsily, and can make long leaps between rocks. Replaces Greek Rock Lizard (p. 174) in dry places as the common climbing lizard of the area.

Males are aggressive and territorial in the breeding season. Females generally lay two clutches of 1–6 (usually 3–4) eggs, about 13–17mm x 8–10mm when first produced. These hatch in about 6 weeks; babies are 3–3.5cm from snout to vent and mature at 5–5.5cm.

Other names. *Podarcis peloponnesiacus.*

MILOS WALL LIZARD *Podarcis milensis*   Pl. 36

Range Milos group and islands to the west of this in the Aegean sea: Milos, Kimolos, Polyaigos, Antimilos, Ananes islands, Falkonera, Velopoula. Map 133.

Identification Adults up to about 7.5cm from snout to vent but usually smaller; tail about twice as long. A robust, deep-headed Wall lizard with a characteristic colour-pattern in the males. Back usually brown, often with a rather weak vertebral stripe; flanks, throat and sides of head black with prominent light green, blue, yellow or whitish spots. Belly also often marked heavily with black, but bright colour absent or limited to second rows of belly-scales from mid-line. Females more nondescript, often with light dorsolateral stripes and a few well-marked blotches on throat.

Variation Populations on Milos, Kimolos and Polyaigos are named *P. m. milensis.* Some of the smaller islands near Milos have populations that are rather dark, e.g. Falkonera (*P. m. gerakuniae*) and Antimilos (*P. m. schweizeri*). On the latter island the throat is often entirely black in males. Another subspecies, *P. m adolfjordansi* occurs on one of the Ananes islands.

Similar Species Balkan Green Lizard (p. 139) occurs within range, but is bigger and has distinctive pattern, even when young. Erhard's Wall Lizard (p. 176) is found on other islands near the Milos group (for example Sifnos, Serifos, Folegandros, and Sikinos) but lacks characteristic male colouring of Milos Wall Lizards.

Habits Mainly lowlands but reaches 685m on Antimilos. Most abundant in areas of cultivation, especially where dry-stone walls, banks and stone piles provide many refuges. Tends to sit and bask in large numbers on these, but hunts mainly on the ground, either in vegetation or in open areas. There may be up to 600 animals to the hectare. Also occurs at lower densities in a wide variety of habitats including scrub-covered hillsides, and even damp marshy land near sea. Often seen on walls with Kotschy's gecko. Interestingly, in sandy places, this Wall lizard will try to escape by diving straight in to loose sand and disappearing in it. Such behaviour is more usual for lizards in really dry regions of the world.

Has a distinctive reproductive pattern compared with other Wall lizards. Females lay repeated small clutches of 1–3 eggs (average 1.73) over a long reproductive period; the size or number of eggs does not increase with the body size of the mother. Adult males may attack young males. Babies often mature within a year at a length of around 4.5–5cm from snout to vent.

SKYROS WALL LIZARD *Podarcis gaigeae*

Range Skyros archipelago and Piperi island in the northern Sporades, north Aegean sea. Map 134.

Identification Up to about 8.5cm from snout to vent; tail about twice as long. A deep-headed Wall lizard, often with a bright green back, dark vertebral and dorsolateral stripes or series of spots, and pale dorsolateral streaks. Flanks usually dark and sometimes reticulated, with a blue spot above the foreleg. Lip scales are frequently dark-edged and underside is whitish, often with dark spots on the throat. Collar may be notched.

Variation Animals on offshore islets tend to be olive or brownish. Population on Piperi island (*P. g. weigandi*) lacks dark vertebral and dark dorsolateral stripes.

Similar Species The only small lacerta in its range. Superficially like some Balkan Wall Lizards (Pl. 34, Fig. 2b) but often has strong dark vertebral stripe, different colouring beneath and back scales that are at most lightly keeled. Erhard's Wall Lizard on neighbouring islands can also be similar.

Habits Populations of large islands live in and around dense, bushy vegetation, but the species occurs in barer situations on offshore islets. On the Diabates islands west of Skyros, the lizard feeds in and around the nests of Eleonora's falcon (*Falco eleonorae*), as Erhard's Wall Lizard does elsewhere.

Other names. *Podarcis milensis gaigeae.* Protein studies indicate that this lizard is closely related to the Milos Wall Lizard and it is now sometimes placed in that species, in spite of its very different appearance and distant range.

CAUCASIAN ROCK LIZARD *Lacerta saxicola*   **Pl. 34**

Range South and south-east Crimea. Also north-east coastal area of the Black sea and the western Caucasus. Map 135.

Identification Up to about 7cm, but usually smaller, tail about twice as long. A distinctly flattened rock lizard with a smooth-edged collar, more or less smooth or feebly keeled back scales, a deep first supratemporal scale and no clear, narrow, light stripes in pattern. Males are grass green to grey above and females usually brown or olive grey; typically back has irregular dark markings and flanks have a dark reticulation that contains lighter spots some of which may be blue and in males become brighter in the breeding season. Underside pale yellow or ochre.

Variation Populations in the Crimea have some distinctive features and are named *L. s. lindholmi.*

Similar Species In Crimea, only likely to be confused with the Balkan Wall Lizard (p. 175), but this species is not flattened, has shallow supratemporal scales (Fig., p. 116) and quite clearly keeled back scales, usually narrow light stripes in pattern and a typically bright orange or red belly in breeding males.

Habits Occurs in mountainous parts of Crimea. A climbing lizard encountered especially on rock faces, boulders, scree, ruins, drystone walls, bridges, tree boles in woods, and even on fences. Prefers rather humid, often semi-shaded places, from sea level to over 1,000m. Takes refuge in rock crevices, beneath bark and even in rodent burrows. May hibernate communally in deep rock fissures.

In breeding season, dominant males defend territories of around 70 square meters in which about three females may also be based. The resident male regularly mates with these, copulation taking about 2–4 minutes, but the females may also mate with males without territories on the periphery of the defended area. Females probably lay two clutches a season, containing 2–5 eggs, 12–16mm x 6–8mm when first laid. These hatch in 5–9 weeks, the babies averaging 2.5cm from snout to vent.

ANATOLIAN ROCK LIZARD *Lacerta anatolica*   **Pl. 36**

Range European islands close to the south-western coast of Anatolia: Samos, Ikaria, Nisiros, Simi, Nisios Strongili, Rhodes, Pentanisios. Also west and south Anatolia. Map 136.

Identification Up to about 7.5cm from snout to vent; tail about twice as long. A rather flattened small lacerta with, two postnasal scales (Fig., p. 112), fine scaling on the temple and the preanal plate in front of the vent relatively small (see for instance

Fig., p. 173). Pattern highly variable; ground colour often fairly dark with two dor-solateral series of light spots along the back that frequently join to form light stripes which are often quite broad and conspicuous. Area between the stripes, together with the flanks and the limbs, frequently with often small light flecks. Underside pale, although throat may be reddish, especially in males, and the sides of the belly may be dark spotted.

Variation European subspecies are *L. anatolica aegea* on Samos, *L. a. oertzeni* on Ikaria and *L. a. pelasgiana* on the remaining islands. Animals from Ikaria tend to have very contrasting colouring.

Similar Species No other small lacertas within range. Other lacertid lizards in the area are Snake-eyed Lacertid (p. 130) and Balkan Green Lizard (p. 139).

Habits A rock-dwelling species found on boulders, cliffs and outcrops and also on drystone walls. Often occurs in wooded regions, and close to water in dry areas. Females lay about 3–8 eggs.

Note Sometimes regarded as two species: *L. anatolica* on Samos and in western and north-west Anatolia, and *L.oertzeni* on the other European islands and in south-west Anatolia.

Canary Island Lizards
(*Gallotia*)

The Canary islands have their own distinctive group of lacertid lizards that look sim-ilar to the lacertas of the European mainland and Mediterranean islands but are most closely related to the species of Psammodromus. Like these they have a voice and often squeak and males grip females by the neck during copulation instead of by the flank as is usual in most lacertid lizards. Canary Island lizards are found on all the islands of the archipelago. They vary from largely insectivorous populations about the same body size as Wall lizards to big herbivorous species with saw-shaped teeth, some of which include the largest lacertid lizards known. The once abundant giant forms in the west Canaries were believed to be extinct but small surviving popula-tions of three species have been discovered in recent years. Their decline followed the colonisation of the Canary archipelago by people, who probably hunted the lizards, but the effects of introduced domestic animals may have been more damaging. For instance cats eat lizards and goats devastate their habitat. These pressures may have made the giant forms more vulnerable to competition with the smaller Canary lizards, especially when young. Only on Gran Canaria, where there are no native small lizards, does a very large form still flourish. Canary lizards appear to have orig-inated in north-west Africa and reached the eastern islands first, spreading west-wards across the archipelago over a period of millions of years. This contrasts with the geckos (*Tarentola*) which have colonised the islands three times and the skinks (*Chalcides*) which have arrived twice, both groups probably being transported from Africa by the south-west running Canary current.

Identification The seven living species of Canary lacertid lizards show clear differences from each other and as no more than two occur naturally on any one island, their iden-tification is easy. Distributions are as follows. Eastern islands: Atlantic Lizard (opposite), Gran Canaria Lizard introduced to Fuertaventura. Gran Canaria: Gran Canaria Lizard (p. 184), Atlantic Lizard introduced around Arinaga. Tenerife: Tenerife Lizard (p. 182), Tenerife Speckled Lizard (p. 183). La Palma: Tenerife Lizard. La Gomera: Boettger's Lizard (p. 182), La Gomera Giant Lizard (p. 185). El Hierro: Boettger's Lizard, El Hierro Giant Lizard (p. 184), Tenerife Lizard introduced at El Matorral.

ATLANTIC LIZARD *Gallotia atlantica*   **Pl. 37**
Range Eastern Canary islands of Lanzarote, Fuertaventura and the smaller nearby islands of Lobos, Graciosa, Alegranza, Monte Clara and Roque del Este. Animals from Fuertaventura have been introduced in a small area of Gran Canaria around Arinaga. Map 137.
Identification Adults up to 10.5cm from snout to vent but usually smaller and often considerably so. Snout pointed, front of ear opening with a toothed edge, no large masseteric scale on cheek, only four upper labial scales in front of eye, scales on back quite large and keeled, and 8–12 large belly scales across mid-body. Throat black or blackish in most animals of all ages; belly whitish. Babies, females and young males usually brownish with two pale stripes on each side, the upper one often bordered with black lines or spots. Males larger and more robust than females, usually with blue spots on each flank
Variation Considerable variation in size and colouring, especially of males. Animals from much of Lanzarote, and Graciosa, Alegranza, Monte Clara and Roque del Este (*G. a. atlantica*) are less than 7.5cm from snout to vent with well developed blue spots on flanks of males that are clearly separated from each other. The malpaís (lava field) of Montaña de la Corona, Lanzarote has very large individuals (*G. a. laurae*) up to 10.5 cm from snout to vent, the males often being dark with blue spots on the flanks that are usually large and may fuse together; these animals are not genetically distinct from neighbouring populations in which animals may sometimes be just as big. Fuertaventura and Lobos are inhabited by small animals (*G. a. maharotae*) with throats that are less dark than in other forms and the blue spots on the sides are small.
Similar Species None occurring naturally within range, although the Gran Canaria Giant Lizard (p. 184) has been introduced on to eastern Fuertaventura. This and other Canary lizards are larger than the Atlantic Lizard, usually with a masseteric scale on the cheek, no toothing of ear opening, usually five or more upper labial scales in front of eye, finer scales on the back, often more than ten large belly scales across the mid-body, and a throat which is not largely black or blackish.
Habits Found in a wide range of habitats ranging from sand dunes, coasts and old lava flows with some vegetation to cultivated and waste land, although it is often absent from very recent, bare lava fields. Occurs up to 800m and is active all year. Frequently very common with up to 1,000 animals per hectare. Especially abundant in modified habitats such as the vicinity of villages, rubbish dumps etc. Often seen sitting on stones, although it frequently forages in vegetation. Old males are generally more conspicuous than females and especially young. Quite territorial and animals are often spaced 3–4m apart. Can dive rapidly into loose earth or sand when disturbed; the pointed snout and partly protected ear opening are probably related to this behaviour. The young are largely insectivorous but adults take some vegetation (fruit and flowers), especially on lava flows where insects are uncommon. Like several other lacertids will also eat from bird remains discarded by nesting Eleonora's falcons (*Falco eleonorae*).

Males approach females while nodding their down-turned heads and with their throats inflated. Copulation takes place without preliminaries and may last about three minutes. Females begin to lay 3–4 weeks after this, producing 2–3 clutches a year of 1–5 eggs, 10–15mm x 7–9mm, which usually hatch in 7–10 weeks. Babies are about 2.5–3.5cm from snout to vent and mature in about a year. Large males may be 5 years old and some individuals live 15 years.

TENERIFE LIZARD *Gallotia galloti*   **Pl. 37**
Range The western Canary islands of Tenerife and La Palma, and the islets of Anaga
and Garachico off northern Tenerife. Also introduced at El Matorral on El Hierro
and to the Funchal botanic gardens on Madeira. Map 138.
Identification Up to 14.5cm from snout to vent with very fine dorsal scales and
12–14 large scales across mid-belly. Adult males are often quite dark with a bright
yellow eye and blue spots on flanks. Females and young are frequently, although not
always, strongly striped with a pair of light lines or rows of spots on each side bordered by often broader darker areas leaving the grey, brown or reddish ground
colour visible along the centre of the back; the flanks often have lighter spots or ocelli between the light stripes. Striping fades in mature males and some females but may
be retained in non-dominant males. The belly may be whitish or pinkish.
Variation Considerable variation in size and male pattern. In south and south-east
Tenerife (*G. g. galloti*) males are up to 12.5cm from snout to vent and largely black
including the throat, with prominent blue spots on flanks the more forward of which
are large and may fuse to form a particularly large patch. On the northern edge of
Tenerife including the Anaga peninsula in the north-east of the island (*G. g. eisentrauti*), males are up to 14cm, with finer scaling and are dark grey with yellow bars
and spots on back; the throat, jowls and chest are intense blue but the blue spots on
the flanks are relatively small. On Anaga islet off north-west Tenerife (*G. g. insulana-
gae*), males are especially large and particularly dark with large blue spots on the
flanks. In contrast animals from La Palma island (*G. g. palmae*) only grow to about
11cm and adult males have largely black bodies and blue throats, jowls and chests.
Similar Species None within range.
Habits Found in all cultivated and wild habitats, except some native woodland and
heath and above 3,000m. Sometimes reaches densities of 2,000 animals per hectare.
Often climbs on drystone walls and may dig burrows among stones or at the base of
bushes. Active all year in lowlands eating a high proportion of plant food but also
insects. Often considered a pest of tomato cultivation as it eats the fruit.
    Males fight in the spring and display to females by flattening the body from side
to side, making their blue flank spots conspicuous, and nodding their heads. Females
lay 1–2 clutches a year of 2–9 eggs, 17–24mm x 9–15mm, which usually hatch in
about 12–13 weeks. Babies are 3–4.5cm from snout to vent.

BOETTGER'S LIZARD *Gallotia caesaris*   **Pl. 37**
Range Western Canary islands of La Gomera and El Hierro. Map 139.
Identification Up to 10 cm from snout to vent, although females smaller, with very
fine scaling on body and 10–12 large scales across mid-belly. Males can be all black
but are often sombre grey or brownish with a darker poorly defined band on the
flank that may bear light spots with one or more in the shoulder region that is blue;
a paler dorsolateral streak may be present at least on the head and neck. Young animals and females are more conspicuously striped, often with two light lines on each
side bordered by broad darker areas which may bear small light spots and sometimes
blue ones above the shoulder; the brown or greyish ground colour usually shows
along the centre of the back. Non-dominant males may be coloured like females.
Variation Animals on La Gomera (*G. c. gomerae*) are rather larger than those of El
Hierro (*G. c. caesaris*) with colouring that is often rather lighter and more contrasting; these trends are especially marked in the highlands where animals also tend to
be bigger.

Similar Species The El Hierro and La Gomera Giant lizards (p. 184, 185) have very restricted ranges within these islands and are much larger than Boettger's Lizard, with different adult patterns, more large belly scales across the mid-body and the parietal scale turned downwards on the side of the head.

Habits Found in all habitats except in some native forest and at very high altitudes, although it reaches 1,800m. Frequently very common in the lowlands where it may reach densities of 1,500 animals per hectare. Apparently more insectivorous than other central and west Canary lacertid lizards.

Females lay up to three clutches in good years when there is adequate rainfall. These consist of 1–5 eggs, 15–20mm x 11–12mm, which hatch in 8–9 weeks, producing babies 3–4cm from snout to vent with tails about twice as long. The young mature in 2–3 years and can survive 7 years in the wild, although such longevity is rare.

Note Two German scientists independently recognised this lizard as distinct: Caesar Boettger and Philip Lehrs. They raced to be first to describe it and Philip Lehrs went to London to enlist the help of the great herpetologist, George Boulenger. Because of this, Lehrs succeeded in publishing his description first, naming the lizard after his rival, but his triumph was short lived. The time was August 1914 and what was to become the First World War was declared. Lehrs tried to return home but was arrested as an enemy alien and spent four years incarcerated on an offshore island. He eventually got back to his country and, perhaps understandably, gave up reptile studies and went into Forestry instead. A collection of preserved reptiles he brought to show Boulenger, including his material of Boettger's Lizard, remained in London but in the 1930s efforts were made to return it to him. However, by the time arrangements were made, the Second World War had broken out, so Philip Lehrs never got his material.

TENERIFE SPECKLED LIZARD *Gallotia intermedia*   Pl. 38

Range North-west Tenerife, in the western Canary islands, where it was discovered in 1996. Map 140.

Identification Up to 15cm from snout to vent; tail about twice as long. 16–18 (rarely 14) large belly scales across mid-body and the parietal scale is turned downwards on to the side of the head. Brown above often with dense small yellow dots and a pattern of pale ocelli of which those forming a line on the upper flank may be blue. Some animals are more uniform above. Throat pale grey

Variation Some in colour.

Similar Species In the El Hierro and La Gomera Giant Lizards and the Gran Canaria Lizard, the parietal scale does not turn downwards on to the side of the head. The Tenerife Lizard also has a different range of colouring.

Habits Found in very rugged country with cliffs and boulders in the extreme north-west of Tenerife where only a few hundred animals survive. They coexist with the Tenerife Lizard, living mainly on small ledges on the cliffs which may have sparse vegetation or be almost entirely bare. In such extreme environments, the lizards live on insects and fragments of vegetation brought by the wind. As with the El Hierro and La Gomera Giant Lizards, to which the Tenerife Speckled Lizard is related, present habitat is probably not typical of the environments once occupied and this species may well have originally occurred in a much wider range of situations.

GRAN CANARIA GIANT LIZARD *Gallotia stehlini*   **Pl. 38**

Range Gran Canaria in the Canary islands; introduced on to Fuertaventura. Map 141.
Identification Up to 27cm from snout to vent, although most animals less than
18cm; tail up to twice as long; largest animals may weigh a kilogram. Back scales
keeled and 16–20 large belly scales across mid-body the collar is fairly smooth.
Reddish to dark grey above with two rows of lighter transverse bars that are stronger
in young males and in females. Sometimes light spots present on flanks. Big males
are very dark with very big heads and jowls and the chin and sides of the head are
often reddish. Babies usually have light dorsolateral bands that may be broken up.
Similar Species None within range. Introduced Atlantic Lizard on Gran Canaria
is much smaller with coarser scales, a fringe at the front of the ear opening, and only
8–12 large scales across the mid-belly.
Habits Found in a wide range of habitats up to 2,300m but especially common in
rocky humid gorges and can reach densities of 1,000 animals per hectare. Sedentary
and males aggressive towards others, threatening them by flattening body and inflating throat. Similar gestures are used towards females in courtship. Young are largely
insectivorous and adults mainly vegetarian with similar special features to the El
Hierro Giant Lizard that help them deal with this diet.
    Females lay one clutch a year of 4–16 eggs, 23–29mm x 14–18mm, which hatch
in 8–10 weeks. Babies are 4–5cm from snout to vent and mature in their fourth or
fifth year at a length of around 11cm. Some animals live 12 years.
Note Like other very large Canary lizard species, maximum size has fallen since the
archipelago was colonised by people. Recent fossils suggest this species once sometimes reached a length of 37cm from snout to vent and a weight of 2 kilograms.

EL HIERRO GIANT LIZARD *Gallotia simonyi*   **Pl. 38**

Range El Hierro island in the west Canaries. Confined to one locality on El Hierro
itself, the cliff, Risco de Tibataje, at Fuga de la Gorreta. Also on the offshore islet of
Roque Chico de Salmor where it became extinct about 1940 but was re-introduced
in 1999. Map 142.
Identification Up to 23cm from snout to vent; tail 1.4–1.7 times as long. A very
large robust lizard with strongly keeled scales across the mid-back and usually 18–20
large belly scales across the mid-body; the collar is serrated. Dark brown or blackish
with one or more rows of yellow spots on flanks; underside dark. Adult males have
much bigger heads and jowls than females. Babies are typically dark-speckled, with
a pair of light stripes on each side and sometimes light spots on the flanks.
Variation The extinct population on Roque Chico de Salmor (*G. s. simonyi*) and
those on Hierro (*G. s. machadoi*) showed slight differences in scaling and proportions.
Similar Species Boettger's Lizard (p. 182) is much smaller and lacks yellow spots
on flanks; it has only 10–12 large belly scales across mid-body.
Habits Originally grew much larger than at present and probably occurred in all
habitats on El Hierro up to 700m or more, but now confined to a small area of cliff
with sparse vegetation where only about 300 animals remain which are outnumbered
14:1 in their habitat by the much smaller Boettger's Lizard. The species was thought
to have become extinct on El Hierro in the Nineteenth century but came to the
attention of scientists again in 1974. Adults are largely vegetarian and have special
features that help them deal with this diet. The teeth are blade-like with extra cusps,
and a cul-de-sac in the gut, opening from the rectum, is occupied by symbiotic

nematode worms which digest cellulose; the salt gland found in the nose of all lizards helps remove excess salts associated with eating plants. Babies initially feed on insects before becoming increasingly vegetarian.

The yellow flank spots of males increase in brightness in spring and they become more aggressive to other males. They display to these and to females, flexing their heads downwards, swelling the throat and nodding. As in many other lacertid lizards, both sexes are promiscuous and may have several partners. Laying begins 3–4 weeks after mating and two clutches of 5–13 eggs may be produced. The eggs vary in size, being 20–40mm x 15–30mm, and hatch after about 9 weeks. Newly hatched lizards are about 4.5–5.5cm from snout to vent with tails 9–11.5cm long that become relatively shorter with age. Most females reach sexual maturity in 3 years but the time varies from 2 to 4 years and some individuals may be as small as 10cm from snout to vent at this time although 16cm is more usual. Some animals live 15–17 years and there is evidence that this species is capable of surviving 45 years in the wild. It is now being bred in captivity on El Hierro for reintroduction to other parts of the island.

Note Similar giant lizards were once present on La Palma and Tenerife. The Tenerife Giant Lizard (*Gallotia goliath*) was extremely large, reaching 45cm from snout to vent with a head length of over 11cm.

LA GOMERA GIANT LIZARD *Gallotia gomerana*   **Pl. 38**
Range One locality on La Gomera island in the western Canary islands. Map 143.
Identification Up to at least 19cm from snout to vent. Similar to the closely related El Hierro Giant Lizard with 16–18 large scales across mid-belly. Very dark above, males blackish and females dark brown, often with inconspicuous small bluish spots on flanks. Underside including throat whitish, this colour extending to the lower jaw and sides of the neck, especially in males. Young animals lack pale colouring on jaw and neck.
Similar Species Boettger's Lizard (p. 182) is much smaller and lacks pale colouring on sides of head and neck; it has only 10–12 large belly scales across the mid-body.
Habits At present known from only one small area, a cliff with large boulders at its base, where there were no more than ten individual lizards when discovered in 1999. Large lizards had not been reported on La Gomera since 1870. Recent fossils indicate that this species was once widespread on La Gomera and probably grew much larger, perhaps to 40cm from snout to vent. It probably mainly occurred in richer habitats with more vegetation than at present. Like other western Canary lacertids this lizard is largely vegetarian when adult. Attempts are being made to breed it in captivity and a clutch of 5 eggs has been produced.

Skinks
(Family *Scincidae*)

A large family of about 1,100 species widely distributed over the warmer parts of the world. Ten occur in Europe where they are small or medium-sized, ground-dwelling lizards with large, smooth, shiny scales which contain a layer of bone. The head is small, the neck relatively thick and the body often elongate. Limbs are small or even absent and there are no femoral pores. Many European skinks have a transparent window in the lower eyelid. Skinks are mainly diurnal, but often rather secretive and feed largely on a wide variety of invertebrates. A fairly high proportion of species, including the European *Chalcides*, give birth to fully formed young but the others lay eggs. Skinks are good colonisers of islands. Those of the Canaries result from two separate invasions from western Morocco, one by the ancestor of the East Canary Skink, the other by the ancestor of the other two species.

SNAKE-EYED SKINK *Ablepharus kitaibellii*   Pl. 39

Range South and east Balkans, north to Serbia and lower Danube valley with an isolated enclave in west Hungary and south Slovakia; many Aegean islands (but not Crete). Also parts of south-west Asia. Map 144.

Identification Up to about 13cm total length, the tail about 1.3–1.6 as long as the body; females larger than males. A slim, diminutive, very glossy lizard with a relatively thick tail; very short legs and small head. At close quarters, easily separated from other skinks by the eye which lacks obvious eyelids and does not close when touched. Bronze-brown to metallic olive above with distinctly darker sides. Back may be plain or with small dark and light flecks, which are often arranged in rows and may form stripes. Males tend to be darker than females. Underside whitish or greenish.

Variation Animals from Greece including the Ionian and many Aegean islands (*A. k. kitaibellii*) are small and slim with small heads and weak limbs, 18–20 scales around mid-body and little patterning. Those from further north (*A. k. stepaneki*) are bigger and robust with wider heads, stronger limbs, 20 scales around the mid-body and 2–4 dark and light stripes or series of spots. The isolated enclave in Hungary and Slovakia is sometimes separated from these as *A. k. fitzingeri* and may have 4 upper labial scales in front of the eye instead of the usual 3, 20–22 scales around the mid-body and 4–6 stripes or series of spots. Snake-eyed Skinks on the small islands of Mikronesi, Amathia, Kasos and Karpathos east of Crete (*A. k. fabichi*) are robust with big relatively deep heads, long strong limbs and hind toes, 20 scales around the mid-body and 2–4 stripes or series of spots.

Similar Species None really similar. The only other skink in range with legs, the Ocellated Skink (opposite), is more robust, even when very young, and has different colouring and normal eye-lids.

Habits Mainly a lowland animal but found up to 800m in places. Over much of range often seen in quite dry places: south facing slopes, meadows, oak, juniper, or box scrub, and edges and clearings of open oak and chestnut woods, where it hides in fallen leaves, bush-bases, under stones etc. May also occur in places with a covering of grass or other low dense vegetation. This is especially so in the south of its range where it is sometimes encountered in quite moist habitats and tends to be active mainly in the spring and autumn and after rain. Often basks but sometimes active in twilight. The limbs may be used when moving about, especially the front ones, but as animals grow they increasingly employ serpentine locomotion. Not particularly agile, but can retreat effectively into its often dense habitat.

The male bites the flank of the female during mating which lasts just 30–60 seconds. 2–4 eggs are produced in a clutch, the average number being lowest in the smaller southern animals. They are buried in soil and are 7–10mm across swelling to as much as 14mm during development which may take around 9 weeks. Hatchlings are about 3–3.5cm from snout to vent and mature in about 2 years. Has been known to live 3.5 years in captivity.

OCELLATED SKINK *Chalcides ocellatus*   **Pl. 39**

Range Sardinia, Sicily, Maltese archipelago, Pantellaria, and Portici near Naples; mainland Greece (Attica region, Peloponnese and Kryoneri), neighbouring Aegean islands of Euboea, Kea, Crete, Chios, Rhodes and Karpathos. Present as well on Linosa, Lampedusa and Conigli islands. Also north and north-east Africa and south-west Asia east to Pakistan. Map 145.

Identification Up to 30 cm total length, of which tail may be about half, but is often much less. A fairly large, glossy, long-bodied skink with a thick neck, small pointed head and short, five-toed limbs. Tail often considerably thinner than body. Loreal scale borders second and third labial scales (see Fig., below); usually 28–38 scales round mid-body. Often yellowish, buff, pale brown or grey above, sometimes with an olive tinge, and with a pattern of dark-edged ocelli, or short pale streaks bordered by dark pigment, the dark areas often joining together to produce irregular cross-bars.

Variation Greek and Aegean animals (*C. o. ocellatus*) are relatively small (rarely much over 20cm total length). European individuals from further west (*C. o. tiligugu*) are usually larger, rather more flattened and have a pale dorsolateral stripe on each side, bordered by a dark streak below it, at least on the front of the body. There is also considerable variation in the amount of dark pigment and animals from Linosa (*C. o. linosae*) and Lampedusa (*C. o. zavattari*) are rather melanistic; some dark individuals also occur in Sardinia.

Similar Species None within European range.

Habits Typically found in sandy places, often in lowlands and frequently near sea, although can occur up to 1,500m. Lives in scrub areas behind beaches, in vineyards, olive groves, field systems, gardens and even towns. Encountered in both dry and rather moist places. Often hides in crevices of dry-stone walls, in holes in the ground and under stones, logs and plant litter. In Europe, largely diurnal, being most active early in morning, at evening, and at other times on warm but overcast days. May also be active at night during very warm weather. Rather timid, retreating when approached, and tends to skulk close to cover. Very fast and agile and burrows swiftly into loose sand or matted plant growth using serpentine locomotion. After grasping prey, may spin on long axis detaching it from the ground and battering it.

Pairs may copulate repeatedly and after around 6–12 weeks females produce 2–20 (usually less than 12) fully formed babies which mature after about 3 years at a length of 9cm from snout to vent. Has lived 13 years in captivity.

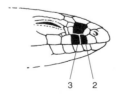   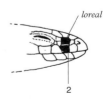

Ocellated Skink   Bedriaga's Skink

Skinks, relationship of loreal and upperlabial scales

BEDRIAGA'S SKINK *Chalcides bedriagai*   Pl. 39

Range Portugal and Spain with the exception of the north; also some offshore islands. Map 146.

Identification Adults up to 17cm total length, of which intact tail may be more than half; females larger than males. Like a small, slender Ocellated Skink. An elongate, small-headed lizard with short, five-toed limbs. Loreal scale borders second upper labial scale only (Fig., p. 187); 24–28 scales around mid-body. Usually buff, brown or grey above with scattered black-edged ocelli and often a pale, dorsolateral band on either side; underside pale. Babies are uniformly dark.

Variation Animals from eastern Spain (*C. b. bedriagai*) have a rounded body section, 3 or 4 upper labial scales in front of the eye and 24 (occasionally 26) scales around the mid-body. Those from the south of Portugal, Andalucía and Murcia (*C. b. cobosi*) are relatively small with four upper labial scales in front of the eye, the ear opening smaller than the nostril, short forelegs that do not reach the ears when turned forwards, 24–26 scales around the mid-body and quite uniform colouring. Populations from mountain ranges in Portugal north west and western Spain (*C. b. pistaceae*) differ from *C. b. bedriagai* in having 3 upper labial scales in front of the eye, long front legs that reach the ears, and 28 scales around the body. This last form may best be regarded as a separate species, *Chalcides pistaceae*.

Similar Species None within range.

Habits Diurnal. In many areas, especially in south of range, lives in open sandy places either with sparse vegetation or with a good cover of low plants. If ground cover extensive, may be quite abundant but secretive, often keeping out of sight and basking under cover. In such situations, is most usually encountered when plant mats are uprooted and stones and logs turned over. Can burrow very quickly in loose sand. May also occur in scrubby areas, woods and clearings, and in hilly regions where it tends to be less secretive than elsewhere and may be found in rocky and grassy places and in leaf litter.

Males are territorial often fighting in the breeding season. Females may mate with several partners and give birth to 1–6 young after a gestation of about 11 weeks; the babies are about 5.5–6cm long.

WESTERN THREE-TOED SKINK *Chalcides striatus*   Pl. 39

Range Portugal, Spain excluding much of east, southern France, north-west Italy (Liguria as far east as 25km south-west of Savona). Map 147.

Identification Up to 43cm total length of which the intact tail may be more than half; females are larger than males. A very elongate, snake-like skink, with tiny, three-toed limbs and 20–26 scales around mid-body. The middle hind toe is about equal to the outer one. Yellow, beige, brown or greyish above with 9–13 narrow dark lines (rarely unmarked) and sometimes a metallic gloss; head scales dark edged; underside white or grey.

Similar Species Italian Three-toed Skink (opposite). Easily separated from snakes and legless lizards by its tiny limbs.

Habits Usually found in damp relatively cool but sunny places with low, dense herbaceous plant-growth, including grassy slopes, water meadows, fields near streams, edges of cultivation, fallow fields, hedges etc. Also occurs in drier places in the west of its range near the Atlantic ocean, such as scrubby areas with scattered low shrubs and sandy places near the sea with a growth of gorse. Less common in mountain areas but found up to 1,800m in the south of its range. May move slowly using its tiny limbs but also employs serpentine locomotion, when it is very swift and agile. While travelling over the surface of vegetation, its body bends from side to side so fast that it can scarcely be seen. Takes refuge in dense herbage, under stones and in holes in the ground in which it also hibernates. Very sensitive to cold, emerging late from hibernation and tending to avoid wind.

Males gather together in the breeding season and may fight. Females produce 1–15 young 2–3 months after copulation which are 8–11.5cm long, and enclosed in a thin membrane at birth like other species of *Chalcides*; they become sexually mature in 3–4 years. Females may eat any still births and sometimes die producing their young.

Note Mistakenly thought to be venomous in parts of its range.

ITALIAN THREE-TOED SKINK *Chalcides chalcides*   Pl. 39
Range Italy, Sicily, Elba and Sardinia. Map 148.
Identification Up to about 48cm but usually smaller; unbroken tail may be more than half of total length; females are larger than males. Very similar to the Western Three-toed Skink (opposite) but limbs even smaller with the middle hind toe longer than the outer one, and colouring is different. There are three main patterns: 1. Some animals have 4–6 dark lines; 2. these may sometimes be combined with two broader pale, dorsolateral stripes; 3. there may be virtually no markings at all.
Variation In Italy and Sicily, all three colour forms are found, whereas only lined ones with two pale stripes occur in Sardinia.
Similar Species Western Three-toed Skink (opposite). Easily separated from snakes and legless lizards by its tiny limbs.
Habits Very like those of the Western Three-toed Skink and is found in a similar range of habitats. Can occur up to 1,270m in mountains and may sometimes be encountered there on arid and degraded slopes. Females give birth to 3–13 young that are 8–8.5cm long.

EAST CANARY SKINK *Chalcides simonyi*   Pl. 40
Range Eastern Canary islands: Fuertaventura, northern Lanzarote (where it is rare) and possibly Lobos. Map 149.
Identification Up to 30cm total length, of which tail may be more than half. A glossy, long-bodied skink with short, five-toed limbs and a round body section. Snout blunt and head relatively large and quite well differentiated, especially in some males. Nostril contacts first upper labial scale, 2 loreal scales present, 30–32 scales round mid-body. Light or dark brown or very dark grey with rows of small paler spots along body; snout often reddish and lips dark spotted; underside lighter than back.
Similar Species None within range; colour and scaling distinguish it from other Canary skinks.
Habits Diurnal and occasionally seen basking, but secretive and poorly known. Found in rocky places with significant moisture and often dense vegetation, including gullies, stone walls and barren areas of recent lava if they hold water. Also sometimes in cultivation, on banks between irrigated fields and around the bases of hedges of prickly pear (*Opuntia*) used to raise cochineal insects for making dye. Occurs up to 400m. Omnivorous and eats some flowers and fruit.

Males are territorial in breeding season and may wave tails at each other; mating is brief but repeated. Females produce a single brood of 4–7 young in summer which are about 10cm long.

GRAN CANARIA SKINK *Chalcides sexlineatus*   **Pl. 40**

Range Gran Canaria in the Canary islands. Map 150.

Identification Up to 18 cm of which the tail may be about half. Generally similar to East Canary Skink but body shorter. 0ne or two loreal scales, 25–38 scales around mid-body. Pattern highly variable but young usually less brightly coloured than adults.

Variation Animals in the dry south-west of Gran Canaria (*C. s. sexlineatus*) are dark with about six narrow light beige stripes on the back and a bright blue or green tail; underside is greyish although the tip of the lower jaw may be orange and the bright tail colouring sometimes extends forwards on to the belly. Individuals from the wetter north and east (*C. s. bistriatus*) may be more robust and are usually less strongly coloured with a speckled back and darker sides and 2–4 paler stripes; the tail is coloured similarly to the back and the belly is often orange or yellow. Animals of this form from La Isleta peninsula are very dark, although sometimes with a pair of lighter stripes, and the belly is black and the throat orange. The two subspecies meet in the north-west without much integradation, but all possible intermediates may occur in the south-east and in some higher parts of the island.

Similar Species None within range.

Habits Found in all habitats up to 1,950m, from humid meadows and valleys in the north-west to dry gullies and sand flats in the south. Occurs in woods and cultivation, especially in stony places and ones with low vegetation and drystone walls. Basks but is rather shy and secretive and often heard rather than seen as it moves among litter and plants. Apparently does not burrow much in loose soil. When escaping often folds limbs back along body and relies on serpentine locomotion.

Females give birth to 2–7 fully formed young which are 7–9cm long. The number of babies tends to be relatively low in the smaller *Chalcides s. sexlineatus*.

WEST CANARY SKINK *Chalcides viridianus*   **Pl.40**

Range Western Canary islands: Tenerife, La Gomera and El Hierro but not La Palma. Map 151.

Identification Usually up to 15cm, occasionally 18cm, total length, of which tail is about half; adult females are bigger than males. A glossy skink with a thick neck, small pointed head and short, five-toed limbs. Only one loreal scale present and 28–30 (rarely 32) scales round mid-body. Brown or dark grey above, often with a coppery or olive tinge; the flanks clearly darker. Rows of small light spots present on back. Usually no light stripes but occasionally one on each side between the back and flanks. Underside black or dark grey.

Variation Some minor differences between animals on each of the main islands where the species is found, and between populations in north and south Tenerife. In adults from La Gomera (*C. v. coeruleopunctata*) the light spots on the back are often blue. Babies from the more southern parts of Tenerife have blue tails.

Similar Species None within range; the dark throat and belly are distinctive.

Habits Very common in many habitats and occurs especially at low and middle altitudes, although on Tenerife it extends to 1,500m and, at low densities, even to 2,800m. Found in meadows and sandy places with stones, wood edges, fields, banana plantations and virtually any relatively open place where there is at least a little moisture. Often found when turning stones or other objects and at the base of drystone walls. Avoids dense natural forest and very recent lava flows. Diurnal and capable of climbing. When escaping through grass will fold limbs back along body and rely on serpentine locomotion.

In spring and summer frequently seen in pairs. The male often follows the female for some hours before copulation which is often very short (10–15 seconds) but repeated. After about 3–3.5 months, females give birth to 1–6 fully-formed young which are 6–8cm long.

LEVANT SKINK *Mabuya aurata*   **Pl. 40**
Range Rhodes and perhaps other Greek islands close to the mainland of Asiatic Turkey. Map 152.
Identification A robust lizard with longer legs than most other European skinks and a more slender tail; rostral scale separated from nostril; 36–38 scales around the mid-body, those on the back, sometimes each with three feeble keels. Usually brownish or greyish above with a two series of dark bars along the back and light dorsolateral and lower lateral stripes on each side of the body, the flanks between them usually have large dark blotches and sometimes small light spots; the underside is pale.
Variation Subspecies in Europe is *M. a. aurata.*
Similar Species Ocellated Skink (p. 187) occurs within range but has shorter limbs, a thicker tail, a different colour pattern and the nostril contacts the rostral scale.
Habits Often found in richly vegetated places such as cultivated land, gardens etc., sometimes near water. Climbs well on tree trunks and drystone walls.

LIMBLESS SKINK *Ophiomorus punctatissimus*   **Pl. 41**
Range Southern and eastern peninsular Greece; Kythera island in the Aegean sea. Also south-west Asia Minor. About 10 related species live in south-west Asia. Map 153.
Identification Up to about 18cm total length, of which the unbroken tail is about half. A small, shiny, snake-like, legless lizard with a rather pointed snout and 18 (occasionally 20) rows of scales around mid-body. Typically cream, buff, or brown above, paler and greyer or flesh-coloured elsewhere. Usually there is a pattern of fine dark lines or rows of flecks which is best developed on the tail and weakest on the back. These markings are more obvious in young animals.
Similar Species Slow worms (p. 192) are larger with a different pattern, blunter snout and usually 23 or more rows of scales round body. Worm Snakes (p. 200) have more uniform colouring, finer scaling and the tail is thicker than the head.
Habits Often occurs in spring on loamy slopes with grass or some other low vegetation and scattered stones under which it is usually found. Prefers stones that are naturally half submerged and appears to go deeper into ground in dry weather. Occurs up to 600m. Burrows swiftly in loose soil and when disturbed, often moves conspicuously striped tail more than head.

Slow Worms and Glass Lizard
(Family *Anguidae*)

A group of about 100 species that are mainly found in the Americas, but a few forms occur in Asia, Europe and north-west Africa. Two kinds are found in Europe: Slow worms and the European Glass Lizard which is the largest anguid lizard. Most anguids are rather long bodied, with either small limbs or no obvious limbs at all, as in the Old World species. These are consequently rather snake-like, although easily separated from real snakes by their closable eyelids and fragile sometimes partly regenerated tails. Unlike most snakes, their bodies are rather stiff. The smooth shiny scales include a bony layer that often persists long after death, and looks like a pale imitation of the living animal. Fossils show anguids have a long history in Europe, extending back at least to the Eocene period about 40 million years ago. Slow worms appear to have evolved from the larger Glass lizards. European anguids are quite secretive and retiring, although males may fight fiercely in the breeding season.

SLOW WORM *Anguis fragilis*   Pl. 41

Range Found over almost whole mainland of Europe but not southern Spain and Portugal, Southern Greece, most Mediterranean islands, Ireland or the extreme north of the continent. Also east to west Siberia, Caucasus, north Asiatic Turkey and north-west Iran. Map 154.

Identification Adults usually up to about 50cm total length, occasionally even larger but typically smaller. Intact tail is longer than the body, but is frequently shorter in adults as it is often broken and scarcely regenerates. A very smooth-scaled, snake-like reptile. Usually 24–30 scales around mid-body. Typically brown, grey or even reddish or coppery above. Females frequently have a vertebral stripe and rather dark sides and belly; males are more uniform. The new born young are strikingly coloured: gold or silvery above with very dark sides, belly and vertebral stripe.

Variation Western animals (*A. f. fragilis*) usually have 24–26 scales round mid-body and blue spotting is uncommon occuring only in some males. Eastern individuals (*A. f. colchicus*) usually have 26–30 scales around mid-body and blue spotting is very frequent in males and sometimes occurs in females as well. The area of contact of these two forms, where some intergrading occurs, runs roughly to the west of Finnland south to central Hungary; they have complex distributions in the Balkan peninsula.

Similar Species In Greece very similar to Peloponnese Slow Worm (opposite). Babies superficially similar to Limbless Skink (p. 191), colouring generally different.

Habits Prefers well vegetated habitats with extensive ground cover in rather damp but not wet situations, tending to disappear in periods of hot dry weather. Found up to 2,000m in the southern parts of its range and sometimes even to 2,400m in Alps. Occurs in pastures, glades in woods and the edges of these, in lush scrub-land, on heaths, hedge-banks, motorway and railway embankments, and in gardens and parks. Spends much of its time in dense vegetation and below the surface among the roots of this and in loose soil. Usually slow-moving and secretive and most likely to be encountered abroad at evening or after rain and may be active in quite cool conditions around 15°C. Occasionally basks directly, usually in small patches of sunlight among plants, but normally gets warm by lying beneath vegetation or sun-warmed objects, such as flat stones, sheets of old iron and discarded rubber mats. Most commonly discovered when turning these over. Can apparently reach densities of 600–2,000 per hectare. May hibernate communally or with other reptiles. Rarely bites when handled but thrashes body and tail which can be shed quite easily. Feeds especially on small slugs and snails and earthworms, but also arthropods and even small reptiles.

In breeding season, males fight fiercely, biting and wrestling with each other. They hold the head or neck region of the females in their jaws and the two animals often twine their hind-bodies and tails. Copulation may last 10 hours. Females often breed every other year, and bask more when gravid. After about 2–3 months they usually give birth to 6–12 live young but the number may vary from 3–26. The babies are usually 6–10cm long and are enclosed in a transparent mebrane which they quickly break. At least in west Europe, males breed at 3 or 4 years and females at 4 or 5 when they approach a length of 30cm. Slow Worms are long-lived, probably sometimes surviving 10–15 years in the wild; one lived 54 years in captivity.

PELOPONNESE SLOW WORM *Anguis cephallonica*

Range Southern Greece: the Peloponnese and the Ionian islands (Cephalonia, Ithaca and Zakynthos). Map 155.

Identification Up to 50cm total length. Very similar to Slow Worm, but adults are often longer and rather slender with narrower heads and relatively longer tails, when these are unbroken. There are also more scales around the mid-body (often 34–36 compared with up to about 30). Back usually coffee brown and underside and flanks black or dark brown, the borders between these areas sharply defined from the tip of the snout to the tip of the tail and markedly wavy on the neck just behind the head. Adults have an often intermittent, dark, short (2–6cm) stripe on midline of neck.

Similar Species Slow Worm (opposite).

Habits Similar to the Slow Worm with which it overlaps in the north Peloponnese. Lives especially in humid localities including open deciduous and coniferous forests and wooded edges of rivers. It occurs up to 1,200m in the Taygetos mountains.

Other names *Anguis peloponnesiaca*

EUROPEAN GLASS LIZARD *Ophisaurus apodus*   Pl. 41

Range Balkans as far north as north-west Croatia, north Greece, South Macedonia and south and east Bulgaria; Crimea. Found on a few islands on the borders of the Aegean Sea and in the Crimea. Also Caucasus and parts of south-west and central Asia. Map 156.

Identification Adults up to 140 cm; unbroken tail is about 1.5 times as long as body. Looks rather like a giant Slow worm (opposite). A heavy-bodied, snake-like animal, easily distinguished from other European reptiles that are more or less limbless by a prominent groove on each side of the body. Most individuals have tiny vestiges of hind legs on either side of the vent. Adults may be as thick as a man's wrist and are usually uniform yellowish brown or warm brown, darkening with age although the head may remain pale. Young have more strongly keeled scales than adults and distinctive colouring: greyish above with well-defined dark bars.

Variation Little variation. European populations are assigned to *O. a. thracius*.

Similar Species Not likely to be confused with any other reptile.

Habits Usually found in fairly dry habitats, often frequenting rocky hill-sides with some cover, light woods, dry-stone walls and embankments, stone piles etc. May be found in cultivated areas and even near human habitations. Diurnal but avoids the hotter parts of the day in very warm weather, occasionally crepuscular and often active after rain. Can move fairly fast when alarmed but lacks stamina. Feeds on larger invertebrates, especially snails, which it crushes whole, and arthropods including beetles, grasshoppers and other insects; occasional vertebrates such as small mice, shrews and lizards are also sometimes taken. The tail breaks with difficulty and usually regenerates very slowly only producing a stump.

Glass lizards lay about 6–10 eggs, approximately 40mm x 20mm; the babies measure 10–12cm long when they hatch.

Other names. Scheltopusik. *Pseudopus apodus*, *Anguis apodus*

Worm Lizards
(Amphisbaenidae)

Worm lizards are a group of about 133 very specialised snake-like reptiles that in nearly all cases are totally limbless. They occur in the west and east Mediterranean area, Arabia, tropical and southern Africa, central and South America and the West Indies. Most are small but the Giant Worm Lizard (*Amphisbaena alba*) of South America grows to 70cm. Worm lizards spend most of their time underground and are only occasionally encountered on the surface. The skull is heavy and modified for burrowing and the skin is very loose. Many elongate reptiles with small limbs or none at all have one lung very reduced in size or absent; usually this is the left one but in Worm lizards it is the right. Most species feed on invertebrates, especially ants and termites. The majority lay eggs.

IBERIAN WORM LIZARD *Blanus cinereus*   Pl. 41
Range Much of Spain and Portugal except the north. Similar species in Morocco and Algeria. Map 157.
Identification Up to about 30cm total length but usually smaller. At first sight, looks like a rather plump earthworm. Head small and pointed with tiny eyes beneath the skin and a marked skin fold where it meets the body. The body has a series of grooves that separate rings of small squarish scales; the tail is very short. Colour variable: may be yellowish, brown or grey, often heavily tinged with pink or violet, sometimes rather speckled; underside paler than back.
Variation Some variation in colour.
Similar Species Only Strauch's Worm Lizard (below) in the eastern Aegean area.
Habits Rarely seen on surface of ground, although may come up occasionally, particularly in heavy rain, at evening or at night. Otherwise entirely subterranean. Spends a lot of time in burrows and is usually found when turning stones and logs, when ground is being dug or ploughed, and when plants are uprooted. Often occurs in rather moist places, both in soils with a lot of humus and in ones that are predominantly sandy; avoids densly packed soils and clay. Frequently found in pine woods and in cultivated areas. Occurs up to 1,800m in south of Spain. Eats ants, insect larvae and other arthropods which it often appears to locate by the sound of their movements. Part of the tail can be shed but does not regenerate. When threatened by a predator, may twist tightly round twigs or other objects.
    Females usually lay a single elongate egg that is often about 30mm x 5mm.

ANATOLIAN WORM LIZARD *Blanus strauchi*   Pl. 41
Range Greek Aegean islands close to Asiatic Turkey. Also Asiatic Turkey to north Iraq and Syria. Map 158.
Identification Up to 20cm total length. Very similar to the Spanish Worm Lizard, but snout overhangs the lower jaw and there is no clear fold in the skin marking the back of the head. Often bluish brown or reddish grey.
Variation Some variation in colour and scaling.
Similar Species Only the Iberian Worm Lizard (above).
Habits Similar to Iberian Worm Lizard. Found under stones and in soil in sparsely vegetated bushy areas. Females lay one or two eggs.

SNAKES

(*Ophidia*)

Possibly as many as 2,900 species of snakes exist, of which only 34 occur in Europe. All are obviously scaly and legless (or at most with tiny vestiges of hind legs), and have eyes that cannot be closed. Legless lizards are superficially similar to snakes but in Europe they differ from all except the Worm Snake (*Typhlops*) in having belly scales that are not greatly enlarged. Most are further distinguished by possessing closable eyelids. Worm lizards lack this feature but have characteristic circular grooves along their length, like earthworms.

Identification Many scale features useful in identifying snakes are illustrated in the key and texts. Head scaling is shown in the illustrations below and opposite. Other helpful characteristics are body-proportions (stout or slender), head-shape, colour and pattern.

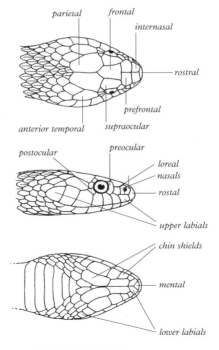

Head scales of Snakes

Counting back (dorsal) scales. It is sometimes necessary to count the small scales on the upper surface of snakes. This is done across the animal, usually at mid-body, that is half-way between the snout tip and the vent. Starting at one side of a belly scale, count diagonally forwards to the middle of the back and then diagonally backwards to the other side of the belly scale. See Fig., below.

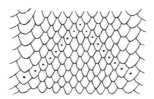

Counting belly (ventral) scales. In a few cases it may be helpful to count the belly scales. The first belly scale counted is the first one bordered by the lowest row of back scales on each side. It is found by following these scale rows forwards along the neck until they break away from the belly scales. A count finishes on the scale in front of the preanal scale See Fig., below).

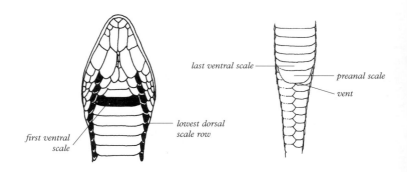

first ventral scale

lowest dorsal scale row

last ventral scale

preanal scale

vent

KEY TO SNAKES

Note Any European snake with clearly keeled scales and a vertical, slit-shaped pupil in bright light is a viper *(Vipera)* and is dangerously venomous.

   1. Balkans and Aegean only. No enlarged scales on belly; adults small (40cm or less) and rather like a dry worm
                        Worm Snake, *Typhlops vermicularis* (p. 200, Pl. 42)
     A row of large scales on belly   **2**

197

2. Balkans and Aegean only. Large scales on belly only up to about a third of body width (Fig., below). Tail very short, rounded at tip, with a single row of broad scales beneath. Eyes very small with slit- shaped pupil.

Sand Boa, *Eryx jaculus* (p. 201, Pl. 42)

Broad scales on belly more than a half of body width (Fig., below); undamaged tail not very short and not rounded at tip unless damaged, a double row of broad scales beneath. Eyes not very small   **3**

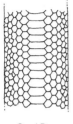

Sand Boa

most other snakes

Snake bellies

3. Dorsal scales clearly and strongly keeled (Fig., below)   **4**

Dorsal scales unkeeled or, if lightly keeled, pupil is round and 23–27 rows of dorsal scales are present. Scales may be grooved in one species   **7**

Keeled

unkeeled

Dorsal scales of snakes

4. Pupil round, preanal scale usually divided (Fig., below)   **5**

Pupil slit-shaped in good light, preanal scale usually undivided (Fig., below)

Vipers, *Vipera* – see separate key on p. 227.

divided

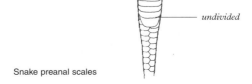

undivided

Snake preanal scales

5. Nostrils not clearly directed upwards; internasal scales broad and rectangular (Fig., p. 217); often a pale collar behind head
Grass Snake, *Natrix natrix* (p. 216, Pl. 46)
Nostrils directed upwards; internasal scales narrow towards front (Fig., p. 216); no obvious pale collar behind head   **6**

6. East and central Europe. Usually eight upper labial scales; fourth and sometimes fifth in contact with eye; three or four (or even more) postocular scales (Fig., p. 216); usually 19 back scales across mid-body
Dice snake, *Natrix tessellata* (p. 220, Pl. 46)
West and south-west Europe. Usually seven upper labial scales, third and fourth in contact with eye; two (rarely three) postocular scales (Fig., p. 216); usually 21 back scales across mid-body
Viperine Snake, *Natrix maura* (p. 218, Pl. 46)

7. Frontal scale very narrow and deep set (Fig., p. 203); a distinct brow ridge over eyes extending forwards on to snout, which gives snake characteristic expression
Montpellier Snake, *Malpolon monspessulanus* (p. 202, Pl. 42)
Frontal not very narrow; no very prominent brow ridge   **8**

8. South-east Europe and Malta only. Pupil vertically slit-shaped in bright light
Cat Snake, *Telescopus fallax* (p. 225, Pl. 47)
Most areas. Pupil always round in south-east Europe and Malta   **9**

9. Iberian peninsula and Balearics only. Rostral scale low, anterior temporal scale usually single, largest upper labial scale (usually the sixth) reaches or approaches parietal scale (Figs, p. 224).
False Smooth Snake, *Macroprotodon cucullatus*, (p. 223, Pl. 47)
Most areas. Rostral scale narrower and higher, often 2 anterior temporal scales, largest upper labial scale usually well separated from parietal scale.   **10**

10. East Aegean islands and European Turkey. Small, under 60cm, with a large sometimes divided dark blotch on head and dark crescent on neck, 17 back scales across mid-body
Dwarf Snake, *Eirenis modestus* (p. 210, Pl. 44)
Not like this. Usually at least 19 back scales across mid-body   **11**

11. Iberian peninsula, Sardinia and Pantellaria. No upper labial scales reach eye; characteristic pattern
Horseshoe Whip Snake, *Coluber hippocrepis* (p. 204, Pl.42)
Some upper labial scales reach eye (only rare exceptions)   **12**

12. Islands off south-east Asia Minor including Rhodes. A pattern of large spots changing into lines on tail
Coin-marked Snake, *Coluber nummifer* (p. 210, Pl. 43)
Combination of range and pattern not like this   **13**

13. Malta only. Usually a series of well-separated, dark bars on back
                     Algerian Whip Snake (*Coluber algirus*) (p. 205, Pl. 43)
     Combination of range and pattern not like this   **14**

14. Balkans only. Extremely slender. Fairly uniform but but with dark bar across
     neck or ocelli on its sides; 19 back scales across body a third of the way along
     its length   **15**
     Not extremely slender; pattern different   **16**

15. South-east Bulgaria and European Turkey. Dark band on neck. Fewer than
     205 belly scales and fewer than 95 pairs of scales under tail
                     Reddish Whip Snake, *Coluber collaris* (p. 206, Pl. 44)
     East Adriatic coast, south Balkans, islands near Asia Minor. One or more
     ocelli on side of neck. More than 204 belly scales and more than 104 pairs
     of scales under tail
                     Dahl's Whip Snake, *Coluber najadum* (p. 205, Pl. 44)

16. Balkans, south Italy, Sicily, Malta, Crimea. Pattern of red or brown, black-
     edged spots or stripes
                     Leopard Snake, *Elaphe situla* (p. 211, Pl. 44)
     Pattern not like this   **17**

17. Not Iberian peninsula or France. Usually with more than two stripes along
     body or large dark blotches, and two precocular scales (Fig., p. 213)   **18**
     Pattern not like this; usually a single preocular scale   **19**

18. Balkans, Ukraine, Italy and Sicily. Back scales lightly but distinctly keeled in
     adults which have either four dark stripes or large blotches, parietal scale
     often meets both postocular scales
                     Four-lined Snake, *Elaphe quatuorlineata* (p. 211, Pl. 45)
     South-east Ukraine. Back scales unkeeled, pattern of dark cross bands and
     light stripes running along body; usually an elongated dark U-shaped mark
     on top of head; parietal scale only meets upper postocular scale
                     Steppe Snake, *Elaphe dione* (p. 213, Pl. 44)

19. Usually 23 or more back scales across mid-body; if 21 then no well-defined
     dark streak from eye to side of neck   **20**
     19 or 21 (rarely 23) back scales across mid-body   **21**

20. Twenty-three (rarely 21) back scales across mid-body; rostral scale not
     clearly pointed behind
                     Aesculapian Snake, *Elaphe longissima* (p. 214, Pl. 45)
                     Italian Aesculapian Snake, *Elaphe lineata* (p. 215, Pl. 45)
     South-west Europe. 27 (rarely 25 to 31) back scales across mid-body; rostral
     scale pointed behind
                     Ladder Snake, *Elaphe scalaris* (p. 215, Pl. 45)

21. Usually a dark streak from eye to side of neck; 19 or 21 (rarely 23) back
     scales across mid-body   **22**

No dark streak from eye to side of neck; usually 19 back scales across mid-body   **23**

22. Belly more or less uniform; usually a dark stripe from nostril to eye; often third and fourth upper labial scales reach eye (Fig. p. 223); usually 19 back scales across mid-body
                           Smooth Snake, *Coronella austriaca* (p. 221, Pl. 47)
    Belly with bold, contrasting pattern; usually no dark stripe from nostril to eye, fourth and fifth upper labial scales reach eye (Fig., p. 223); usually 21 back scales across mid-body
                      Southern Smooth Snake, *Coronella girondica* (p. 222, Pl. 47)

23. East Adriatic coastal area and Greece. Generally under 100cm. Pale, often with small dark blotches and pale flecks. Belly yellowish or whitish, usually with some dark spots. 160–187 belly scales
                      Balkan Whip Snake, *Coluber gemonenesis* (p. 208, Pl. 43)
    Not like this. 187–227 belly scales   **24**

24. South-east Europe. Adults finely streaked with yellow or orange above without dark pigment. Belly unspotted yellow or orange-red. 189–211 belly scales, 80–110 pairs of scales beneath tail
                           Large Whip Snake, *Coluber caspius* (p. 209, Pl. 43)
    Not most of Balkan mainland. Adults black and yellow above or almost entirely black. 187–227 belly scales, 94–125 pairs of scales beneath tail   **25**

25. France, Italy and neighbouring areas, Corsica and Sardinia. Colour variable; up to 160cm or more
                      Western Whip Snake, *Coluber viridiflavus* (p. 207, Pl. 43)
    Gyaros island in west Aegean sea. Black above, less than 100cm
                        Gyaros Whip Snake, *Coluber gyrosensis* (p. 208, Pl. 43)

Worm Snakes
(Family *Typhlopidae*)

There are about 200 species of Worm snakes, all very similar in appearance, which occur throughout the warmer parts of the world. Typically they are small, most being under 30cm long but some reaching nearly a meter, and are slender with an inconspicuous head and reduced eyes. There are no enlarged belly scales and the body is completely covered with small, shiny scales. Worm snakes are usually secretive burrowers.

WORM SNAKE *Typhlops vermicularis*   **Pl. 42**
Range South Balkans north to Dugi Otok (island) in Croatia, south-west Montenegro, west Albania, central Macedonia (Vardar valley), southern Bulgaria and Turkey. Present on a few Greek islands such as Corfu, Euboa, Skyros?, Andros, Naxos, Kithera, Crete, Lesbos, Chios, Patmos, Leros, Kalymnos, Rhodes and Megisti. Also south-west Asia and Caucasus. Map 159.

Identification Adults occasionally up to 40cm, including tail, but usually smaller. Quite different from other European snakes and more like a dry shiny worm. The only snake in our area without large belly scales. Very slender and cylindrical but slightly thicker towards tail. Head inconspicuous and rather flattened with a rounded snout and not distinct from body. Eyes placed on top of head, but underneath scales, and appear as two tiny dark spots. Tail rounded, very short (about as wide as long) with a distinct spine at tip, rather longer in males than females with about 10 instead of about 7 scales in a row along the underside. Scales smooth, 21–24 rows around mid-body Colour usually brownish, often tinged yellow, pinkish, or purple. Belly slightly paler than upper surface.

Similar Species Unlikely to be confused with any other European snake. In Greece, sometimes found in same habitats as the Greek Limbless Skink (p. 191), which can appear superficially similar, but the skink is relatively thicker, has a much longer tail, a pointed head with eyes on the sides, and usually a pattern of stripes or rows of spots that increase in intensity towards the tail-tip.

Habits Mainly subterranean in fairly dry, open habitats without a dense covering of high vegetation. Grassy fields and slopes with scattered stones are often favoured, but Worm Snakes are sometimes encountered in more barren areas, and even fairly close to the sea. In spring, can often be found by turning half sunken stones, but in summer retreats deeper into ground. Occupies narrow burrows like those of earthworms, down which it retreats very quickly when disturbed. Occasionally seen on the surface, especially at twilight or in wet weather, and more rarely during the day. Pointed tail is used to gain purchase when travelling over the ground. Feeds mainly on small invertebrates especially ants and their larvae and pupae, but sometimes also takes very small spiders, centipedes, beetles and crickets.

Females lay 4–8 yellowish-white eggs that are elongate with pointed ends and 11–25mm long.

Sand Boas
(Family *Boidae*)

Boid snakes constitute a rather primitive family of about 80 species. Most of these are confined to the tropics although a few are found in more temperate areas. The family contains the largest known snakes (pythons and anacondas), but many species are quite small. This is true of the European Sand Boa, one of a group of ten species adapted to dry, or even desert conditions and found mainly in central and south-west Asia and North Africa. Like many other boids, Sand Boas give birth to living young.

SAND BOA *Eryx jaculus*   **Pl. 42**

Range South Balkans north to extreme south Albania, Macedonia, southern Bulgaria and Turkey, with isolated populations in north Bulgaria and south-east Roumania (where now possibly extinct); occurs on many Aegean islands, especially the Cyclades and those close to Asiatic Turkey. Also west of Caspian sea, and in south-west Asia and North Africa. Map 160.

Identification Adults up to 80cm, including tail, rarely more and usually less. A very stout snake with a short, blunt tail usually less than one-tenth of its total length, and a poorly defined, rather pointed head. Snout is chisel-haped and overhanging and the eyes are very small with slit-shaped pupils in good light. Belly scales are extremely narrow and about a third of the body width or less (Fig., p. 197). Dorsal scales, including those of most of head are small, rather glossy, and not obviously keeled and there are 40–50 or more across the mid-body. Colouring very variable: usually pale greyish, buffish, brownish, or even reddish with darker bars and blotches that often join forming an irregular network enclosing isolated areas of ground colour. Frequently a dark ^-shaped mark on back of head, and a dark streak from the eye to the corner of the mouth. Belly pale yellowish or whitish.

Similar Species None within European range. Adults have general build of large viper, but the head is narrower, and even the undamaged tail has a blunt, rounded tip.

Habits Found principally in dry habitats, usually with a good covering of light soil or sand, such as arable land, dry beaches or even sandy soil at the bottom of rocky hollows; sometimes also occurs in quite stony places, Spends much of time in rodent galleries but may burrow actively in loose soil. Largely nocturnal, but may be partly diurnal in spring and autumn, otherwise usually encountered during day only when dug or ploughed up, or when stones, logs, etc. are turned over. May come to surface, especially at night or twilight, when it is fairly swift (animals uncovered during day tend to move slowly). Feeds largely on small rodents, but lizards, nestling birds, big insects and even slugs may be eaten occasionally. Large animals are constricted before being swallowed. Apparently hunts mainly in mammal burrows, but sometimes lies beneath surface of soft soil and catches passing prey.

Females produce 6–20 live young, about 12–15cm long which are coloured like the adults. Takes at least three years to reach sexual maturity.

Typical Snakes
(Family *Colubridae*)

This very large family of more than 1,700 species is extremely widely distributed, being absent only from much of Australia, from very cold areas and from some islands, such as Ireland and New Zealand. It contains the majority of European snake species. All European colubrids have the head covered by large scales. Most are essentially diurnal, the main exceptions being the crepuscular Dwarf snake (Gitenis modestus), Southern Smooth Snake (*Coronella girondica*), False Smooth Snake (*Macroprotodon cucullatus*) and Cat Snake (*Telescopus fallax*); These two last species and the Montpellier snake (*Malpolon monspessulanus*) are the only European colubrid snakes with fangs, which are at the back of the upper jaw. Mating in colubrids is often prolonged and males may leave a plug in females afterwards, which probably makes it harder for other males to mate with them. Most species lay eggs, but the Smooth Snake (*Coronella austriaca*) produces fully formed young.

MONTPELLIER SNAKE *Malpolon monspessulanus*   Pl. 42
Range Iberian peninsula (except parts of north), France (Mediterranean coast and an isolated population in the south-west), Italy (Liguria and Trentino only), east Adriatic coast, south Balkan peninsula north to Albania, Macedonia and southern Bulgaria, and a few Greek islands, including Corfu, Cephalonia, Zakynthos, Thasos, Samothraki, Skopelos and Euboa, and some close to the coast of Asintic Turkey; Lampedusa. Also North Africa, west Caspian area and south-west Asia. Map 161.

Identification Adults up to 240cm, including tail, but usually considerably less. Can reach a weight of 3kg. A large, formidable, often uniformly coloured snake with rather stiff, slender body and narrow characteristically shaped head: eyes very large and 'eye-brows' raised and overhanging, extending forwards onto snout as two strong ridges with a hollow between them. Snout overhangs lower jaw and the frontal scale is very narrow. The combination of large eyes and strong 'brows' give this snake a very penetrating expression. Dorsal scales large, in 17 or 19 rows at mid-body, and are unkeeled but may be grooved becoming more so with age. Ground colour grey, reddish-brown, olive, greenish, or blackish. Many adults are more or less uniform, but may have scattered light or dark spots, or both. Some have flanks darker than back and others retain a weak version of blotched juvenile colouring. Belly is often yellowish, mottled or suffused with dark pigment. Some hatchlings are fairly uniform, others have a series of vague dark blotches on back and rows of small darker spots on sides. Irregular, light spots may also be present and the head and throat often bear regular, well-defined light and dark markings.

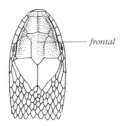

frontal

Montpellier Snake

Variation Considerable variation in colouring. Western animals (_M. m. monspessulanus_) usually have 19 rows of dorsal scales at mid-body, while eastern ones (_M. m. fuscus_, formerly assigned to _M. m. insignitus_) tend to have 17. The name _M. m. insignitus_ is now used for Levant and many north African populations including that on Lampedusa.

Similar Species The Montpellier Snake is unmistakable if the head can be seen.

Habits An aggressive, agile and mainly terrestrial snake. Most usually found in warm, dry Mediterranean habitats, nearly always with some plant cover in which it often hides. Prefers open rocky or sandy country with bushy vegetation, but is also encountered on arable land, in open woods and even in salt-marsh and sand-dune vegetation near sea. May sometimes occur on edge of moist woods (for instance in Italy) and on river banks and irrigated land, and occasionally even swims. Largely diurnal but can be partly crepuscular in hot weather. Occurs up to 2,150m in southern parts of range. Able to climb in vegetation and often basks on roads where it is frequently run over by traffic. Montpellier Snakes have a large gland on each side of the snout, beneath the skin between the nostril and the eye. This produces a liquid secretion that exits through the nostril and is wiped over the body, especially after the skin has been shed. The secretion appears to 'waterproof' the snake, so that loss of water through the skin is greatly reduced. When threatened, hisses loudly and for long periods; may also flatten and inflate the front of the body and spread neck. Temperament varies but provoked animals often bite. Prey is hunted largely by sight.

Food includes a large proportion of lizards (adults even take fully grown Ocellated Lizards); other snakes and small mammals (up to size of small rabbits) are also eaten, as well as occasional birds (especially the young of ground-dwelling species) and, in the case of juveniles, invertebrates.

Males take part in combats with each other in the breeding season, sometimes even during mating. Females lay 4–20 (usually 4–14) elongate eggs, about 27–55mm x 12–40mm, which are deposited under stones, in cracks in timber, under dead leaves and in rabbit and bee-eater burrows where large communal clutches are sometimes found. Eggs hatch in about 2 months producing babies 20–36 cm long which become sexually mature in 3–5 years. Males have been known to live 25 years and females 15 years in the wild.

Venom Prey animals are killed within minutes by the venom. However, the fangs are at the back of the upper jaw, so the snake must take a very secure grip before they can function. Because of this the fangs are only liable to be used effectively on human beings if the snake is actually picked up and a severe bite from a free snake is unlikely. In people, prolonged bites to the hand produce numbing and stiffness in the arm, as well as swelling and even fever. This usually passes in a few hours.

HORSESHOE WHIP SNAKE *Coluber hippocrepis*   **Pl. 42**

Range Iberian peninsula (particularly south, west, centre and east), south-west Sardinia and Pantellaria. Also north-west Africa. Map 162.

Identification Adults up to 180cm, but usually smaller. A fairly slender snake with a characteristic pattern. The only European snake with broad belly scales (apart from vipers) that regularly has a complete row of small scales below the eye. Head well defined, eyes fairly large with round pupils and back scales smooth. A series of large, closely spaced spots on back, which are often blackish or dark brown with a black edge; flanks also have one or more rows of smaller dark spots. Ground colour olive, greyish, yellowish, reddish or brown but frequently largely hidden by dark pigment so that snake appears mainly blackish. Often a ^ or V or horshoe shaped mark on the back of the head which may be joined to first spot on neck. Belly yellow, orange, or red, usually with dark spots, especially at sides and towards tail. Young are generally like adults but patterning is more contrasting: ground colour not overlaid by dark pigment and spots on back are yellowish, grey or greenish. 25–29 rows of back scales across mid-body, 214–258 belly scales (214–242 in males and 222–258 in females), and 72–109 pairs of scales beneath tail (72–106 in males and 81–109 in females).

Variation The population on Pantellaria is assigned to the subspecies *C. h. nigrescens*.

Similar Species None within range. Young may resemble juvenile Leopard Snakes (Pl. 44) but the two species do not occur in the same places and the Leopard Snake lacks a row of small scales below the eye. See also Algerian Whip Snake (opposite).

Habits A fast, shy, mainly diurnal, though occasionally crepuscular, snake found in dry, often rocky places, scrub-covered hillsides, coastal plains with low vegetation and arable areas often with almond, olive and carob trees. Also occurs in and around human habitations, buildings and drystone walls. Found up to 1,800m in south of range. Largely ground-dwelling but can climb well in bushes, on shelving rock faces and on stone walls. Usually flees but if cornered will flatten head, inflate body and hiss loudly and bites very readily if handled. May explore holes and crevices when hunting. Main food of adults is small mammals up to the size of rats including even occasionally bats. A smaller proportion of lizards, sometimes as large as Ocellated

Lizards, is also taken, as well as a few birds. Young eat mainly lizards and some invertebrates.

Females lay up to 29 elongate white eggs, about 60–80mm x 20mm, although usually far fewer and often only 4–11. Clutches are deposited under stones or logs or in rodent burrows and hatching occurs after 1 or 2 months producing young about 15–35cm long. In captivity males may mature in 5 years and females in 8, at lengths of about 50cm and 70 cm respectively.

ALGERIAN WHIP SNAKE *Coluber algirus*   **Pl. 43**

Range In Europe, confined to Malta. Also north-west Africa from where it may have been introduced to Europe. Map 163.

Identification Adults up to about 115cm including tail, possibly even larger but usually smaller. A slender snake with a well-defined, rather narrow head, and prominent eyes with roundish pupils. Scales smooth. Adult usually grey, grey-yellow or grey-brown, with a darker head bordered behind by a yet darker collar. Usually a series of widely separated dark bars on back and two rows of dark spots on each flank, the upper out of phase with the bars on back, and the lower reaching the edge of the belly scales. Pattern may fade with age. Belly whitish. Young are a more intense yellowish, pinkish or red-brown with more contrasting markings including an often blackish head and collar. 25 rows back scales across mid-body, 210–237 belly scales, and 90–110 pairs of scales beneath tail. May occasionally have a complete row of small scales below the eye.

Similar Species None on Malta. See also Horseshoe Whip Snake (opposite).

Habits Apparently lives in dry stony places often with scattered bushes; also ruins, rock piles, arable land, dry ditches and roadside banks. Not common on Malta. Agressive and bites readily. Feeds on lizards, small mammals, birds, snakes and insects. Like other whip snakes, lays eggs.

DAHL'S WHIP SNAKE *Coluber najadum*   **Pl. 44**

Range Mainly south Balkans: from mainland Greece, north to coastal Croatia, Macedonia and extreme south Bulgaria. Absent from many Aegean islands but found on the ones that lie close to Asiatic Turkey. Also Caucasus region and southwest Asia. Map 164.

Identification Adults up to 135cm but usually less than 100cm. Easily identified by build and pattern. A very slender snake with a narrow well defined head and a tail that makes up a quarter or even a third of the total length; belly scales with a distinct keel on either side (Fig., p. 206). Foreparts usually grey-green to olive brown often becoming browner or reddish towards tail. A row of large black to olive spots on sides of neck each of which may be surrounded by concentric dark and outer light rings. These spots are more prominent and extend further along body in young animals. Skin in front of and behind eye light coloured. Belly yellow or whitish, unspotted. Body scales smooth, 19 (rarely 17) back scales across body, about 205–223 belly scales, and 104–138 pairs of scales under tail.

Variation Some variation in colour and extent of neck markings. European populations are assigned to the subspecies *Coluber najadum dahlii*. The population on Kalymnos off the coast of Asiatic Turkey (*C. n. kalymnensis*) is melanistic.

Similar Species Reddish Whip Snake (p. 206). Differs from other whip snakes in range in its slender build and pattern and in having a well developed keel along each side of the belly scales.

Habits Diurnal and mainly terrestrial in dry, often stony habitats, usually with bushes and some dense grassy vegetation in which it climbs. Occurs in open areas in woods, by overgrown walls, stony banks, boulder-choked gullies, ruins, path edges etc. Found up to 2,000m in southern parts of range. Extremely fast, even compared to other Whip snakes, running down prey by speed and sometimes pursuing it for considerable distances. Small prey items are simply grabbed in the jaws, while larger ones may be pressed against the ground with the body, or this may be wrapped two or three times round them while they are swallowed. May bite when handled; unlike most snakes, does not appear to hiss. Feeds mainly on small lizards, especially lacertids but known to also take skinks and even small agamas; additionally eats grasshoppers and other invertebrates and, rarely, small mammals. Like many colubrid snakes, the saliva of this species appears to be toxic to its usual prey. Although such snakes may lack proper venom-injecting fangs, the saliva enters through the punctures made in the prey by the ordinary teeth and can cause death quite quickly.

Females lay 3–16 very elongate eggs, 21–44mm x 5–14mm, producing hatchlings less than 30cm long.

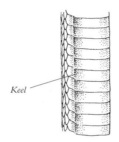

Keel

Underside of snake showing keeled belly scales

REDDISH WHIP SNAKE *Coluber collaris*   **Pl. 44**

Range Coastal south-eastern Bulgaria and Bosphorus area of European Turkey; also north-west and south Asiatic Turkey, Syria, Lebanon, Israel and Jordan. Map 165.

Identification Up to 70cm including tail in Europe, but to 100cm elsewhere. A very slender snake with a long tail, generally like Dahl's Whip Snake, but head tends to be smaller and flatter and colouring differs. Reddish grey to red-brown above with a dark, light-bordered band across the neck and scattered dark, light-bordered spots on the foreparts that are more widely spaced than in Dahl's Whip Snake. Skin in front of and behind the eye is usually dark. Belly yellowish white. 19 back scales across body; in Europe animals have 209 or fewer belly scales and 95 or fewer pairs of scales beneath tail.

Variation Animals from Europe and West Asiatic Turkey have lower numbers of belly and tail scales than elsewhere.

Similar Species Dahl's Whip Snake.

Habits Dry coastal areas, especially stony or rocky plains with bushes, scrub and other low vegetation. Eats mainly lizards but also large insects. A diurnal, extremely fast ground-dwelling snake.

Lays 3–5 cylindrical eggs, 25–30mm x 8–9mm.

Other names *Coluber rubriceps*.

WESTERN WHIP SNAKE *Coluber viridiflavus*   **Pl. 43**

Range France (except north), Luxembourg (?), southern Switzerland, and north-west Croatia (as far south as Senj), and south to extreme north-eastern Spain (Pyrenees), Sicily and Malta. Occurs on Corsica, Sardinia, and various smaller islands such as Krk, in the Adriatic, and Elba, Montecristo and the Pontini and Aeolian islands in the Tyrrhenian Sea. Sporadic in north of range. Map 166.

Identification Adults up to about 150cm, including tail, but may occasionally grow larger. A rather slender snake with a smallish but fairly well-defined head, smooth scales and fairly prominent eyes with round pupils. Over most of range has greenish-yellow ground colour, largely obscured by black or dark green pigment that forms indistinct crossbars on foreparts and reduces the ground colour on rest of body to yellowish streaks or rows of spots. These characteristic markings extend on to the tail and are often visible as the snake retreats into cover. Belly yellowish or greyish (rarely reddish), sometimes with small, dark spots. Most adults from the north-east of the range, southern Italy, Sicily, and some islands are almost entirely black. Young are pale grey or olive with a bold head pattern: usually a light bar between the eyes, round spots on the parietal scales and a light V- or W-shaped mark behind these; there may also be dark marks on foreparts. Full adult colouring is developed in about the fourth year. Usually 19 back scales across the mid-body (rarely 17 or 21), 187–227 belly scales (187–212 in males and 197–227 in females). and 95–125 pairs of scales beneath the tail.

Variation Marked variation in colouring (see above).

Similar Species Most likely to be confused with Balkan Whip Snake (p. 208) and Large Whip Snake (p. 209). Horshoe and Algerian Whip Snakes (p. 204) and Rat snakes (*Elaphe* species, p. 211–215) have characteristic patterns and higher numbers of dorsal scale rows. Montpellier Snake (p. 202) has a characteristic head-shape

Habits Found in wide variety of mainly dry, open but well vegetated habitats, such as sunny, rocky hillsides, bushy areas, maccia and scrub, open woods (both deciduous and evergreen) and their edges, ruins and gardens. Occasionally also in damp meadows and other humid places; found on heaths in parts of France and in Pyrenees occurs in the vicinity of warm-water spas. Ranges up to 1,500m (rarely to 2,000m) in south of range. Frequently very abundant and largely diurnal and terrestrial, although it also climbs well among rocks and in bushes. Hibernates in rock cracks, mammal burrows and outbuildings, sometimes communally. May move quite long distances from these hibernating places in the spring to its summer quarters. Here may spend most of its time in a relatively limited area (averaging around 3,000 sq. m in parts of Italy). Within this home range, snakes move to feed and bask, spending the hotter part of the day in the shade. As with the Adder (p. 230), very dark colouring is believed to increase heat absorption in cool weather and black animals tend to grow slightly larger than others and produce rather more eggs. However these advantages are balanced by such animals often being poorly camouflaged and consequently probably more prone to predation. This snake is very fast and agile, hunting by sight and extremely aggressive when captured, biting hard and persistently. May sometimes constrict prey. Food of adults includes many lizards and a varying proportion of mammals; also some nestling birds, other snakes (even vipers and members of its own species) and frogs. Juveniles take mainly small lizards and their eggs and large grasshoppers.

   Males fight in the breeding season, lashing each other with their tails and will exceptionally travel up to 3km searching for mates. Females lay 4–15 elongate eggs (often 4–7 in Italy) that are about 30–40mm x 15–22mm with star-shaped concretions on the shell. These hatch in around 6–8 weeks producing babies about 20–25cm long.

BALKAN WHIP SNAKE *Coluber gemonensis*   Pl. 43

Range East Adriatic coastal area from extreme north-east Italy through Croatia and Albania including many offshore islands, even distant ones; also mainland Greece and the Ionian islands, and Euboa, Kythera, Crete and nearby islets and Karpathos, but otherwise absent from the Aegean sea. Map 167.

Identification Adults normally under l00cm including tail, occasionally up to 130cm or even longer. A rather slender snake with a fairly well-defined head, smooth scales, fairly prominent eyes and round pupils. Olive-grey, grey-brown, or yellowish-brown with dark spots on foreparts that are often divided by light streaks and may form very irregular bars. Rest of body tends to have regular narrow light and dark stripes. Small white spots frequently present on edges of some back scales. Underside yellowish or whitish with dark spots typically present at least on sides of neck. Young very similar to those of Western Whip Snake. Usually 19 dorsal scales across mid-body (very rarely 17), 160–187 belly scales and 80–116 pairs of scales under tail.

Variation Some minor variation in the boldness and extent of dark markings.

Similar Species Overlaps with the Western Whip Snake in north of range, but this species grows larger, and where it occurs with the Balkan Whip Snake, is black with no small white spots on the scales; the number of ventral scales is also higher in this area (188 or more) and this feature will usually distinguish juveniles. Large Whip Snake (opposite) is also similar. Rat snakes (*Elaphe* species, pp. 211–215) have different patterns and more back scales across mid-body. Montpellier Snake (p. 202) has characteristic head shape.

Habits Very similar to Western Whip Snake; occurs in dry stony places, scrub areas, vineyards, overgrown ruins, open woods and low macchia, and on road banks, bare scree etc. Sometimes extremely common. On Crete is found in marshes by streams. Occurs up to 800m in Croatia and 1,400m on Crete but is most abundant close to sea level. Diurnal and terrestrial although sometimes climbs in bushes etc. Bites fiercely when handled. Feeds on lizards and large insects including grasshoppers; also small mammals and nestling birds.

Females lay clutches of 4–10 eggs which are about 25–40mm x 15–20mm.

GYAROS WHIP SNAKE *Coluber gyarosensis*   Pl. 43

Range Known only from Gyaros island in the Aegean Sea, 90 kilometers south-east of Athens. Map 168.

Identification Adults up to 91cm, including tail. A slender snake with a fairly well defined head, smooth scales and a round pupil. Uniform black above and smoke grey below; Lips and chin area white with the edges the lip scales dark, belly scales clouded black at front, scales under tail with scattered black spots. Young coloured like juvenile Balkan Whip Snakes. Back scales in 19 rows at mid-body, 194–210 belly scales (194–199 in males, 201–210 in females), 94–103 pairs of scales under tail.

Similar Species Very similar to Western Whip Snake but nearest populations of this species are 600km away in southern Italy. Gyaros whip snake is distinguished from the Balkan Whip Snake by its dark colouring and higher number of belly scales.

Other names. Originally thought to be a subspecies of the Balkan Whip Snake, *C. gemonensis gyarosensis*

Habits Very little information available. The few specimens known were found in broken, densely vegetated scrub areas where Wall Lizard is common. Behaviour is presumably like Balkan and Western Whip Snakes.

LARGE WHIP SNAKE *Coluber caspius*   Pl. 43

Range Mainly south and east Balkans, north to Hungary, extreme south and east Romania and the Ukraine; absent from south Greece (Peloponnese) but on many Aegean islands (not Gyaros, Crete or Rhodes). Also south-west Asia. Map 169.

Identification Adults up to about 200cm including tail, rarely to 250cm; males larger than females. One of the largest European snakes, with a fairly well-defined but smallish head, smooth scales, and quite prominent eyes with round pupils. Body colour yellow-brown, olive-brown, or reddish above, with a pattern of, often rather weak, narrow stripes that extend all over body and run along each scale; no obvious dark blotches. Belly light yellow, orange, or orange-red, without dark spots in adults, even at sides of neck (very rare exceptions). Young animals are greyish or brownish, usually with well spaced, short bars on back and no bold light markings on head; a dark streak often present on midline of crown and belly may have a few dark spots at sides. Usually 19 rows of dorsal scales at mid- body, 189–211 belly scales with a weak keel on each side (189–204 in males and 198–211 in females) and 80–110 scales under tail (80–110 in males and 91–104 in females).

Variation Members of some Aegean populations (on Andros, Kythnos, Tinos, Karpathos) are small, usually under 100cm long, and often retain traces of juvenile colouring. Around Corfu the Large Whip Snake may have dark bars on the foreparts.

Similar Species Balkan Whip Snake (opposite) is rather similar, but is smaller with dark blotches and often small light spots and streaks on forepart of body, the belly is paler with some dark spots, there are fewer belly scales (pp. 160–187), and the young have different colouring. Rat snakes (pp. 211–215) occurring in range have different patterns and more rows of dorsal scales. Montpellier Snake (p. 202) has characteristic head shape.

Habits Generally like Western and Balkan Whip Snakes. A diurnal, very swift and largely terrestrial species living in dry, open habitats usually with some vegetation: rocky hillsides, embankments, vineyards, gardens, scrub, dry-stone walls, open woods, wooded steppe and even semi-desert. A lowland form in north but extends up to 1,600m in other areas. Often basks on roads and is killed by traffic. Highly aggressive and frequently not very inclined to retreat; strikes repeatedly when approached and bites readily and fiercely when handled; large animals can jump forwards up to 1m when striking. May climb 5–7m up bushes and trees when hunting. Adults eat mainly small mammals and lizards plus occasional small snakes and birds. Young take a high proportion of lizards and grasshoppers etc.

Males grasp mates by neck during copulation. Females lay 5–18 eggs, around 45mm x 22mm, which develop into babies about 30cm long that become sexually mature at 65–70cm when about 3 years old. May live 8–10 years in the wild.

Other names. Previously regarded as a subspecies of the Asian *Coluber jugularis*, as *C. jugularis caspius*.

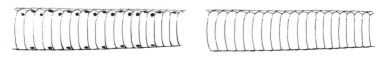

Balkan Whip Snake   Large Whip Snake

Whip snakes, undersides of necks

COIN-MARKED SNAKE *Coluber nummifer*   Pl. 43

Range Greek islands in the south-eastern Aegean sea: Leros, Kalymnos, Kos, Symi, and Rhodes. Also Asiatic Turkey, Cyprus, south-west and central Asia and north-east Egypt. Map 170.

Identification Up to about 100cm including tail, often less. A strong, fairly robust snake with well defined head and relatively small eyes with a round pupil. Back scales keeled. Brown, grey or olive grey above, often with a row of about 57–65 large, rounded dark spots along the back that is replaced by a continuous line on the tail; smaller spots present on each side that alternate with the back spots. Often a dark bar between the eyes and irregular markings on the back of the head. Belly grey white, frequently with fine stippling and the edges of the scales often spotted. Young animals may have a more contrasting pattern than adults. 23–25 back scales across mid-body; 195–221 belly scales (195–206 in males, 208–221 in females), 79–102 pairs of scales under the tail (82–102 in males, 79–100 in females).

Variation Little in Europe but very variable in pattern over its large range.

Similar Species None within European range.

Habits Lives in dry, sunny, rocky places with bushy vegetation and is commonest in lowlands. Diurnal and very fast and active. Feeds on small mammals, lizards and small birds which are killed by constriction. Appears to gain protection from predators by evolving colouring similar to local viper populations, the similarity being enhanced by behaviour, the snake often flattening its head and hissing loudly when approached.

Females lay clutches of 4–10 eggs.

DWARF SNAKE *Eirenis modestus*   Pl. 44

Range Edge of the Sea of Marmara in European Turkey, and islands along the coast of Asiatic Turkey including Lesbos, Chios, Samos, Alzonisi in the Fournoi group, Leros, Kalymnos and Symi. Also Asia Minor, the Caucasus area and possibly western Syria. Map 171.

Identification Adults up to 60cm, including tail, but usually smaller. A small, graceful pencil-thin snake with the head poorly differentiated from the body and round pupils. Back scales smooth. Upper parts uniform grey, grey-yellow to sandy brown, the centre of the scales lighter than their edges. Top of head with a large dark blotch enclosing a pair of light spots or an often zig-zag light transverse stripe just behind the eyes. A dark crescent-shaped band on the neck that is separated from the head blotch by a light band and has an often light border behind. Belly scales whitish to light grey with a strong gloss. Back scales in 17 rows at mid-body; belly scales 146–177 (146–176 in males, 154–177 in females), scales under tail 61–81 (61–81 in males, 63–72 in females).

Similar Species Cat Snake has collar but not heavy head markings and its pupils are vertical slits in good light.

Habits Found in dry places with sparse vegetation including both fallow and cultivated land and also wooded places, where it tends to occur in the drier areas. Shelters under stones and in rock crevices and several may sometimes be found together. Active especially at twilight but may occasionally be seen in shady places, such as in thickets, during the day. Quite slow and not very shy. Hunts actively for invertebrates and eats especially spiders, but also centipedes, insects, scorpions and small lizards.

Lays 3–8 elongated relatively large eggs, 31–38mm long. Newly hatched young are up to 12cm long.

LEOPARD SNAKE *Elaphe situla* **Pl. 44**

Range Southern Italy, E and S Sicily, Malta, Adriatic coast and islands and S Balkan peninsula including Albania, Macedonia, Greece, Turkey and southern Bulgaria; Crete and most Aegean islands; south-east Crimea. Also Asiatic Turkey. Map 172.

Identification Up to about 116cm but usually smaller; females rather larger than males. A medium-sized fairly slender snake with a characteristic range of patterns. Head rather narrow but well-defined, pupils round and scales smooth. Unlike most other rat snakes (*Elaphe* species), adults retain a juvenile pattern which often consists of a row of black-edged brown to red spots on back and a row of smaller spots on each flank. Sometimes back spots are dumb-bell shaped or divided in two, or replaced by two dark-edged stripes. Ground colour is yellowish, greyish or buff. Underside is yellowish-buff near head but becomes heavily marked towards tail so that middle and hind belly is often largely black. Head is boldly marked with dark stripes. 27 (occasionally 25) back scales across mid-body, 215–255 belly scales, and 54–92 pairs of scales beneath tail.

Variation Considerable variation in pattern.

Similar Species Young Horshoe Whip Snake (p. 204) may have similar markings, but does not occur in same areas. Leopard Snake is not likely to be confused with other species within its range, although Smooth Snakes (p. 221) may be vaguely similar as may young Four-lined snakes. The latter usually have very dark spots on back and two preocular scales are present (Fig., p. 213). Steppe Snake (p. 213) usually has 25 back scales across mid-body, fewer belly scales and a different pattern.

Habits A largely ground-dwelling snake characteristic of Mediterranean maccia and usually found below 500m, but up to 1,600m in the southern Balkans. Usually in sunny habitats, especially those including numerous rocks and stones and some plant cover: field-edges, road banks, stone-piles, screes, dry-stone walls; also sometimes marshes and stream edges. May be encountered in human habitats including gardens, vineyards, olive groves, cemeteries and around barns and houses which it sometimes enters. Active by day but also sometimes seen at dusk. Rather slow moving but climbs quite well on stone-piles, walls and bushes where often seen draped over branches avoiding high ground temperatures. Frequently bites when handled but may remain still when disturbed. Sometimes vibrates tip of tail rapidly among dry leaves making a rattling noise. Larger prey may be constricted but small food items are pressed against the ground before being swallowed. Food of adults consists almost entirely of mammals and their nestlings, especially small rodents and occasionally shrews; reptiles and birds form a minor part of the diet; the young eat lizards.

Mating may take some hours during which the males bite the neck of their mates. Females may possibly breed only every other year when they lay 2–8 eggs (usually only 2 in the Crimea) which are large and cylindrical, 35–70mm x 10–22mm. The eggs hatch in 6–9 weeks and babies are 29–36cm long and probably take 3 or more years to reach maturity. Has lived 25 years in captivity.

FOUR-LINED SNAKE *Elaphe quatuorlineata* **Pl. 45**

Range South-eastern Europe north to western Slovenia, and adjoining north-east Italy, Macedonia, Bulgaria, parts of Romania and southern Ukraine; Ionian islands and Aegean islands; north-west, central and south Italy and Sicily. Also parts of south-west Asia. Map 173.

Identification Adults up to about 260cm, including tail, but most animals under 150cm. A large, moderately built snake that is more robust than other big snakes within its range. Head well defined, rather long and somewhat pointed, pupil round and back scales lightly but distinctly keeled in adults giving the snake a rather rough appearance. Young have a row of dark, often black-edged, broad spots or bars on back and one or two series of smaller spots on each flank; their head is boldly marked and the belly has dark markings that may form two streaks. Adults may be marked more or less like this or with four dark stripes along the back. Two preocular scales are present (see Fig., opposite). 25 (rarely 23, 26 or 27) back scales across mid-body, 187–234 belly scales, and 56–90 pairs of scales beneath tail.

Variation Adult size and colouring varies geographically.

Western Four-lined Snake (*E. q. quatuorlineata*) is found in Italy, Sicily, the west Balkans, south-west Bulgaria, mainland Greece and the Aegean islands of Skiathos, Skopelos, Euboa, Kea, Tinos and Spetsai. It is large with 25, rarely 23, back scales across the mid-body; adults are yellowish, pale brown or grey with four dark stripes along the back and a dark streak on the side of the head from the eye to the corner of mouth; underside is usually yellowish, often with some darker markings especially on the tail. The adult pattern is developed after about 2–3 years.

Dwarf Four-lined Snake (*E. q. muenteri*) occurs on many Aegean islands including Mykonos, Paros, Naxos, Iraklia, Ios, Amorgos, Schinoussa, Milos, Antimilos, and possibly Thiera. Skyros animals have been distinguished as *E. q. scyrensis* but are generally similar. A small form, only up to about 90–180cm long, that becomes striped at lengths of 50–60cm. The stripes are very narrow being only about back scale 1–1.5 back scales wide and the snout is rather long. Some individuals from Amorgos in the Cyclades are quite variable and can be dark with poorly developed stripes, while young may be rather pale. Such animals have been regarded as a separate species, Rechinger's Snake (*Elaphe rechingeri*), but are now known to be atypical Dwarf Four-lined Snakes.

Blotched Snake (*E. q. sauromates*) occurs in eastern areas including north-east Greece, European Turkey, Bulgaria, Romania and southern Ukraine. A moderately large form rarely longer than 180cm with 25, rarely 27, back scales across mid-body that are less distinctly keeled in adults than in Western Four-lined Snakes. The adults lack stripes and tend to retain the juvenile markings, although they often become more uniform through darkening of the ground colour.

Similar Species Aesculapean Snake (p. 214) is much more slender, lacks keels on the dorsal scales at all ages and has only one preocular scale. In Ukraine see also Steppe Snake (opposite).

Habits Often found along wood-edges and hedges, in open woods, on rocky overgrown hillsides etc. and occurs to over 1,400m in south of range. Prefers some shade and likes warm, rather humid environments and may be encountered in marshy areas and near pools and streams. Often hunts in warm, cloudy conditions and at dusk, and climbs and swims well. The Western Four-lined Snake is a rather phlegmatic, slow-moving form whereas the more eastern Blotched Snake is faster and may be more aggressive. This form vibrates the tail tip and bites when disturbed and also raises the foreparts, flattens the head and lunges forwards while hissing. As in other European Rat snakes (*Elaphe*), large prey is constricted. Food of adult Four-lined Snakes consists mainly of small mammals up to the size of rats and young rabbits, but birds, especially nestlings, are taken particularly in the spring, and eggs and occasional lizards may also be eaten; sometimes raids poultry yards. The young take a relatively high proportion of lizards

Females breed most years, at least in some areas, and lay 4–16 eggs 30–70mm in length that hatch in about 7–9 weeks. Hatchlings are 20–40cm long in the Western

Four-lined Snake and 15–25cm in the Blotched Snake; they mature in 3–4 years.
Note The Western Four-lined Snake and the Blotched Snake meet in extreme north-east Greece and south-west Bulgaria, yet apparently remain separate. This, together with differences in behaviour, suggest they are separate species, *E. quatuorlineata* and *E. sauromates*. This has recently been confirmed by studies of their DNA.

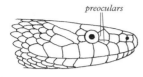

Four-lined Snake

Aesculapian Snake

STEPPE SNAKE *Elaphe dione*   **Pl. 44**

Range South-east Ukraine as far west as eastern Sea of Azov. Also eastwards to Pacific coasts of Russian Federation, North China and Korea. Map 174.

Identification Up to about 100cm, females rather larger than males. General form rather like Four-lined Snake and head quite short and robust. Usually a dark U-shaped mark on top of head that is open at the back and may sometimes be broken up; also a dark stripe extending back from the eye. Body with series of short dark transverse spots or bars, each about three scale rows wide. This is combined with four vague dark stripes alternating with three light ones running along the snake to the tail tip. General colouring usually yellowish or dirty white and brown, occasionally more olive. Underside light grey or yellowish, sometimes reddish in vent area, with numerous dark spots. 25 (occasionally 23–26) unkeeled back scales across midbody, 183–205 belly scales (183–194 in males, 197–205 in females), and 58–70 pairs of scales beneath tail (65–70 in males and 58–63 in females).

Similar Species Distinguished from Four-lined Snake by general pattern, consistently spotted belly, often lower numbers of belly and tail scales and in having the parietal scale only contacting the upper postocular scale, instead of the upper and lower one as is usual in the Four-lined Snake.

Habits Ground-dwelling at low altitudes in dry areas, especially steppe on stony or rocky ground or clay, with low often sparse vegetation or bushes; also in open woodland. Occurs mainly below 500m. Rarely bites, but may vibrate tail tip rapidly when threatened making a rustling noise when it is in contact with dry leaves. This behaviour is similar to that of venomous Pit vipers (*Agkistrodon*), with which it coexists in the more eastern parts of its range, and might possibly be partial mimicry of these. Eastern animals may hibernate communally below ground, sometimes forming assemblages of thousands of animals with other snake species. Takes refuge in marmot and other rodent burrows and also hunts in them. Diet consists of small mammals and birds, especially nestlings and eggs, and lizards.

Lays cylindrical eggs 30–55mm long in clutches of 3–24, the number of eggs varying in different parts of the large geographical distribution of this species. In at least some populations, development of the young snakes is well advanced at laying and they hatch in just a few days.

214 SNAKES

AESCULAPIAN SNAKE *Elaphe longissima* **Pl. 45**

Range France, except north, west and south Switzerland, south and east Austria, south-east Czech Republic, Slovakia, south-east Poland, and Ukraine; south to north-west Spain (as far west as Santander province), central Italy and southern Greece although absent from the Aegean islands; also a few isolated localities in Germany near Heidelberg and one in the north-west Czech Republic; possibly west Sardinia. Outside Europe, occurs in Turkey, the Caucasus and north Iran. Map 175.

Identification Adults up to 200cm including tail but usually under 140cm; males bigger than females. A large, slender snake with a rather narrow, poorly defined head and a round pupil. Belly scales have a slight keel on each side (Fig., p. 206). Adults are usually fairly uniform grey-buff, olive-brown or brown, the foreparts often lighter; frequently with small white spots on the scale edges, especially at mid-body. In some animals, there is also a vague pattern of dark or light stripes along the body (especially in Italy). Usually a dark streak on temple and a vague yellow patch behind this; upper lip is yellow and the eye brown or grey. Underside pale yellowish or whitish. Young with 4–7 rows of small dark spots on body and a boldly marked head: often a dark A- or ?-shaped mark on neck, a bar across snout in front of the eyes and another on the temple which is followed by brightish yellow blotch like a collar. 23 (rarely 21) back scales across mid-body, 211–250 belly scales, and 60–91 pairs of scales under tail.

Variation Some variation in pattern. Black or very dark animals occur occasionally.

Similar Species Italian Aesculapian Snake (opposite), Ladder Snake (south-west Europe only, opposite), Four-lined Snake (Italy and Balkans, p. 211). Whip Snakes (p. 204–211) and Montpellier Snake (p. 202) differ in general appearance and lower number of back scales across mid-body. Young may look rather like Grass Snake (p. 216).

Habits Encountered in often dry habitats, especially sunny woods, shrubby vegetation, field borders etc., but also on old walls, ruins, stony banks and even hay stacks. In north of range confined to favourable localities such as sheltered, south facing slopes on light soils and river valleys. May be found in quite humid places in the south. Occurs up to about 2,000m in some areas. Diurnal, although occasionally active on hot evenings; often rather inconspicuous. Enjoys sun and may bask on paths and tracks but retreats from excessive heat. Moves rather deliberately but is a very adept climber in bushes and trees, even ascending tall vertical trunks. Often flees upwards when disturbed, but may hold ground when approached and make chewing movements with jaws; frequently bites and voids the malodorous contents of cloacal glands when handled. Food is often constricted and consists mainly of small mammals (especially mice and voles but even squirrels); lizards and birds, especially nestlings, are also taken; the young often eat lizards. Adults may take a prey item every 3 days in summer.

Males can travel quite long distances in breeding season, sometimes up to 2km. They often wrestle with each other for up to half an hour, twining their bodies and tails. Most females appear to reproduce every year, producing a single clutch of 2–18 (often 5–11) elongate or pear-shaped eggs, 35–60mm x 17–25mm, with longitudinal grooves on the shell. These may be laid in holes including ones in trees, in soil, and sometimes communally in fermenting material often with the eggs of the Grass Snake. Length of hatchlings appears very variable, from 12cm to 37cm. Males mature at around 100cm and females first lay at about 85cm. May live 25–30 years.

Note This snake was associated with the classical god of healing (the Greek

Asclepius and Roman Aesculapius) and encouraged around temples dedicated to him. The god's staff, entwined with a snake, is often used as a symbol of the medical profession. The Aesculapian and other snakes are carried in a yearly religious procession in Cucullo, central Italy.

ITALIAN AESCULAPIAN SNAKE *Elaphe lineata*   Pl. 45
Range Southern Italy, as far north as Naples and Monte Gargano; also Sicily. Map 176.
Identification Up to 140cm, males larger than females. Very similar to the Aesculapian Snake which occurs immediately north of its range. Differs in its generally lighter colour and grey underside. If dark stripes present on back, these are narrower than the spaces between them; any white speckles on the scale edges are confined to these stripes. There is no light blotch on the side of the neck, either in adults or young; the eye is reddish. Usually 23 back scales across mid-body, 225–238 belly scales, and 72–82 pairs of scales under tail.
Similar Species Aesculapian Snake (opposite).
Habits Apparently similar to Aesculapian Snake. Found up to 1,600m.

LADDER SNAKE *Elaphe scalaris*   Pl. 45
Range Spain and Portugal, Mediterranean littoral of France extending just over Italian border, Iles d'Hyères, Minorca. Map 177.
Identification Adults up to about 160cm, including tail, but most individuals less than 120cm; males are larger than females. A large, moderately built snake with a pointed, overhanging snout and short tail. Pupil round, scales smooth; rostral shield very pointed behind and projecting between internasal scales. Preocular scale usually single, occasionally double. Adults fairly uniform in colour: yellow-grey to mid-brown with a pair of dark brown stripes on back that extend from the neck to the tail tip. Usually a dark stripe from the eye towards the angle of the mouth; belly unmarked or with a few dark spots, eye dark brown to black. Young boldly marked with 'H'-shaped blotches on back which may join to form a 'ladder'; head spotted and irregular spots, streaks or bars present on sides. Belly of young yellowish or whitish, marked with black that sometimes covers whole surface. 27 (rarely 25–31) back scales across mid-body, 198–228 belly scales, and 48–68 pairs of scales beneath tail.
Variation Considerable variation in pattern with age but otherwise fairly constant, although dark body stripes may ocasionally be poorly developed or interrupted; melanistic animals occur.
Similar Species In north and north-east of range poorly marked adults could be confused with Aesculapian Snake (opposite) but this species is more slender, has a more rounded snout (with rostral scale less pointed behind), a 'keel' on each side of the belly scales, and only 23 back scales across mid-body. Montpellier Snake (p. 211) lacks stripes on back, has more rounded snout, narrow frontal scale, and only 17 or 19 back scales across mid-body. Young Ladder Snakes may look superficially like Southern Smooth Snakes (p. 222), but their pattern is rather different and they have more back scales across the mid-body.
Habits Usually found in sunny, often stony Mediterranean habitats, typically with some bushy vegetation: hedges, vineyards, field-edges, scrub, open woods, overgrown dry-stone walls etc. Ranges up to 2,100m in south of range but usually under 700m. Largely diurnal and sometimes active in the hottest part of the day, but may be crepuscular in spring and nocturnal in hot weather. Occurs mainly on ground, but also climbs well on rock piles, banks and outcrops and in bushes and even trees.

May travel 100m in the course of the day and have an average home range of 4,500 sq. m but often considerably larger. Aggressive, sometimes lunging forwards with open mouth, and often hissing and biting if captured, when it may also void the pungent contents of its cloacal glands. May sometimes hibernate communally. Appears to often hunt by scent. Adults eat warm-blooded prey, constricting larger items. Mammals (rodents, small rabbits etc.) predominate but a smaller proportion of birds (principally nestlings and eggs which it often climbs to obtain) are taken and trivial quantities of lizards and their eggs; males often take larger prey than females. Very young snakes eat nestlings of small rodents, grasshoppers and possibly small lizards and similar sized animals. Adults sometimes hunt rodents in buildings.

Mating lasts up to an hour and, about 3–6 weeks after this, females lay 4–15 (even up to 24) eggs, 45–70mm x 14–33mm, that stick together and are deposited under stones or dead vegetation, in burrows of other animals or ones dug by the snake. The eggs can make up 30–45 per cent of the weight of a pregnant snake and the mother may stop with them for some days after they are laid. Not all adult females breed in a particular year. Eggs hatch in around 7–11 weeks and babies are about 25–35cm long. Females mature at around 5 years at a length of about 65cm while males take less time, maturing at about 50cm. Has been known to live 19 years in captivity and at least as long in the wild.

GRASS SNAKE *Natrix natrix*   **Pl. 46**
Range Nearly all Europe; north to southern Norway and Sweden (with isolated populations on the coast of the Gulf of Bothnia and old records as far north as 67°N), southern Finland and Russia; absent from some islands, such as Ireland, the Balearics, Malta, Crete, and some Cyclades. Also occurs in north-west Africa and Asia east to Lake Baikal, and north-west China and Mongolia. Map 178.
Identification Usually up to 120cm, including tail, often less but occasionally up to 200cm; females grow larger than males and sometimes twice as long. An often rather large, usually quite thick-bodied snake, with a well-defined rounded head especially when adult, round pupils, and 19 keeled back scales across the mid-body. Colour very variable, but many specimens have characteristic yellow (or, less commonly, white, orange or red), black bordered collar just behind the head. Body usually olive-grey, greenish, olive-brown or even steel-grey with various dark blotches and sometimes light stripes. Underside whitish or grey, chequered with black and occasionally entirely this colour.
Variation Occasional black individuals occur quite widely but vary in incidence, for instance being very rare in Britain. There is considerable geographical variation in colouring. Commonest patterns are as follows.

Britain, France, Belgium, Netherlands, much of Switzerland and north Italy (*N. n. helvetica*). Robust with a light yellow or whitish collar, dark vertical bars on sides, and sometimes small spots on back.

Central and eastern Europe from the Rhine area, east Switzerland and north-east Italy eastwards (*N. n. natrix*). Fairly slender, with a yellow or even orange collar and no obvious bars on sides, but often dark spotted.

Gotland island, Sweden (*N. n. gotlandica*). Small, collar often orange, sides may be barred; often entirely dark.

Spain and Portugal (*N. n. astreptophora*). Adults very robust with a large head and sometimes a blood red eye, collar pale whitish or yellowish and absent in old females, body grey or grey green, often quite uniform.

Sardinia (*N. n. cetti*). Slender with a short tail and a variable collar, broad dark bars present on flanks and often reaching midline of back where they alternate on each side of the body.

Corsica (*N. n. corsa*). Similar to Sardinian animals but pale collar usually absent and the dark bars are narrower, more numerous and do not usually reach the midline.

Italian peninsula (*N. n. lanzai*). Collar pale and often absent in old females, edges of head scales dark, bars present on flanks and alternating with each other, sometimes additional bars present along the mid-back.

Sicily (*N. n. sicula*). Fairly robust, no pale collar, snout orange, bars present on sides and also along mid-back.

East Adriatic coast, southern Balkan peninsula eastwards to the Crimea, and many Greek islands (*N. n. persa*). Head narrow and tail long, light collar broadly divided on mid- neck, often two pale stripes along back.

Kea island in south-west Aegean sea (*N. n. fusca*). Often three or four rows of dark round spots.

Milos, Kimolos and Polyagos in the south-west Aegean sea (*N. n. schweizeri*). Robust, without a light collar but with three rows of large black spots on a silvery grey ground; entirely dark individuals also occur and ones that are flecked light yellow. Animals from Aegean islands further to the north tend to be intermediate between this form and *N. n. persa*.

All western mainland animals, including those from Spain and Portugal, and much of Italy are sometimes assigned to *N. n. helvetica* and all eastern ones as *N. n. natrix*, the only other forms recognised as separate in this arrangement being those from Corsica and Sardinia.

Similar Species Young Aesculapian Snakes (p. 214) may look similar but Grass snake is more likely to be confused with Viperine and Dice Snakes (p. 218, p. 220). Pattern is often useful in identification. Apart from this, the following characters are helpful.

1. Nostrils at sides of snout (not directed upwards).
2. Broad rectangular internasal scales.
3. Typically 7 upper labial scales with 3rd and 4th entering eye.
4. Usually 19 back scales across mid-body.
5. Usually a single preocular scale.
6. Two to four postocular scales (rarely one).
7. Keeling of back scales is not especially strong and does not always extend onto the tail.

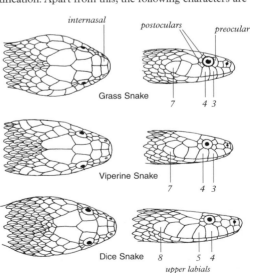

In north-east coastal regions of Black Sea, just outside the area of this book, the Large-headed Grass Snake (*Natrix megalocephala*) is found: the borders of the scales on the massive head are not clear and the scales themselves are lumpy.

Habits Over most of its range, a snake of damp places, such as moist fields and woods but in the south, where it may reach altitudes of 2,400m, usually occurs near water. In north Europe, the Grass Snake is more of a lowland animal, but is less restricted in habitats and may sometimes be found in quite dry woods, hedgrows, and meadows. It may also occur on sea coasts. Often quite common and on some stretches of river there may be one for every couple of metres of bank. Largely diurnal, although crepuscular in south in hot weather. Sardinian population is said to be largely nocturnal. Swims well and often hunts actively in water, but is less aquatic than both Viperine and Dice snakes, although it can remain submerged for 30 minutes. Individual snakes are often quite mobile, frequently with home ranges of 3–120 hectares within which they may travel 10–300m in a day. Some snakes cover 4km in a year. Adult males may shed the skin twice a year and females once. When disturbed often hisses and strikes with mouth closed, but rarely bites. Frequently voids foul-smelling contents of anal glands when handled, and may feign death, lying on back with mouth open and tongue hanging out. The frequency of this behaviour varies geographically, and in some places a third of animals may show it. In most areas, food consists predominantly of frogs and toads, but newts, tadpoles and fish are also occasionally taken and even small mammals, nestling birds, other snakes and slugs. In the Mediterranean area females eat the very large Common Toads found there but males tend to take smaller prey. Animals may eat the equivalent of 5–8 toads a year in Britain; babies take tadpoles and invertebrates. In Cyclades, the Grass Snake eats mainly geckos, lacertid lizards and small mammals.

Males rub potential mates with their chins and several males (sometimes as many as 22) may form a ball around a female, wrestling each other with their tails as they attempt to mate with her. Copulation may last up to 3 hours, females often only mating once. 2–5 weeks afterwards they lay 2–105 white eggs, around 30 being usual for a mature female. These tend to stick together and also subsequently swell, the total range of lengths being about 20mm to 40mm. Eggs are sometimes hidden in holes and crevices, mammal burrows and under stones and logs but the Grass snake also characteristically often lays eggs in compost and dung heaps and piles of leaves and other vegetation including sea weed. In such situations, fermentation raises the temperature and so helps incubate the eggs. Artificial heat from ovens in house walls may also be exploited. Sometimes as many as 3,000–4,000 eggs are laid communally in particularly favourable places, occasionally with those of the Viperine, Dice or Aesculapian snakes. This habit may partly explain why the Grass snake is able to extend further north than any other European egg-laying snake. Incubation lasts 6–10 weeks in south and hatchlings are about 14–22cm long. Males mature in about three years at around 40–50cm (occasionally 30cm) while females mature at about 5 years and a size of about 60cm. Has been known to live 28 years in wild.

VIPERINE SNAKE *Natrix maura*   **Pl. 46**

Range Spain and Portugal, France (except north of about Paris), south-west Switzerland, north-west Italy (east to west Emilia Romagna), Balearic islands (Majorca and Menorca), Iles d'Hyères and Sardinia; reported from Corsica. Also north-west Africa east to north-east Libya. Map 179.

Identification Adults occasionally up to about l00cm, including tail, but most animals less than 70cm; females grow larger than males. A medium-sized snake with well defined, fairly broad head, and rather thickset body in adults. Pupil round and dorsal scales strongly keeled. Colour varies: usually brown or greyish but may be tinged with olive yellow or red. Typically two rows of staggered dark blotches down mid-back, which often merge to produce bars or a well-defined zig-zag stripe. Flanks have dark blotches or, more usually, large light-centred ocelli. Some animals have two narrow, light yellow or reddish stripes running along the back. Head is typically boldly patterned, with one or two ^-shaped marks on crown and neck that may be joined by a central blotch; the light upper lip scales have conspicuous dark borders. Belly whitish, yellow, red, or brown chequered with black or dark brown. 21 (rarely 17,19 or 23) back scales across mid-body, 142–163 belly scales, and 44–73 pairs of scales beneath tail (44–73 in males, 44–59 in females).

Variation Varies considerably in colour and pattern; melanistic animals may sometimes occur and proportion of striped individuals varies geographically. Mediterranean animals often have light bellies, while those from the north Iberian peninsula and north and west France tend to be sombre with little contrast in their colouring. Sardinian Viperine Snakes resemble ones from Tunisia and were perhaps introduced from there. The number of scales across the mid-back is often reduced to 19 in north-west France. Populations on the north shore of Lake Leman in Switzerland also have 19 scales across the mid-body, but their pattern is also reduced and they have long heads, these features making them look like Dice Snakes.

Similar Species Animals with a zig-zag stripe on back can look like vipers (p. 226) and this impression is enhanced by behaviour (below), but true vipers have vertical pupils and smaller head scales. Most likely to be confused with other Water snakes (*Natrix*). For distinction from Grass Snake, see this species. Viperine Snake can be separated from Dice Snake, in the small area where their ranges meet, by the following characters (see also Fig., p. 217).

1. Usually 7 upper labial scales, 3rd and 4th often entering eye (typically 8, 4th and often 5th entering eye in Dice Snake).
2. 2, rarely 3, postocular scales (3 or 4, even 5 or 6 in Dice Snake).
3. Usually 21 rows of back scales across mid-body (19 in most Dice Snakes).
4. About 147–164 belly scales (160–190 in Dice Snake).
5. Head broader and snout more rounded in most Viperine Snakes and body often thicker.
6. Pattern usually bolder in Viperine Snake and ocelli often present on flanks.

Habits Diurnal, although can be crepuscular and even nocturnal in hot weather. Usually found in or near water, both still and flowing. Often very common in weedy ponds and river pools, but also encountered in brackish conditions and by mountain streams, occurring up to more than 1,800m in the south of its range. Also found in very damp woods and meadows and sometimes in drier places, especially old animals. Individuals may range over 0.5 hectare. Often seen when swimming and diving, or basking on water's edge where it may rest on low bushes, rocks etc. Frequently occurs with Grass Snake but is more aquatic, especially in warm conditions when it may be almost completely so. A very good swimmer that can remain submerged for more than 15 minutes. Often hunts in water, using scent and touch to explore for hidden prey under stones and in weed. May take food to bank to swallow and sometimes strikes prey in water from there; also forages on land. When disturbed often retreats into water and dives, but if cornered hisses fiercely and flattens body and head, striking repeatedly, although usually with mouth closed. Such behaviour, especially combined with some of the colour patterns usual in this species, produces a very viper-like impression. When picked up, the snake often voids smelly contents of cloacal glands and may sometimes feign death. Hibernates in holes on land. The skin is shed 2–5 times a year. Feeds mainly on frogs, toads, newts, tadpoles, and a varying proportion of fish and aquatic invertebrates. Earthworms and even small mammals are also eaten occasionally.

Copulation may continue for an hour; females lay 3–16 (exceptionally up to 24) eggs, although usually less than 10 are produced and the average is around 7. The eggs are about 28–40mm x 14–19mm and are laid in clumps in holes near banks of water bodies. Hatching takes place in 6–13 weeks and babies are about 14–22cm long; males mature in three years at 30–40cm and females in 4–5 years when they are considerably larger. May live 20 years in wild.

DICE SNAKE *Natrix tessellata*   **Pl.46**
Range Most of Balkan peninsula and much of Italy, north to southern Switzerland, south and east Austria, the Czech Republic, Slovakia, and south-west Ukraine. Occurs on the Ionian islands and on Serifos, Tinos, Crete and Rhodes in the Aegean sea. Isolated colonies exist in the Rhine valley of Germany but the species is now extinct on the river Elbe. Occurrences north of the Alps in Switzerland are probably introductions. Also found eastwards to central Asia, north-west China and Pakistan. Map 180.

Identification Adults up to 130cm or more, including tail, but usually less than 80cm. Females grow larger than males. A medium-sized snake usually with a rather small, narrow, pointed head, round pupils, and very strongly keeled dorsal scales. Colouring variable: most frequently greyish or brownish but sometimes yellowish or greenish; often with a pattern of regular dark spots evenly dispersed over body. These spots may be large, small, or sometimes completely absent, or they may fuse to form dark bars on back and flanks. Those on flanks often alternate with narrower light bars. Sometimes a ^-shaped mark on nape, but head markings are often obscure. Underside whitish, yellow, pink, or red, dark chequered or with one or two irregular dark stripes, or almost entirely black. In most cases, 19 (occasionally 17 or 21) back scales across mid-body; usually 160–187 belly scales but number can range from 148 to198 and even down to 130–140 in some Swiss populations; there are 47–89 pairs of scales beneath the tail.

Variation Varies greatly in colour and pattern, often from place to place. Occasional animals lack markings or have red and black flanks, or broken stripes, or are completely black. For instance black animals occur on Serpilor (= Snake) Island, off the Danube Delta, and here tend to have 21 rows of dorsal scales and 6 postocular scales.

Similar Species Only likely to be confused with other water snakes (*Natrix*). See Grass Snake (p. 216) and Viperine Snake (p. 218) for separation.

Habits Generally very like Viperine snake and found in similar habitats including rivers and lakes, but even more aquatic, regularly spending much of its time in water. Usually found below 1,000m but can extend 2,200m. Often remains beneath surface for long periods, even hours. It forages actively, searching under stones and weed for hidden prey but also occasionally hunts from ambush, lying with the body and tail largely buried among stones and weed and striking at passing prey. Often abundant and along rivers and streams there may be one animal for every 2–3m of bank. Like other Water snakes, rarely bites but may empty contents of cloacal glands if handled.

Diet consists almost entirely of fish, but amphibians are also taken occasionally.

May court communally like other European Water snakes and eggs are laid 7–10 weeks after mating. These are 30–45mm x 20–25mm and number 5–37, being laid in crevices, under stones and sometimes in fermenting material occasionally with Grass Snake eggs. Newly hatched young are 14–25cm long and males mature after three years at about 40cm.

SMOOTH SNAKE *Coronella austriaca*   **Pl. 47**

Range Southern England, France and north and central Iberian peninsula (isolated records from further south), east to south Scandinavia and Russian Federation and south to Italy, Sicily, and Greece. Also north Asiatic Turkey to north Iran. Map 181.

Identification Adults usually up to about 70cm including tail, occasionally over 80cm; females often rather bigger than males. A moderately small snake with a cylindrical body, poorly defined neck, and rather small head with a fairly pointed snout. Eyes small with round pupils; scales smooth. Colouring variable: usually greyish, brownish, pinkish, or even reddish, sometimes more intense on each side of mid-line giving the effect of two often vague streaks; females tend to be greyer than males and pattern shows some correlation with habitat. Usually small dark spots or blotches present on back; these are clearest on neck (where there are often two short dark stripes) and often form irregular transverse bars or are arranged in two lines. Nearly always a dark stripe from side of neck through eye to nostril and sometimes a vague 'bridle' on snout as well. Belly usually darkish: red, orange, grey, or blackish, generally with some mottling or fine spotting. Commonly 19 back scales across mid-body (rarely 17 or 21), 150–200 belly scales (150–182 in males, 170–200 in females), and 41–70 pairs of scales beneath tail. For head scaling see Fig., p. 223.

Variation Occasional melanistic animals occur. Southern animals often have the rostral scale extending more obviously between the internasals. Size is variable: most although not all animals from England are quite small, as are those from southern Italy and Sicily (*C. a. fitzingeri*) which also tend to be quite uniform in pattern. Populations of small body size in extreme north-west Spain and north Portugal (*C. a. acutirostris*) have a more pointed snout than is usual elsewhere and more scales on the head and neck

Similar Species Southern Smooth Snake (p. 222); has also been confused with vipers, but lacks keeled scales and vertical pupil.

Habits In England and other northern areas associated with sandy heathland with stands of old heather, but elsewhere occurs in hedgerows, wood-edges and open woods, and on bushy and rocky slopes, embankments etc. In south is found in more open places, often with only sparse vegetation, such as screes, stone piles and even cliffs and rock-cuttings, where it lives partly in crevices. Also occasionally encountered in moist habitats, especially in the south. Occurs down to sea level in northern areas but tends to be montane in the south of its range, extending to over 1,800m in some places and reaching 2,600m in southern Spain. A diurnal, although rather secretive, snake found in a variety of usually dry sunny habitats. It avoids extreme heat and is often active in the cooler parts of the day, in warm cloudy conditions and even at night during warm weather. Often basks under cover, beneath vegetation or discarded pieces of metal. Home range in England may be about 0.5–3 hectares and snakes often do not move far in a day (13–100m). Largely ground dwelling but capable of climbing in bushy vegetation such as heather. Said to be intelligent (for a snake!). May shed skin 4–6 times a year. Rather slow-moving and phlegmatic but bites readily when handled and voids smelly contents of anal glands. Does not run down prey by speed but often hunts by scent taking food items from their hiding places. Holds larger prey in coils of body. Food consists largely of lizards (often 70 per cent of diet), especially lacertids up to the size of a half-grown Green lizard; also Slow worms and skinks in the south. Rest of diet is made up of small mammals and their young, small snakes (even young vipers), reptile eggs and nestling birds; females are more likely to take these non-lizard items. Young especially may eat insects as well as lizards.

Males may fight and mating often takes place in the spring and can last some hours. After 4–5 months in northern Europe, 2–16 (very exceptionally 19) fully formed young are produced, although most broods are made up of about 3–9 babies. In south may mate again in summer, the young being carried through hibernation to be borne after the mother emerges from it. In the north, females breed every two or three years. Young are 12–21cm long and coloured like adults, growing to about 30–40cm in third year. Males mature in two and a half to three years in south of range and four years in north, while females take longer. Known to occasionally live 18 years in the wild.

SOUTHERN SMOOTH SNAKE *Coronella girondica*   Pl. 47
Range Spain, Portugal, southern France, Italy, Sicily. Also north-west Africa. Map 182.
Identification Adults up to 95cm or more including tail, but usually well under 70cm; males smaller than females. Similar to Smooth Snake but tends to be more slender with a more rounded snout. Colour above brownish, greyish, ochre, or pinkish, with irregular darker bars that are usually bolder than in the Smooth Snake. Belly often yellow, orange or red overlaid with black in a bold diced pattern, sometimes forming two lines. Typically no dark stripe from nostril to eye, but in most cases has a clear 'bridle' over snout. Young marked similarly to adults but rather more brightly coloured especially beneath. There are usually eight (not seven) upper labial scales (see Fig., opposite), the fourth and fifth reaching the eye (not third and fourth as in many Smooth Snakes). The rostral scale is not large and does not extend between supranasal scales (see Fig., opposite). Usually 21 (rarely 19 or 23) back scales across mid-body (not usually 19 as in the Smooth Snake), 172–198 belly scales (172–197 in males, 177–198 in females), and 52–85 pairs of scales under tail (61–85 in males, 52–65 in females).
Similar Species Smooth Snake (p. 223); False Smooth Snake (opposite). Young Ladder Snakes (p. 215) are more boldly marked with finer scaling (25–31 scales across mid-body). Vipers have vertical pupils and keeled scales.
Habits Generally a more lowland species than Smooth Snakes within its range and encountered in warmer, drier places, although it may sometimes be found quite close to them. Does however also occur in mountain regions, occasionally even up to 2,000m in south, although usually under 1,000m. In contrast to Smooth Snake, generally comes out to hunt in the evening, and at night when conditions are warm enough, being especially active in rainy periods with mild nights in the spring. May occasionally be active in the day during moist weather. Found in a wide variety of dry habitats, especially in hedgerows and open woods (such as of various oak species and pine and cultivations of almond, olive and carob). Often seen around piles of old vegetation, but also sometimes frequents rocky places, stone piles, dry-stone walls etc., where it may spend the day under sun-warmed stones. Much more docile than Smooth Snake and rarely bites, but may flatten and spread head, making it look more

viper-like, and often voids contents of cloacal glands. Feeds mainly on lizards which it hunts by scent; lacertids up to the size of Green Lizards are the commonest food items but geckos, slow worms and skinks are also taken, as well as small snakes and mammals and even some insects. Most prey is constricted but small items may be simply swallowed, usually head first.

Females reproduce annually and lay 1–16 (often 4–10) elongate eggs, 20–45mm x 13–16mm, that tend to stick together and take 6–9 weeks to hatch, producing babies about 11–20cm long. May live 15 years.

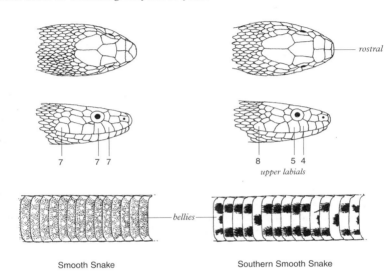

upper labials

bellies

Smooth Snake   Southern Smooth Snake

FALSE SMOOTH SNAKE *Macroprotodon cucullatus*   Pl. 47

Range Portugal, Spain except much of north, Majorca, Minorca and Lampedusa Also North Africa eastwards to Israel. Majorcan and Minorcan populations are probably an early accidental introduction from northern Africa. Map 183.

Identification Adults up to 65cm, including tail, but more usually under 45cm; females larger than males. A small snake with a fairly well defined head that is distinctively flattened, especially the snout. Eyes small, with a round pupil, and placed fairly near front of head. Upper parts rather pallid grey, brown or reddish with small dark markings that often form vague bars or even streaks. In many cases a blackish collar present on neck that may extend on to top of head and usually a dark streak from nostril, through eye, and often on to lower cheek. Belly whitish, greyish, yellow, pink, or reddish, with weak dark markings or a bold black 'diced' pattern, which may consist of a central band or two stripes. The rostral scale is low and scarcely visible from above, and there is usually only a single anterior temporal scale (see Fig. p 224 below). Usually 19–21 (occasionally up to 23) back scales across mid-body, 158–189 belly scales, and 37–53 pairs of scales under tail.

Variation Animals from Spain and Portugal (*Macroprotodon cucullatus ibericus*) often have a complete dark collar, the dark streak from the eye on to the cheek is long and extends back to the end of the upper labial scales, the belly is boldly patterned, and there are often 21 back scales across mid-body (occasionally 19 or 23), and eight upper labial scales of which the sixth is usually in contact with the parietal scale (Figs., below). In south of Iberian peninsula, some animals have an entirely black head and these may be especially large. Snakes from Majorca and Minorca and from Algiers to north Tunisia (*Macroprotodon cucullatus mauritanicus*) often have the collar divided into three parts, the dark streak from the eye on to the cheek is short, or if long, curling up on to the side of the head, the belly is plain or sparsely spotted, there are 19 rows of dorsal scales at mid-body and often nine upper labial scales of which the sixth is usually separated from the parietal scale. Balearic animals are also large. Snakes from Lampedusa belong to the subspecies *M. c. cucullatus*.

Similar Species In Spain and Portugal may be mistaken for either of the Smooth Snake species (p. 221, p. 222).

Habits Mainly a lowland species found in a variety of warm, fairly dry or rather humid habitats including sandy open woods, for instance of holm and cork oak and pine, scrub areas and cultivation. Occurs up to 1,500m in Andalucía. Encountered most often in stony places, rock piles, old walls, and ruins where it is found during the day either under stones, in burrows of other animals or buried in light soils into which it tunnels easily. In Spain and Portugal not usually active by day and hunts at dusk and at night, but is more diurnal on Majorca and Minorca. Generally slow-moving but reasonably swift when disturbed and bites when handled. When alarmed may strike with closed mouth, throw head back to expose underside of neck, or hide the head under the body; the end of the tail may also be curled. Some prey animals are caught while resting in their hiding places. Food is mainly small lizards in Spain and Portugal, including worm lizards, small lacertid lizards, skinks and geckos; a significant proportion of small mammals and bird nestlings are taken on Majorca and Minorca.

Females may breed every 2 years and, about 4–8 weeks after mating, lay 2–6 eggs, about 27–42mm x 10–13mm. These are placed in damp situations under stones or vegetation or in sandy soil and hatch in about 6–9 weeks. Babies are 12–16cm from snout to vent and mature at lengths of 24–28cm, females being larger than males.

Venom Grooved fangs are present at the back of the upper jaws; the small size of this snake prevents their effective use on people.

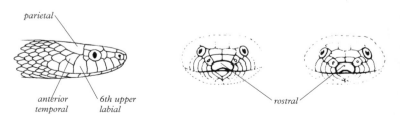

False Smooth Snake   Smooth Snake

CAT SNAKE *Telescopus fallax* **Pl. 47**

Range East Adriatic coast and islands southwards from extreme north-eastern Italy; south Balkans southward from Macedonia and south-east Bulgaria, as well as many Agean islands including Crete and Rhodes; Malta. Also Caucasus and south-west Asia. Map 184.

Identification Adults usually up to about 75cm, including tail, but sometimes over 100cm and even reaching 130cm. A slender snake with a broad, flat head. Body often deeper than wide; snout tapering but blunt, eyes small with vertical cat-like pupils in good light. Head scales large, dorsal scales smooth in 19 (sometimes 17 to 23) rows at mid-body. Grey, beige, or brownish above with a conspicuous dark spot or collar just behind head, and series of dark transverse bars or blotches on back. These are often oblique, especially on neck where they are also best developed. Weaker bars present on flanks and often a dark stripe from each eye to the angle of the mouth. Belly pale yellowish, whitish, or even pinkish, sometimes heavily suffused with grey or brown, which may form irregular streaks or spots.

Variation Markings vary in intensity. Some populations on the southern edge of the range, on Crete and nearby islands and on Rhodes, tend to be fairly uniform, either pallid or with very weak markings above, or finely and evenly speckled with dark brown; animals in the Crete area also tend to have 21–22 back scales across mid-body. Several subspecies have been named.

Similar Species Only other snakes in range with vertical pupils are the Sand Boa (p. 201), which has a blunt tail, plump body and small head scales, and vipers which have plump bodies, more head scales and keeled body scales.

Habits Usually found in stony places, rocky degraded woodland, old walls, rock piles, ruins, etc., but also occasionally in heaps of old vegetation and sandy areas with bushy plant cover. Sometimes found close to human habitation. Mainly a lowland snake, but can occur up to 1,300m in mountains in south of range. Hunts largely at twilight, but sometimes active at night in summer, and by day in cooler parts of year. If molested, may behave in a viper-like way, hissing quite loudly, flattening the body and head and striking. Cat snakes vary in temperament and some animals bite when handled. Feeds substantially on lizards including especially geckos and lacertids but also occasional agamas and slow worms; may additionally take small snakes and less commonly small mammals and birds; also insects, especially the young. Large Cat snakes take prey up to size of sub-adult Balkan Green Lizards. Victims may be dragged from their refuges, but are also stalked in the open, the snake moving forwards cautiously from behind cover (this is probably origin of the name 'Cat Snake'). Once caught, lizards are held in the jaws while the venom takes effect.

Lays 5–9 elongate eggs, about 25–40mm x 10–15mm, that are finely grooved along their length and produce babies measuring 14–20cm.

Venom Has grooved fangs at back of upper jaws which inject venom into prey, sometimes causing death of small lizards in 1–5 minutes. Unlikely to be dangerous to people, as the mouth is too small to allow fangs to be used effectively.

Note This snake is used in religious ceremonies on Cephalonia.

round

vertical, cat-like

Pupil shapes in snakes

Vipers
(Family *Viperidae*)

In all, there are about 43 species of Typical vipers (subfamily *Viperinae*). These are distributed throughout Europe, Asia, and Africa, and are quite closely related to the Asian and American Pit vipers (subfamily *Crotalinae*), which include the rattlesnakes. Pit vipers differ from Typical vipers in having a conspicuous depression (the 'pit' of their name) in front of each eye. This is a heat sensitive organ and enables them to locate warm-blooded prey, even in the dark.

All vipers possess a characteristic venom apparatus that is more sophisticated than that of most other snakes. Each of the two venom glands leads directly into the base of a long, hollow fang, like a hypdermic syringe. When the snake strikes, the fangs are embedded in the prey and venom is expelled from the fang tip, which means that it is very efficiently transmitted deep into the tissues of the victim. The fangs are also mobile: when in use they project nearly at right angles to the upper jaw, but at other times the fangs are rotated backwards so that they lie against the roof of the mouth, enclosed in a sheath of soft tissue.

All eight European vipers belong to the same genus, *Vipera*. They are the only dangerously venomous snakes in the area covered by this book. In most parts of the European mainland, at least one species is present and sometimes two, or even three, are found close together. They are absent from many, but not all, islands including quite large ones such as Crete, Malta, Corsica, Sardinia, the Balearics, and Ireland.

European vipers are fairly heavy-bodied snakes with short tails and well defined, frequently triangular, heads. The dorsal body scales are strongly keeled and the shields on top of the head sometimes nearly all fragmented. The eye is rather small and the pupil vertical. Unlike most other European snakes, vipers have an undivided preanal scale (Fig., p. 197). They vary considerably in head profile and some species have a soft, scaly 'nose-horn' on the tip of the snout.

Vipers are mainly ground dwelling, although some species will occasionally climb on stone piles and on bushes, and they may also sometimes swim. All species are often active by day, but most of them become partly nocturnal wherever the night temperature is high enough. European vipers feed on small mammals (mice, voles, shrews etc), but some also take birds or lizards, and one, Orsini's Viper, often eats a high proportion of large insects. Most feeding takes place during the day and prey is often ambushed.

All European vipers are relatively slow-moving unless disturbed and frequently hunt from cover, striking mammalian prey as it passes and then tracking it by scent after the venom has taken effect. In contrast, birds, lizards and insects are usually held, and swallowed as soon as they stop moving. Vipers will also hunt in the burrows of mammals for prey.

Most European species give birth to live young, but one, the Milos Viper, lays eggs in Europe. Before mating, rival males may take part in a combat 'dance' in which they rear up and press against each other, the weaker animal eventually retiring from the contest.

In areas where the climate makes it necessary, vipers hibernate, often communally. Some species (at least the Adder and Asp Viper and perhaps the Nose-horned Viper) may migrate from their winter refuges and breeding grounds to other areas for feeding; the distances involved are usually not large, typically being a few hundred metres.

Vipers are not aggressive unless disturbed or molested, when most species will bite fiercely.

Identification Some species of vipers have very distinctive features and are easily identified. Others are extremely variable and this can cause difficulties, especially as it is inadvisable to handle them. Principal features to look for are general appearance, snout profile (flat, turned up, or with a nose-horn), pattern, and, if possible, size of scales on top of the head and other minor features of head scaling (see key). If the snake is not disturbed, many of these features can often be seen with binoculars.

KEY TO VIPERS

Because some European vipers are variable, it is very difficult to produce a simple, yet entirely effective key for them. All identifications must, therefore, be checked with the relevant species texts.

1. Only in West Cyclades islands in the Aegean sea. No large scales on top of head, not even over eyes

    Milos Viper, *Vipera schweizeri* (p. 237, Pl. 49)

    All other areas. Some large scales on top of head, at least over eyes   **2**

2. Only European Turkey and north-east Greece. Large scales over eyes, but rest of head scales small; snout rounded with no nose-horn; characteristic bold pattern on head

    Ottoman Viper, *Vipera xanthina* (p. 236, Pl. 49)

    Not like above   **3**

3. A very distinct nose-horn present (Fig. a, below). Central head scales small   **4**
   No distinct nose-horn present, but snout-tip upturned in profile (Fig. b, below). Central head scales often small   **5**
   No distinct nose-horn and snout not upturned in profile (Fig. c below). Central head scales usually large (frontal and parietal scales well developed, Fig, p. 230)   **6**

a
Nose-horned Viper
(*V. ammodytes*)

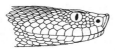

b
Asp Viper
(*V. aspis*)

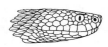

c
Adder
(*V. berus*)

4. East Europe only. Rostral scale does not clearly extend onto front of nose-horn, which is covered with 9–20 small scales

    Nose-horned Viper, *Vipera ammodytes* (p. 235, Pl. 49)

    Iberian peninsula only. Rostral scale clearly extends onto front of nose-horn, which is usually covered by less than 9 scales

    Lataste's Viper, *Vipera latasti* (p. 234, Pl. 49)

5. North-east Spain, France (except north), Italy, Sicily, Elba, Switzerland and
   south-west Germany. Snout nearly always clearly upturned in profile (raised
   section covered behind by 2–3 scales). Typically has two rows of scales
   beneath eye. Pattern variable
   Asp Viper, Vipera aspis (p. 232, Pls.48 and 49)
   North-west Iberia. Snout may be weakly turned up. But usually only one row
   of scales beneath eye
   Seoane's Viper, Vipera seoanei, (p. 231, Pl. 48)
   Iberia, except extreme north. Snout always distinctly up-turned; raised
   section usually covered behind by 4 or more scales
   some Lataste's Viper, Vipera latasti (p. 234, Pl. 49)

6. **Note** Rare individuals of Asp Viper which lack a clearly upturned snout may
   key out here. See Asp Viper texts for eliminating such animals, p. 232.
   Widespread but not northwest Iberia. Snout rather blunt (viewed from
   above). Nostril large and in centre of nasal scale. Two apical scales in contact
   with rostral scale. Scales on top of snout numerous (more than 13). Upper
   preocular scale normally not in contact with nasal scale. (See Fig., p. 230 for
   details of head scaling.) Typically has simple dark zig-zag stripe on back
   Adder, Vipera berus, (p. 230, Pl. 48)
   Northwest Iberia only. Similar to Adder but pattern very variable.
   Seoanei's Viper, Vipera seoanei (p. 234, Pl. 49).
   Restricted range. Small, usually under 50cm. Scaling often appears rougher
   than Adder; head narrower and snout more tapering. Nostril small and
   towards lower edge of nasal scale. One apical scale in contact with rostral
   scale. Normally less than 12 scales on top of snout. Upper preocular scale
   normally in contact with nasal scale (see Fig., p. 230 for details of head
   scaling). Zig-zag stripe tends to have clear, dark edge
   Orsini's Viper, Vipera ursinii (below, Pl. 48)

ORSINI'S VIPER Vipera ursinii   Pl. 48

Range Discontinuous. In Europe occurs in isolated, often quite small populations in
south-east. France (Vaucluse, Alpes-de-Haute-Provence and Alpes Maritimes), central Italy (Marche, Umbria and the Abruzzo), west Balkans (west Croatia, west
Bosnia, north Albania and adjoining Montenegro and Macedonia), north Greece
(Pindus mountains), Hungary, Romania, Moldavia, Ukraine and southern Russia.
Although formerly abundant, populations in extreme eastern Austria are now extinct
and this may be true of Bulgarian ones. Extends eastwards to Central Asia and also
occurs in Asiatic Turkey and north Iran. Map 185.

Identification Adults usually less than 50cm, including tail; rarely over 60cm,
females tend to be larger than males. The smallest European viper: a small, thick-
bodied, narrow headed species that often has a rough appearance. Only likely to be
confused with the Asp Viper or Adder. Differs from Asp Viper in lacking an obvious
upturned snout and always having several large scales on top of head (frontal and
parietal scales well developed) and a low number of dorsal scales (nearly always 19
rows across midbody but see **Variation** below). Differs from Adder in smaller adult
size, in having a narrower head with more tapering snout, and in several features of
head scaling: only a single apical scale in contact with rostral, fewer scales on top of
snout (not more than 12), nostril small and near bottom of nasal scale, upper

preocular scale nearly always in contact with nasal scale (see Fig., p. 230 for details of head scaling). Orsini's Viper also tends to have fewer back scales across mid-body than the Adder (19 instead of usually 21). The scales of Orsini's Viper are often wavy in cross-section, have a more pronounced keel, and may be rather short so that the dark skin between them sometimes shows through; these features produce the rough 'texture' of this species.

Pattern is often not very variable: tends to be greyish, pale brown, or yellowish with a dark zig-zag dorsal stripe that is usually edged with black and may be occasionally broken into spots. Flanks often rather dark, underside may be blackish, whitish, or dark grey, or even rosy, with or without spots. Underside of tail tip sometimes dark or with yellow markings.

Variation. There is some variation in snout shape, eye size, number of back scales across mid-body and colouring (especially in darkness of flanks, continuity of vertebral stripe, and ventral pigmentation). On the basis of these differences several subspecies have been described. Italian animals are assigned to *V. u. ursinii* as are French ones although they were previously distinguished as *V. u. wettsteini*. West Balkan animals are named *V. u. macrops* and the isolated population in Greece is *V. u. graeca*; this has the first 3–4 upper labial scales enlarged and the vertebral stripe may be reduced to a narrow line.. Animals from Hungary and formerly in adjoining Austria and Romania are *V. u. rakosiensis*. Those from north-east Romania and Moldavia are *V. u. moldavica* and those from the Ukraine and Russia are *V. u. renardi* which also extends eastwards to China and may be a separate species. This form is distinguished by having 21 back scales across the mid-body and like *V. u. moldavica* often has the canthi raised so that the snout is slightly concave above. Melanistic individuals occur, at last in *V. u. macrops*.

Similar Species Adder and Asp Viper (see above).

Habits *V. u. ursinii* , *V. u. macrops* and *V. u. graeca* are mountain forms, occurring above 1,000m and exending in some cases to 2,700m. They live on well drained hillsides with some vegetation, or more commonly on high, often dry, meadows. In contrast, the other subspecies are essentially lowland forms, occurring in steppe and both dry and moist meadows, and even occasionally in marshy conditions. These kinds of habitats have been largely destroyed by cultivation in many parts of the range resulting in the extinction of Orsini's Viper in Austria and west Romania and its great reduction in Hungary, north-east Romania and Moldavia. It only rarely occurs at the same localities as the Adder and in such cases the Adder tends to occupy the damper situations.

Unlike other vipers, Orsini's Viper is usually quite docile and almost never bites. In undisturbed areas it can be quite common, *V. u. ursinii* sometimes existing at densities of 20–30 per hectare. Food varies: some populations eat many lizards and others a high proportion of grasshoppers; small mammals may also be taken. Diet may also vary through activity period: French and Italian *V. u. ursinii* and Ukrainian *V. u. renardi* eat mammals, lizards and fledglings early in the season but take predominantly grasshoppers in late summer when they are abundant. The west Balkan *V. u. macrops* also takes a high proportion of grasshoppers, up to a hundred being found in a single snake. Prey is often swallowed live rather than being envenomated but this does occur.

Clutch size varies overall from 2–22 but numbers vary between regions, for instance French and Italian *V. u. ursinii* produce an average of 4 young while *V. u. rakosiensis* averages over 11. Babies range from 12–15cm in length.

Venom The venom is weaker than that of the other vipers. This, together with its placid disposition, makes Orsini's Viper the least dangerous of European vipers. It should still be treated with respect, especially as misidentification is possible.
Other names. Meadow Viper.

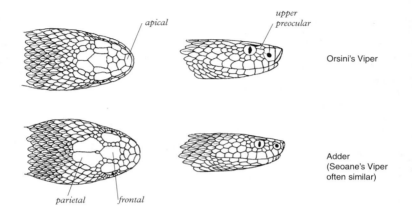

apical

upper preocular

Orsini's Viper

parietal   frontal

Adder
(Seoane's Viper
often similar)

ADDER *Vipera berus*   Pl. 48

Range Occurs over much of Europe extending north to beyond the Arctic Circle and south to northern France (with a southern isolate in the Massif Centrale), north Italy, north Albania, northern Greece and west European Turkey. Rather sporadic in central Europe and southern parts of its range. Also extends across Russia to Sakhalin island in the Pacific ocean. Map 186.

Identification Adults usually up to 65cm, including tail, exceptionally almost 90cm; females tend to be larger than males. An often thick-bodied viper with a flat (not upturned) snout, nearly always several large scales (including frontal and parietals) on top of head, a single row of small scales (suboculars) beneath eye, and usually 21 (occasionally 20–22) back scales across mid-body.

Most Adders have a clearly marked dark, zig-zag vertebral stripe. This is usually without a very distinct paler, central band (as in some other similar vipers). In rare cases, the vertebral stripe is straight-edged, broken up, faint, or even absent. Colouring varies according to sex: many males are very contrasting especially in spring, often being whitish or pale grey with intense black markings; females are frequently brownish or reddish with dark brown markings. Other colour combinations exist and entirely black specimens may be common (up to 50 per cent in some populations). Young animals are often reddish. Belly is grey, grey-brown, or black, sometimes with white spots; tail tip yellow, orange, or even red beneath.

Variation In the south-east of its range, as far north as southern Slovenia and southern Hungary, the Adder is very variable and more extreme animals may look like Asp Vipers, with dark cross bars on the back and sometimes two rows of scales beneath the eye. However the central head scales are not fragmented and the snout is not upturned. Populations containing such animals are named *V. berus bosniensis*.

Similar Species Seoane's Viper (below) of northern Spain is often very similar but lives outside the range of the Adder which is most likely to be confused with Orsini's Viper (see p. 228 for distinction) and the Asp Viper. The latter can usually be distinguished by its clearly upturned snout and, in rare cases where this is poorly developed, there are usually other features that are uncommon in the Adder. For instance, one of the characteristic Asp Viper colour patterns, fragmented head scales, and two rows of small scales between the eye and the large scales forming the upper lip. The unrelated Viperine Snake (p. 218) can also be similar.

Habits Occurs in wide variety of habitats particularly in north of range. Here it is found on moors, heaths and dunes, and in bogs, open woods, field-edges, hedgerows, marshy meadows, and even salt marshes. In the south is more restricted, and usually encountered in mountain areas; where it occurs in lowlands, for instance in parts of northern Italy, the Adder occupies moist habitats. Found up to 2,600m or more in Alps. Largely diurnal, especially in north of range. Flattens body when basking. Black animals absorb heat more effectively (see Western Whip Snake, p. 207) and may be especially common on moors and at high altitudes. Adders may travel 0.5–2km, from the areas where they hibernate and often mate, to their feeding grounds, and males may travel up to 200m in a day during the breeding season. In Britain, densities are often 1–12 individuals per hectare but are occasionally much higher, the numbers reflecting recent prey abundance. Adults feed mainly on small mammals, although, birds, lizards and frogs may also be taken; babies eat nestling rodents etc. and small lizards and frogs. In Britain, adults may take 9 voles or their equivalent per year. Once prey is struck and killed other Adders may try to take it, the animals competing in the way courting males do (see below).

Males leave hibernation earlier than females and go to breeding areas first. When courting they will attempt to drive off rivals by hissing and lunging, after which combatants may both rear up and press their forebodies against each other, and then often wrestle on the ground, the larger animal usually winning. The sexes press their bodies together in foreplay, before mating with their tails raised, a process that can last 2 hours. A male often remains with a particular female for some time, copulating repeatedly, but multiple paternity of clutches still sometimes occurs. 3–18 young are produced which may mature in 3–4 years. Female Adders breed every other year in southern Britain but only once in three years in some montane parts of Italy. In Sweden males only survive 3 breeding seasons on average and females 2, although occasional animals may live 10 years in wild.

Venom Quite potent, though bites are probably not as dangerous as those of Asp Viper and certainly those of exclusively southeastern European species. Human fatalities do occasionally occur but are rare.

Other names. Common Viper, Northern Viper.

SEOANE'S VIPER *Vipera seoanei*   **Pl. 48**

Range North-west Spain and small adjoining areas of northern Portugal and extreme south-west France. Map 187.

Identification Adults up to 65cm but usually 40–50cm. Females may be larger than males. Similar to Adder but central head scales are fragmented in some animals and the snout may sometimes be lightly upturned. Usually 2 (rarely 3) apical scales in contact with rostral scale on snout tip (see Fig., opposite). 21 (rarely 19 or 23) back scales across mid-body.

Often beige, light brown or grey above with a variable pattern. The central stripe may have wavy edges or form a zig zag, or be broken into transverse bars. Alternatively, a dark stripe along the back that may be straight-sided and bordered by lighter streaks (the *bilineata* form). Some animals are like Asp Vipers from south-west France with a straight, wavy, or zig-zag stripe along the back which has a paler centre that may reduce the dark areas to a series of spots on each side of it. Other animals are entirely uniform black or dark greyish. Colours of males tend to be more contrasting than females. Grey or blackish below, usually with some yellow or orange under tail tip.

Variation Very varied in appearance and the presence and proportion of the different colour patterns varies from place to place. Black animals are more common in mountain areas. Animals from north León, south-west of the Picos de Europa, the mountains of eastern Galicia and south-west of Asturias (*V. s. cantabrica*) are grey or brownish grey with a narrow zig-zag along the back which may be divided into cross-bars; they may also have the head scales more broken up. In the Picos de Europa area, animals with the *bilineata* pattern or uniform colouring are common.

Similar Species Adder (p. 230) is outside range. Confusion with Asp Viper (below) is likely only in small area of overlap in the Basque country. In this region, Seoane's Viper can usually be separated by lack of strongly upturned snout, nearly always a single row of small scales between the eye and the scales forming the upper lip, and usually a lower number of belly scales (in most cases under 143 in Seoane's Viper and over 140 in Spanish Asp Vipers, with extremes of about 150 and 136 respectively).

Habits Often found in areas of low scrub, among bramble patches, at wood edges and in clearings, avoiding very dry habitats and not being closely associated with rocky ones. Occurs from sea level to 1,900m and is one of the commonest snakes in its range. It is usually diurnal but may be nocturnal in hot weather. Ground dwelling but quite often climbs in low plants and basks on them. Most adults shed their skins twice a year. Eats mainly small mammals, but also reptiles, amphibians and small birds, the young taking a larger proportion of reptiles and amphibians.

Females often breed in alternate years but sometimes in consecutive ones if food is abundant. They produce 3–10 young, 14–19cm long which become sexually mature after 3–5 years. Males mature in 3–4 years at 30–40cm. May sometimes live at least 13 years in wild.

Venom Potency of the venom varies from area to area, being strongest in *V. s. cantabrica*. In people, it often causes swelling and pain but rarely death.

Other names. Was originally considered a subspecies of the Adder, *Vipera berus seoanei*.

ASP VIPER *Vipera aspis*   **Pls. 48 and 49**

Range West and central Europe, from north-east Spain through France (except north), south-west Germany (Schwarzwald) and Switzerland (west, centre and south) to Italy and Sicily; also occurs on Elba and Montecristo islands. Balkan records exist but most appear to result from misidentified Adders; others may be based on rare hybrids between the Adder and the Nose-horned Viper. Map 188.

Identification Adults usually up to 60cm, including tail, occasionally to 90cm; males tend to grow larger than females. A widespread and extremely variable viper that can usually be identified by its characteristic head profile, in which the snout is distinctly upturned but lacks a real nose-horn (Fig., p. 227). Has typical viper appearance with broad triangular head, but body is rather slender. Usually 21 or 23 back scales across mid-body. Very variable in colouring; underside of the tail tends to be yellow or red.

Variation Principal geographical variants are described below; they appear to gradually merge into each other.

In France (except south-west), north-east Spain (except the central Pyrenees), the Alpine region and north and central Italy, animals typically have a characteristic pattern of dark transverse bars on the back; the bars are often staggered on each side of the body and may be joined by a narrow vertebral streak. Completely black specimens are fairly common especially in highland areas. Apart from the upturned snout, these populations nearly always have the scales on top of head fragmented (except the scale over each eye and sometimes frontal), and two rows of small scales between the eye and the large scales forming the upper lip. All were previously referred to the subspecies *V. a. aspis*, but those from south-east France, Switzerland and north-west Italy, which include a high proportion of black individuals, are now sometimes separated as *V. a. atra*, while other Italian populations, which often have the head swollen behind the eyes and light spots on the belly scales, are named *V. a. francisciredi*.

Populations from southern Italy and Sicily (*V. a. hugyi*) have similar scaling but the pattern consists of a broad, wavy, dorsal stripe that is often rich brown and has a darker edge and may be broken up into series of oval blotches. Animals from Montecristo island between Italy and Corsica (*V. a. montecristi*) are similarly marked.

In snakes from south-west France and the central Pyrenees *(V. a. zinnikeri)*, the dark vertebral stripe may be wavy, zig-zag or almost straight, its centre distinctly pale and often greyish (such a pale centre is rare in the Adder). In some cases, really dark pigment on back may be limited to a row of blackish spots on each side of paler central band. Scaling is more variable than elsewhere and in some populations there are often several large scales on top of head, and sometimes only a single row of small scales between the eye and the large scales forming the upper lip.

Hybrids between the Asp viper and the Adder are believed to occur but only very rarely.

Similar Species Lataste's Viper (234) and Nose-horned Viper (p. 235) both have a large projection on the snout, and only overlap with the Asp Viper in very restricted areas (north-east Spain and north-east Italy respectively). Rare Asp Vipers with little upturn of the snout could be confused with the Adder (p. 230) or Orsini's Viper (p. 228), but pattern and usual presence of two rows of small scales below the eye normally distinguish them. The unrelated Viperine Snake (p. 218) can also be superficially similar.

Habits Often found in dry habitats, especially rocky hillsides with scrub, but also open woods and their edges, hedge bottoms, and drystone walls. Extends into high wet mountainous regions, reaching 2,900m in Pyrenees and 3,000m in the Alps, and is mainly a hill form in the south of its range. Where it overlaps with the Adder, the Asp Viper tends to be limited to warmer areas, and in hilly country generally occurs at lower altitudes. Diurnal, but also partly nocturnal where climate allows. Males may have home ranges of 3,000 sq. m but females are more sedentary. Mature animals shed the skin 2–3 times a year. In most areas, adults feed predominantly on small mammals but also take some reptiles and birds. In parts of Switzerland, a high proportion of lizards and frogs are eaten. Babies feed especially on lizards. Adults often eat 2–3 times their body weight in a year, although breeding females take less.

Females give birth to 5–22 (often about 6–8) young, some 2–3 months after mating and may reproduce every 2 or 3 years (rarely every 1 or 4). Males mature in their third or fourth year, females in their fifth or sixth. May sometimes live at least 18 years in wild.

Venom This is generally more potent than that of the Adder but varies geographically, being more neurotoxic in south-west France. Human deaths have occurred, especially when bites were untreated.

LATASTE'S VIPER *Vipera latasti*   **Pl. 49**

Range Spain and Portugal, except extreme north. Also extreme north Morocco and adjoining Algeria. Map 189.

Identification Adults usually under 60cm, including tail, occasionally up to 75cm; males larger than females. The only viper in Spain and Portugal that usually has a distinct soft nose-horn. Occasionally the snout-tip is just turned upwards as in the Asp Viper; in such cases range is useful in identification as the two species overlap only slightly. Also dorsal pattern tends to differ, and Lataste's Viper has a narrower rostral scale (height 1.5–2 times width, compared with 1.5 times or less in Asp viper); the raised portion of the snout in Lataste's Viper also tends to have more scales (usually 4 or more compared with 2 or 3 in Asp Viper). Body relatively stout with broad, triangular head covered by small scales, except for a large one over each eye and sometimes a frontal scale. Usually two rows of small scales between eye and the large scales forming the upper lip, but sometimes three. Normally 21 back scales across mid-body (sometimes 19, rarely 23).

Ground colour usually greyish, brownish, occasionally with an orange or reddish tinge. Pattern fairly constant; typically a lobed or zig-zag dorsal stripe with darker edges; vertical bars or rounded spots often present on flanks but frequently weakly developed. Belly generally greyish or blackish, usually with lighter or darker spots and sometimes tinged red; often some yellow or red on underside of tail.

Variation Animals in south-west of Iberian peninsula (*V. l. gaditana*) and nearby north-west Africa are rather small with fewer belly scales (average around 130 instead of 140) and no obvious frontal scale. Elsewhere (*V. l. latasti*), individuals from the west and centre of the peninsula tend to be relatively dark and the stripe along the back consists of a series of connected spots or lozenges, while in the east and south-east animals tend to be lighter with a more obviously zigzag stripe.

Similar Species Only likely to be confused with Asp (p. 232) or Seoane's Viper (p. 231). Nose-horned Viper (opposite) similar but does not occur in same area.

Habits Prefers dry, sunny places and is found up to about 2,800m in south of European range. Is commoner in hilly areas above 800m, perhaps because it has been persecuted more in the lowlands and coastal areas (for instance populations on the Columbretes islands are now extinct). Occurs in rocky areas, open woods and their edges, clearings in woodland, hedge rows, drystone walls; also sometimes in sandy places including coastal dunes in the south of range Generally like other vipers in being diurnal, but nocturnal as well in warm conditions. Main prey is often small mammals but also takes a varying proportion of lizards including lacertids, slow worms, and skinks, as well as small snakes and occasional birds, amphibians and invertebrates (centipedes, scorpions etc). Young eat lizards and invertebrates. Females, which do not generally breed every year, give birth to 4–9 young, about 15–21cm long, which become sexually mature at lengths of around 30–40cm

Venom Quite irascible, but bite not considered serious for most healthy people.

NOSE-HORNED VIPER *Vipera ammodytes*   Pl. 49

Range Mainly Balkan peninsula; north to north-east Italy, southern Austria, Bosnia, central Serbia and south-west and south-east Romania; extends south to southern Greece and the Cyclades, but absent from islands occupied by the Milos Viper. Also north-east Turkey and Transcaucasia. Map 190.

Identification Adults usually under 65 cm, including tail, but occasionally up to 90cm; males bigger than females. The only eastern European snake with a distinct soft nose-horn. Body fairly stout with well-defined triangular head which is covered with small scales, except for a single large scale over each eye. Two rows of small scales separate the eye from the large scales forming the upper lip. Rostral scale does not extend upwards on to nose-horn, which is covered by about 9–20 small scales. Usually 21 or 23 back scales across mid-body.

Males often light grey and females greyish, brownish, or red-brown (rarely yellowish or pinkish). Nearly always a clearly defined, dark-edged stripe along centre of back which may form a zig-zag or consist of a series of connected lozenges (occasionally these may not be joined). Very rarely, the vertebral stripe is a continuous straight-edged band or is very faint. Belly greyish, or pinkish with darker spots or mottling; underside of tail may be red, yellowish, or greenish. Melanistic animals rarely occur.

Variation Great variation in colouring and some in head shape and scaling. In the south Balkan peninsula (north to parts of Albania and Montenegro, Macedonia, south Serbia, north Greece and west European Turkey) and the Cyclades, animals are small with an often rather forwardly directed horn covered by numerous scales, and the underside of the tail tip is yellowish or greenish. They have been named *V. a. meridionalis*. In east Romania and Bulgaria and adjoining European Turkey (*V. a. montandoni*), snakes are similar but the horn is smaller with fewer scales and the rostral scale at the tip of the snout is high. Over most of northern part of range, most animals are large with a big upright horn and often a red (less commonly yellow) underside to the tail. These have previously been called *V. a. ammodytes*, but this unit is sometimes broken up with *V. a. gregorwallneri* in Austria, *V. aruffoi* in the Alto Adige of north-east Italy, *V. a. illyrica* in Slovenia and *V. a. ammodytes* in the rest of the northern distribution of the species.

Similar Species None within range.

Habits Found in a wide variety of habitats, but most favoured ones are on dry rocky slopes, with some vegetation and good exposure to sun. Also occurs in light woods and on screes, rock-piles, embankments and dry-stone walls in cultivated areas including gardens and vineyards. Occasionally climbs in bushes, and also on rock faces where it may be seen working its way along crevices looking for lizards. Occurs from sea level up to about 2,500m in south of its range but usually under 1,700m. Generally encountered by day, but said to be sometimes nocturnal in warmer parts of range. A slow rather phlegmatic snake that is not very irascible, although when disturbed it hisses loudly. Feeds mainly on small mammals, but also birds, other snakes, and lizards which are the predominant food on some Aegean islands. Young feed on small lizards and sometimes large insects.

Females produce 4–15 (occasionally 20) young about 15–23cm long.

Venom Highly poisonous, the venom being delivered by fangs that may be 1cm long. Bites to extremities often produce dramatic swelling of the limb concerned. This snake is more dangerous than any other widespread European viper and caused regular human deaths before modern treatment was available.

OTTOMAN VIPER *Vipera xanthina*   Pl. 49

Range Extreme north-east Greece (around Makri and Alexandropolis) and European Turkey (around Kilitbahir and Istanbul); east Aegean islands of Lesbos, Chios, Samos, Patmos, Leipsoi, Leros, Kalymnos, Kos, Symi, Chalki and Oenousses. Also Asiatic Turkey. Map 191.

Identification Adults up to 120cm, including tail, but usually smaller. A large, thick-bodied viper, easily separated from other species within its European range by its lack of nose-horn and characteristic, often vivid, head pattern. Scales on top of head small, except for a larger one over each eye; usually two or three rows of small scales between the eye and the large scales forming the upper lip. 23–25 back scales across the mid-body.

Usually a stripe running from the eye to the angle of mouth and a vertical bar or blotch below the eye. Also, two forwardly converging stripes present on top of head; these may extend between eyes or, if not, there may be one or two clear small spots in front of them. All these head markings are usually very bold. Body grey, yellowish, olive, or brownish, with a dark vertebral stripe that is often darker edged. This may be wavy, or made up of connected lozenges, or sometimes broken into a series of blotches. Flanks may sometimes be darker than back, with darker spots or vertical bars, or both. Underside often greyish with darker spots and tail tip frequently yellow or orange beneath. In general, males are more contrastingly coloured than females.

Variation No obvious geographical variation in Europe, but considerable individual differences in colouring.

Similar Species No really similar species within European range. See Milos Viper (opposite).

Habits Occurs in wide variety of habitats including open woods, stony hillsides, pastures and damp meadows especially those with rock outcrops, and even swamps. Also found in cultivated areas where it appears to be attracted by water. Frequent in hilly and mountain areas, extending up to 2,500m in Asiatic Turkey. Diurnal but active in evening and at night in hotter parts of the year. Fairly sluggish but strikes quickly and also hisses loudly when threatened. Feeds mainly on mammals and birds when adult, and lizards when young. Females give birth to 2–15 (rarely up to 20) young.

Venom A dangerous snake: with cytotoxic venom that destroys tissue. Its bite may be fatal to people and domestic animals if not treated promptly.

Ottoman Viper

MILOS VIPER *Vipera schweizeri*   **Pl. 49**

Range In Europe, confined to west Cyclades islands in the Aegean sea: Milos, Kimolos, Polyaigos, Sifnos. Previously reported from Kythnos and Antimilos. Map 189.

Identification Adults up to 80cm. Head fairly rounded, with no large scales on its upper surface (not even over the eyes), and two or three rows of small scales between the eyes and the large scales forming the upper lip. Usually 23 (rarely 21 or 25) scale rows at mid-body.

Very variable in colouring but pattern usually not very vivid. Females often brownish and males greyer and lighter, but both sexes brighter in spring, when they are frequently straw-yellow and light grey respectively. Often four rows of blotches on body, the two central ones joining on the mid-line in adults. Some specimens are a uniform brick red. Juveniles are typically blue-grey with four series of dark olive markings, the two central ones not usually meeting across the back. Belly pale, speckled with dark pigment; underside of tail-tip-sometimes yellow. In general, males are more contrastingly coloured than females.

Variation Considerable variation in colour.

Similar Species Easily identified as it is the only viper species in its range. Lack of large scales above eyes and absence of strongly contrasting head markings differentiate it from Ottoman Viper.

Habits Found below 400m in a wide variety of habitats, including dry sunny hillsides and cultivated land. Occurs in densely vegetated places with rock outcrops and said to prefer sheltered sunny situations along water courses; may also occur in marshes. Largely nocturnal in hotter part of year, but diurnal at other times, although is often rather sluggish during the day. Estimates suggest 7,000–8,000 animals survive on Milos, 100 on Siphnos and less than 10,000 in all. Feeds mainly on mammals but takes birds and other snakes. Lizards (particularly the Milos Wall Lizard) and beetles are also eaten especially by the young. Birds are caught particularly during their migration periods, the snakes perching in bushes and striking them as they land in these. The birds are held until they die from the venom and then swallowed. This is presumably because, if released, such prey could not be tracked across the ground by scent.

The only European viper to lay eggs, which are 35–47mm long and produced in clutches of 4–11. Even at this stage they have well-developed embryos and hatch in about 5–7 weeks.

Venom Bites of this snake are dangerous and it should be treated with extreme caution.

Other names. *Macrovipera schweizeri*. Was previously regarded as a subspecies of the Blunt-nosed Viper (as *Vipera lebetina schweizeri*), a species which is largely distributed in south-west and central Asia.

IDENTIFICATION OF AMPHIBIAN EGGS

In nearly all cases, the eggs of European amphibians are laid in water. This usually takes place in the spring, but may be earlier in the South, and some species lay again later in the year (for instance discoglossids, p. 59, Parsley frogs, p. 71, Natterjack, p. 74). Eggs are usually enclosed in a gelatinous capsule. In some species this is very small when the eggs are laid but rapidly swells in the next day or two.

Important recognition features are size of eggs and capsules, shape of capsule (round or elongated), colouring of eggs, and whether they are laid separately, in small groups, or in large clumps, strings or bands. In some species eggs are simply broad-cast but in others they are carefully attached to plants or stones. The place used for depositing eggs varies: for instance, some forms choose cold mountain streams, while others prefer still, lowland pools. Species distribution is a very useful means of checking identifications. Another clue is the presence of adult animals in the vicinity.

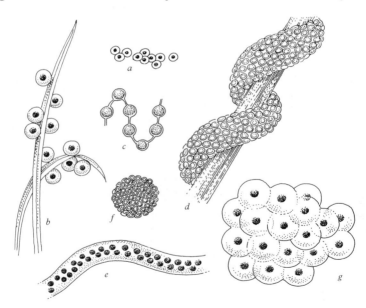

Eggs of frogs and toads (not to same scale)

a. Painted frogs   e. Typical toads
b. Fire-bellied toads   f. Tree Frogs
c. Midwife toads   g. Typical frogs
d. Spadefoots

Eggs can be most easily examined if placed with water in a transparent plastic bag. A lens is also very helpful. Attached eggs should not be separated from their moorings, or kept too long in plastic bags. After examination, they should be carefully returned to the place where they were found.

KEY TO AMPHIBIAN EGGS

Some salamanders habitually give birth to fully developed young or to larvae and consequently have no free eggs. These are the Fire salamander (*Salamandra salamandra*, p. 31), Alpine Salamander (*Salamandra atra*, p. 33), Lanza's salamander (*Salamandra lanzai*, p. 34) and Luschan's salamander (*Mertensiella luschani*, p. 34).

Eggs of the Olm (*Proteus anguinus*, p. 53) are unlikely to be encountered as they are laid in subterranean waters, 12-70 eggs being attached to rocks. They are white and about 4.5mm in diameter, with a capsule about 8-12mm.

1. Eggs laid singly or in small clumps (maximum of 30 eggs)   **2**
   Eggs laid in strings or clumps or bands in most cases containing 30 to more than 10,000 eggs.   **6**

2. South-east France, north Italy, Sardinia. 5-10 (occasionally up to 14) eggs about 5-6.5mm in diameter laid out of water, usually in deep moist crevices, often in caves; usually guarded by mother
                              Cave salamanders (*Speleomantes*, p. 50)
   Eggs laid in water   **3**

3. Iberian peninsula, Tyrrhenian area, Sicily, Malta. Eggs black above, paler below, 1-1.5mm, capsule about 3 or 4mm. Laid singly and often forming a layer one egg thick on bottom, or are attached to plants. Laid in still or slow flowing water
                              Painted frogs, *Discoglossus* (p. 62–64)
   Eggs brown, grey or whitish   **4**

4. Not Iberia or most Mediterranean islands except north-east Sicily. Eggs brown above, paler below, 1.5-2mm, capsule about 5-8mm. Laid singly or in small loose clumps of up to about 30 eggs, which are usually attached to weeds but sometimes deposited free on bottom of pools. In still or slow flowing water
                              Fire-bellied Toads, *Bombina* (p. 60–61)
   Portugal and Spain but not much of north. Eggs dark above, whitish below, about 1.5-2mm, capsule about 7mm. Often attached to plants or stones in clumps of 10-20, usually in still water
                              Sharp-ribbed Salamander, *Pleurodeles waltl* (p. 36)
   Peninsular Italy (mainly west). Eggs brown above, paler below, about 1.5-2mm, capsule about 5mm. Laid in gelatinous clumps of up to 20 on twigs, stones etc., in clean, running water
                              Spectacled Salamander, *Salamandrina terdigitata* (p. 35)
   Pyrenees, Corsica and Sardinia. Eggs whitish or pale buff, about 2-3mm, capsule about 3-7 mm. Laid separately but often placed close to each other among stones or attached to their surfaces, in cool, running water
                              Brook newts, *Euproctus* (p. 37-39)

North-west Iberian peninsula. Eggs light yellow, 2-5mm, capsule 3-7mm. Females each produce 12-20 eggs but these are sometimes laid communally and may form thick hanging bunches. Usually placed in shallow flowing water or on surfaces over which water runs.

Golden-striped salamander, *Chioglossa lusitanica* (p. 42)
All areas. Eggs brown or whitish above, capsule typically longer than broad. Usually laid singly, often carefully wrapped in leaves of submerged plants in still water.

Pond newts, *Triturus*,   **5**

5. South-west Europe. Eggs pale, yellowish white or greenish- white. Relatively large, about 2mm, capsule about 4.5 mm maximum diameter. Eggs more frequently greenish white in Marbled newt.

Marbled newt, *Triturus marmoratus* (p. 42)
Southern Marbled Newt, Triturus pygmaeus (p. 42)
Crested newts, *Triturus cristatus, T. carnifex, T. dobrogicus, T. karelinii* (p. 43-44)
Eggs pale brownish, grey-brown, or greyish above. Relatively small, about 1.5mm or a little more, capsule up to 4mm maximum diameter but usually smaller

Other, smaller Pond newts, *Triturus* (p. 44-49)

6. West Europe only. Eggs large and pale without an obvious gelatinous capsule, about 2.5-7mm. Laid in rosary-like strings, each egg attached to the next by a narrow connector. Usually seen wrapped round hind legs of adult males, or in water when empty. Eggs become darker as the embryos develop.

Midwife toads, *Alytes* (p. 64-67)
Eggs dark, 1-2mm. Laid in long gelatinous strings.

Typical toads, *Bufo*,   **7**
Eggs about l.5-2mm. Laid in a relatively thick band 3mm- 2cm across with eggs irregularly arranged. Often wrapped around the stems of water plants; may smell fishy and bands sometimes break into sections   **8**
Eggs pale brownish above, yellowish-white below, about 1.5-2mm, capsule 3-4mm. Laid in small clumps, about the size of a walnut containing 2-60 (occasionally up to 125) eggs, usually in submerged vegetation. Embryos are light yellowish.

Tree frogs, *Hyla* (p. 77-79)
Eggs brown or black above, about 1.5-3.5 mm, capsule about 6-12mm. Laid in large, often amorphous clumps

Typical frogs, *Rana*   **9**

7. Eggs black all over, 1.5-2mm, capsule 5-8mm. Often in three or four rows when strings are free in water, two rows when these are gently stretched. Usually laid among stems of aquatic vegetation or twigs.

Common toad, *Bufo bufo* (p. 73)
Not most of western Europe. Eggs black-brown, l-1.5mm, capsule 4-6mm. Often in three or four rows when strings are free in water, two rows when these are gently stretched. Frequently laid on the bottom of pools with little aquatic vegetation.

Green Toad, *Bufo viridis* (p. 76)

Not Italy or south-east Europe. Eggs black above, paler below, 1-1.7mm, capsule 4-6mm. Often in two rows when strings are free in water, one row when these are gently stretched. Frequently laid on the bottom of pools with little aquatic vegetation

Natterjack, *Bufo calamita* (p. 74)

8. Eggs dark brown to black, about 1.5mm, in bands about 3- 4mm across and 20cm long (occasionally to 50cm), containing 40-150 (occasionally 360) eggs
Parsley frogs, *Pelodytes* (p. 71-72)

Eggs grey or brown with a light spot below when first laid, about 1.7-2mm, in bands 1.3-2cm thick and 40-100cm long, containing up to 7000 eggs.
Spadefoots, *Pelobates* (p. 68-70)

9. Typical Frogs, *Rana*

The eggs of these species are often difficult to identify with certainty. Range and habitat are often helpful.

Eggs laid in a flat, floating layer about one egg thick and 30-150cm across. Either attached to weeds or free

American Bullfrog, *Rana catesbeiana* (p. 97)

Eggs laid in irregular clumps several eggs thick **10**

10. Eggs brown above, yellowish below, about 1-2.5mm diameter, capsule 6-8mm. Produced in balls of a few hundred eggs. Those of Edible Frog may vary considerably in size (1- 2.5mm see p. 92), others usually 2mm or less.
Water frogs (p. 88)

Eggs black with small, pale spot on lower surface, about 2-3mm, capsule 8-10mm. Produced in clumps of up to 4500 eggs, often many females laying together in shallow water (up to about 30cm deep), clumps soon float to surface
Common frog, *Rana temporaria*, (p. 80).

Western central Pyrenees only. Eggs black, sometimes over 3mm. Produced in clumps of up to about 150 attached to stones or placed beneath them or in cavities protected from the current. In mountain streams and torrents above 1100m.

Pyrenean frog, *Rana pyrenaica* (p. 81).

Eggs blackish to grey brown above, lower third or half whitish when first laid with a poorly defined border, 1.5-2mm, capsule 6-8mm. Laid in clumps of up to 4000 eggs. Placed in shallow water with rich plant growth. Often occurring with spawn of Common Frog and may be mixed up with it.
Moor Frog, *Rana arvalis* (p. 82).

Eggs dark brown to black above, a clearly defined pale spot below when first laid, 1.5-3mm, capsule 9-12mm. Laid in clumps of up to 4000 eggs placed in shallow (less than 40cm) water arranged around twigs or stems, so clumps appear to be impaled. These later float to surface.
Agile Frog, *Rana dalmatina* (p. 83).

North Italy etc. Eggs dark brown above, lower third whitish, 1.5-2mm, capsule 6-7mm. Clumps more compact than Agile frog often with 300-400 (sometimes up to up to 900) eggs and usually laid in water up to 30cm deep, arranged around twigs or stems and later floating to the surface.
Italian Agile Frog, *Rana latastei* (p.84).

Balkan peninsula and peninsular Italy. Eggs brown above, 2-3.5mm, capsule 4-7mm. Laid in small clumps like bunches of grapes containing 600-800 (sometimes 200-2000) eggs. Often placed beneath stones or in hollows of bank in quiet stretches of streams.

Balkan stream frog, *Rana graeca* (p. 85).
Italian stream frog, *Rana italica* (p. 86).

West and central Iberian peninsula. Eggs 2-4mm, capsules 4- 7mm. In small clumps of 100-450 eggs. Often in pools in streams and may be placed between or under stones, in aquatic vegetation, or on bottom.

Iberian frog, *Rana iberica* (p. 87).

Identification of Amphibian Larvae

Most European amphibians have free-living aquatic larvae, which are often called tadpoles (strictly, this word applies only to the larvae of frogs and toads but it is sometimes used for those of newts and salamanders as well). Newt and salamander larvae are long-bodied, with feathery gills and they develop their forelegs before the hind ones. In contrast, the tadpoles of frogs and toads have rounded bodies, the gills are not visible externally, except in the early stages of development, and their hindlegs develop before the forelimbs. They also have rather specialised mouths with a horny beak bordered by several rows of labial (lip) teeth. Because the gills are enclosed, they are aerated by a stream of water that enters via the mouth and leaves through an opening in the body-wall, the spiracle.

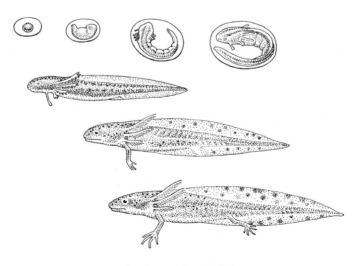

Development of newt tadpole

The main features that are helpful in the identification of larvae are shown in the Figures on p. 244 and p. 248; the mouth of a typical frog or toad larva is also illustrated (Fig. p. 248). Tadpoles are best examined in a small transparent plastic bag, although they should not be kept in this for very long. A lens is very useful in checking most characters but, in the few cases where mouth parts need to be examined (which should only be attempted on dead larvae), a dissecting microscope is needed. As with amphibian eggs, habitat, species range and presence of adult animals nearby can be helpful clues in identification. The tadpoles of some groups (e.g. Brown frogs and the smaller newts) are all similar yet rather variable. They are therefore sometimes difficult to identify. Like most adult amphibians, larvae can change the intensity of their colouring quite rapidly. Stages in the development of a newt and a frog larva are shown in Figs., opposite and below).

Neoteny. Sometimes tadpoles do not change (metamorphose) into the adult form. Instead they retain larval characteristics and, in the case of some newts and salamanders, they may even breed in this condition. In some species, only the neotenous larval form is known and normal adults do not exist. The Olm (*Proteus anguinus*) is one of these permanent larvae. Neoteny occurs occasionally in many species of amphibians, and is most frequent in highland areas where the water contains relatively little nutriment. However it is also found in other situations and is sometimes associated with albinism. The condition is relatively common in the Alpine Newt (*Triturus alpestris*), which has some completely neotenous populations in Slovenia, Bosnia and Montenegro. Neotenous tadpoles usually grow much larger then normal ones and they may also take on some adult characters. Both these features are useful in recognising them.

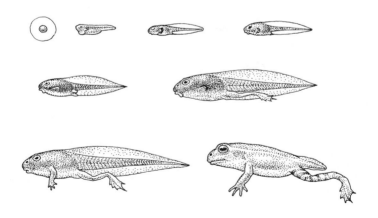

Development of frog tadpole

Identification of Salamander and Newt Larvae

A few species do not have free-living larvae. These are the Alpine Salamander (*Salamandra atra*), Lanza's salamander (*Salamandra lanzai*), Luschan's Salamander (*Mertensiella luschani*) and the Cave Salamanders (*Speleomantes*). Sizes given are for large larvae, just before metamorphosis; the illustrations depict animals at this stage. The figure below shows some of the features important for identification; to make this easier, newt and salamander tadpoles are dealt with in three geographical groups. Tail and crest shape are often quite variable within species and young tadpoles may not have their full complement of toes.

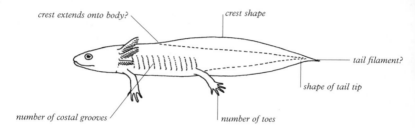

Some features to check when identifying salamander and newt tadpoles

KEY TO SALAMANDER AND NEWT LARVAE
ALL AREAS EXCEPT IBERIA, THE PYRENEES, CORSICA AND SARDINIA

1. East Adriatic coastal area only. Very long-bodied and very pale, with not more than two toes on the hind feet. Usually in subterranean waters (or more rarely in their surface outlets).

Olm, *Proteus anguinus* (p. 53)

Italy only (mainly in west). Up to 3cm. Tail tip rounded, sometimes with a small projection. Crest low although may reach middle of body. Four toes on hind feet, even in well developed larvae (not five as in other species). Usually in cool, flowing water

Spectacled Salamander, *Salamandrina terdigitata* (p. 35 and Fig., p. 246)

Up to 7cm. Tail tip blunt or rounded, upper crest usually extends a short distance on to body. Head broad and rounded when viewed from above. Distance between nostrils much greater than distance from nostril to eye. Often a light spot at the base of each leg. Older animals may show yellow or orange markings of adults. Found in clean, cool, often flowing water

Fire Salamander, *Salamandra salamandra*, (p. 31 and Fig., p. 241)

Not south-west and south France. Up to 8cm (usually smaller but rarely less than 5cm when fully developed). Undamaged tail tapers gradually to a long filament. Crest extends far forwards on back, even to back of head. Toes long. 15 or 16 vertial costal grooves between fore and hind-limbs. Brownish with no green colouring; large dark spots often present, including some on the upper and lower borders of the tail. Usually in still or slow-flowing water

Crested newts, *Triturus cristatus* group (pp. 43-44 and Fig., p. 246) West, south-west and central France. Up to 8cm. Larvae like those of crested newts but only 12 or 13 costal grooves between fore and hindlimbs and usually at least a tinge of green in coloration of flanks. Heavy dark markings may be confined to spots on the upper and lower edges of the tail, where they may alternate with bluish or silvery markings.

Marbled Newt, *Triturus marmoratus* (p. 42 and Fig., p. 246) Up to 4.5cm (usually less). Tail tapering steadily to a point although without a filament; its upper and lower edges not roughly parallel.

Not southern France or southern Italy

Common Newt, *Triturus vulgaris* (p. 46)

South and central Italy only

Italian Newt, *Triturus italicus*, (p. 49)

Not eastern areas or Italy.

Palmate Newt, *Triturus helveticus* (p. 48 and Fig., p. 246) Common and Italian newts are said to have a greater distance between the nostrils than between the nostril and the eye, and this last distance equal to the eye diameter. In Palmate newts the distance between the nostrils is said to equal to that between the nostril and the eye, both these distances being smaller than the eye diameter. However, the constancy of these differences is disputed.

Up to 5cm, rarely more. Tail usually ends quite abruptly, being rounded or only obliquely pointed, although there may be a short projection beyond this; upper and lower edges of the tail roughly parallel.   **2**

2. Up to 5cm (rarely to 8cm). Tail ends in a short projection. Often quite dark with a fine dark reticulation visible on the crests when viewed against the light with a lens.

Alpine Newt, *Triturus alpestris* (p. 44 and Fig., p. 246) Carpathian and Tatras mountains. Up to 4cm, usually less. Tail tip broadly rounded or obtusely pointed; no obvious projection. Just before metamorphosis, there may be one or two rows of light spots on each side of the back

Montandon's Newt, *Triturus montandoni* (p. 45)

IBERIAN PENINSULA AND THE PYRENEES

North-west Iberian peninsula only. Up to 5cm (exceptionally 7cm). Body long and slender, tail-tip rounded, crest does not extend on to body. Found in clear, flowing water.

Golden-striped Salamander, *Chioglossa lusitanica* (p. 35 and Fig., p. 246) Pyrenees only. Up to 6cm (exceptionally 9cm). Tail-tip blunt or rounded, crest only extends a short distance onto body. Head narrow, and not

Spectacled Salamander

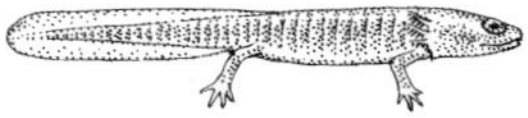

Fire Salamander

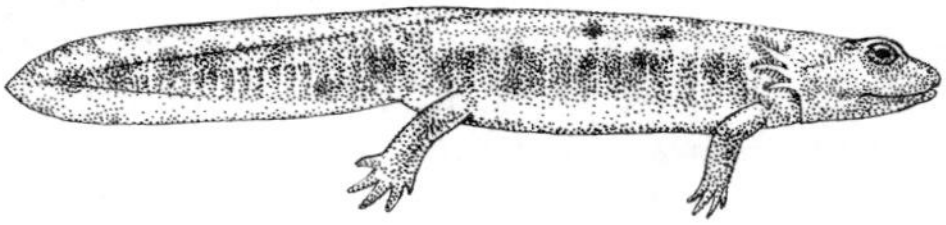

Warty Newt

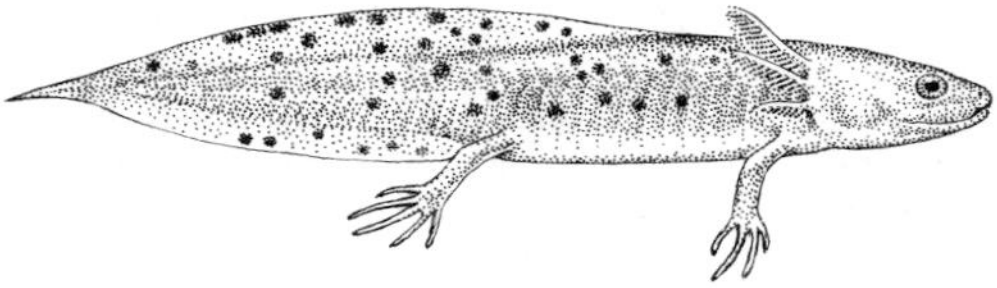

Marbled Newt

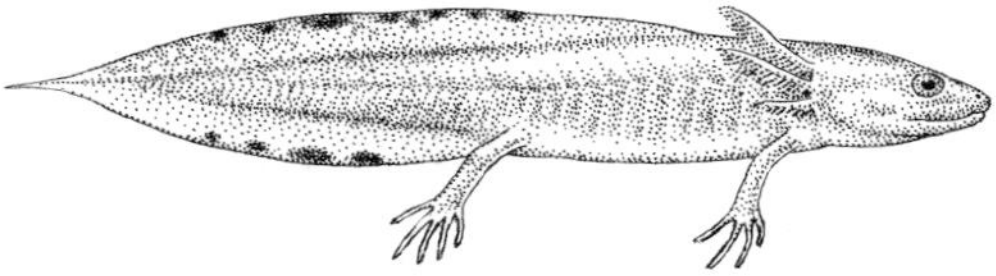

Alpine Newt

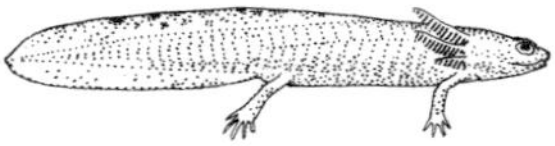

Palmate Newt

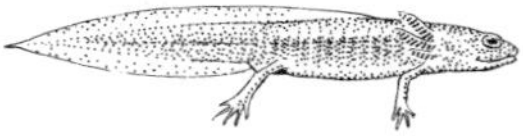

Golden-striped Salamander

Pyrenean Brook Salamander,
other Brook Salamander
species similar

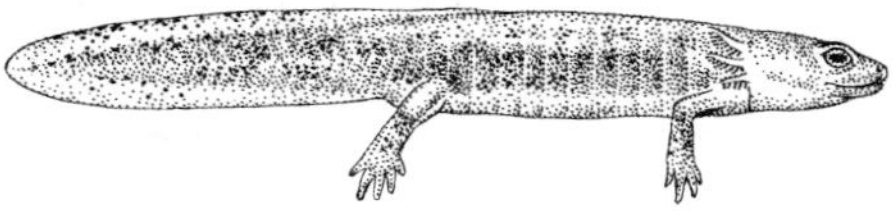

Sharp-ribbed Salamander

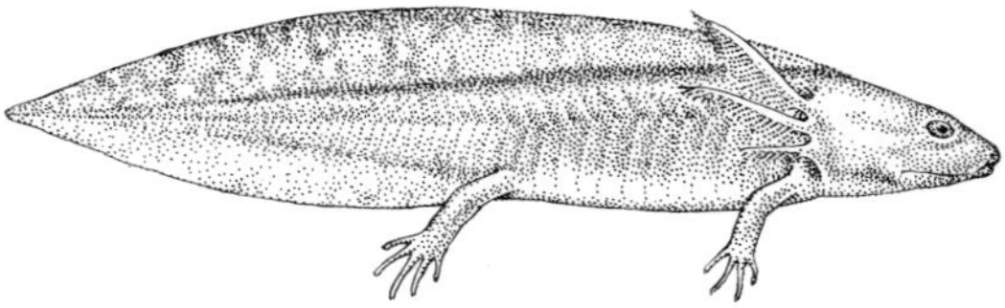

rounded when viewed from above, over 1½ times longer than wide and not much wider than body. Found in cold, often flowing water.

Pyrenean Brook Newt, *Euproctus asper* (p. 37 and Fig., opposite) Mainly in mountain areas. Up to 7. 5cm (9.5cm, high in Sierra de Gredos). Tail-tip blunt or rounded, upper crest usually extends at least a short distance onto body. Head broad and rounded and clearly wider than body when viewed from above. Distance between nostrils much greater than distance from nostril to eye. Often, a light spot at the base of each leg. Older animals may show yellow or orange markings of adults. Found in clean, cool often flowing water

Fire Salamander, *Salamandra salamandra* (p. 31 and Fig., opposite) Not north Spain. Up to 8cm (occasionally 10cm). Tail tapers steadily to a rounded point, crest high and may extend to head. Head quite broad and snout rounded when viewed from above. Gills feathery. Eyes small. Often found in standing waters in lowlands and hilly country. May be dark spotted.

Sharp-ribbed Newt, *Pleurodeles waltl* (p. 36 and Fig., opposite) Usually up to 8cm (generally smaller but rarely less than 5cm when fully developed). Undamaged tail tapers gradually to a long filament. Crest extends far forwards on back. Toes long. Some green in coloration of flanks. Heavy dark markings may be confined to spots on upper and lower edges of tail, where they may alternate with silvery or bluish markings. Found in still or slow-flowing water.

Marbled Newt, *Triturus marmoratus* (p. 47 and Fig., opposite)
Southern Marbled Newt, *Triturus pygmaeus* (p. 42)
North Iberian peninsula and Pyrenees only. Up to 4.5cm but usually less. Tail tapers gradually to a point without a filament, its upper and lower borders not roughly parallel).

Palmate Newt *Triturus helveticus* (p. 48 and Fig., opposite) North-west, west and central Iberian peninsula. Up to 5cm. Tail ends abruptly being rounded or ending in an oblique point, frequently with a short projection. 2

2. Cantabrian mountains and mountains near Madrid. Up to 5cm. Tail ends in a short filament. Often quite dark with a fine dark reticulation visible on the crests when viewed against the light with a lens.

Alpine Newt, *Triturus alpestris* (p. 44 and Fig., opposite)

North-west, west and central Iberian peninsula. Up to 3.5cm. Sometimes pale; without a fine dark reticulation on crests.

Bosca's Newt, *Triturus boscai* (p. 48)

CORSICA AND SARDINIA

Corsica only. Up to about 6.5cm. Tail tip blunt or rounded, crest usually extends at least a short distance on to body. Head broad and rounded when viewed from above; distance between nostrils much greater than distance from eye to nostril. Often a light blotch at the base of each leg. Older animals may have adult pattern of yellow or orange markings.

Corsican Newt Salamander, *Salamandra corsica* (p. 31)

Corsica only. Up to 5cm (seldom to 6cm). Tail-tip rounded, no crest on body. Head relatively narrow.

Corsican Brook Newt, *Euproctus montanus* (p. 38 and Fig., p. 246)

Sardinia only. Up to 5cm (rarely to 7cm). Tail-tip often pointed, crest usually extends at least a short distance onto body. Head relatively narrow. The only free-living salamander or newt larva on Sardinia

Sardinian Brook Newt, *Euproctus platycephalus*
(p. 39 and Fig., p. 246)

Identification of
Frog and Toad Larvae

Sizes given are for large larvae with well-developed hind legs. The illustrations depict the tadpoles at this stage. Figs., below show some of the features important for identification. Tail and crest shape are often quite variable within species. The number of upper and lower teeth rows is sometimes given as a formula, for instance 2/3, or if they are variable, 2-3, 3-4.

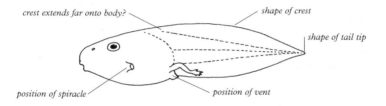

Some features to check when identifying frog and toad tadpoles

Mouth of frog or toad tadpole

KEY

1. Spiracle on mid-line of belly   2
   Spiracle on left flank   5

2. Iberian peninsula etc., Corsica and Sardinia, Sicily, Malta. Up to about 3.5cm. Spiracle centrally placed on underside. Tail over 1½ times body length with a rounded tip. Dark; a polygonal network of fine black lines on crests often visible with lens   Painted frogs, *Discoglossus* (p. 62-64 and Fig., 251)

Not Iberian peninsula or most Mediterranean islands (except north-east Sicily). Up to about 5.5cm. Spiracle placed slightly towards back of body. Upper crest extends on to body. Tail less than 1½ times body length, deep; tail tip bluntly pointed or rounded. Numerous intersecting dark lines often visible on crests and sometimes body.

Fire-bellied toads, *Bombina* **3**

Western Europe. Up to 7- 9cm. Spiracle placed slightly towards front of body. Upper crest barely extends on to body. Tail 1½ times body length or more, often fairly deep, tip bluntly pointed or rounded. No fine black lines on crest or body but dark spots or blotches often present.   **3**

Midwife toads, *Alytes* **4**

3. East Europe. Mouth more or less triangular, crest high, extending far forwards, sometimes to level of eye. Tail moderately long, longer than body. Third lower tooth row often with a gap. Rarely in small, shallow water bodies with little vegetation.

Fire-bellied Toad, *Bombina bombina* (p. 61 and Fig. p. 251)
Range includes Italy and south-east Europe. Mouth rounded, crest less high, extending no more than half way along body. Tail short, sometimes less than length of body. Frequently in small, shallow water bodies with little vegetation.

Yellow-bellied Toad, *Bombina variegata* (p. 60).

4. Western Europe but not much of southern Iberian peninsula. Up to 9cm. Nostril closer to tip of snout than to eye, body broadest behind level of spiracle. Dark spots on muscular part of tail not in obvious rows, crests with small spots. In mature larvae with forelimbs, 3 tubercles present on palm of hand.

Common Midwife Toad, *Alytes obstetricans* (p. 64 and Fig., p. 251),
Southern Iberian Peninsula. Up to 7cm Nostril closer to tip of snout than to eye, body broadest behind level of spiracle. Dark markings on muscular part of tail arranged in one or two rows, crests with big dark spots. In mature larvae with forelimbs, 2 tubercles present on palm of hand.

Iberian Midwife Toad, *Alytes cisternasii* (p. 66)
South Spain only. Dark markings on upper muscular part of tail form a line above on each side and these join each other on the body at the tail base. In mature larvae with forelimbs, 3 tubercles present on palm of hand.

Southern Midwife Toad, *Alytes dickhelleni* (p. 67)
Majorca only. Up to 7.5cm. Snout long, nostril closer to eye than to tip of snout, body broadest at level of spiracle. Crest low, not as high as body. Mouth broad, tail relatively short. In mature larvae with forelimbs, 3 tubercles on hand.

Majorcan Midwife Toad, *Alytes muletensis* (p. 67)

5. Small (up to 5cm); spiracle points straight backwards; tail tip clearly rounded; no papillae below lower lip

Typical toads, *Bufo*, **6**

Small to large; spiracle points backwards and at least slightly upwards; tail
tip in most cases at least bluntly pointed; papillae below lower lip   7

6. Up to about 4cm. Blackish above, belly very dark grey. Mouth as wide as
   space between eyes which is about twice the distance between the nostrils.
   Second row of upper labial teeth almost continuous, or with short break
                          Common Toad *Bufo bufo* (p. 73 and Fig., p. 251)
   Up to about 5cm. Brown or olive-grey above, belly greyish white. Mouth
   nearly as wide as space between eyes which is about 1½ the distance between
   the nostrils. Second row of upper labial teeth often broken in middle
                          Green Toad, *Bufo viridis* (p. 76 and Fig., p. 252)
   Up to about 3.5cm. Blackish above, belly dark grey, often with bronzy spots;
   may have light stripe on back. Mouth often only half as wide as space
   between eyes which is about twice the distance between the nostrils. Second
   row of upper labial teeth with wide gap in middle
                          Natterjack, *Bufo calamita* (p. 74)

7. Upper crest extremely high on sides of head and can be seen from below. Tail
   tip slender. Vent opens on right of lower tail fin and well above its lower edge.
   Usually golden-olive above and white below (though young may only show
   green below). Tooth rows 2/3. Very fast and fish-like
                          Tree frogs, *Hyla*   **8**
   Upper crest not extremely high, eyes tend to be towards top of head   **9**

8. Much of Europe. Crest may extend between eyes. Muscular part of tail plain
   or spotted, and there may be a single line of dark pigment. Second row of
   upper labial teeth with a narrow gap in middle
                          Common Tree Frog, *Hyla arborea* (p. 77 and Fig., p. 252)
                          Italian Tree Frog, *Hyla intermedia* (p. 78)
   Corsica, Sardinia etc. Similar but crest extends to about level of spiracle
                          Tyrrehnian Tree Frog, *Hyla sarda* (p. 78)
   South-west Europe. Crest extends to about level of spiracle. Muscular part
   of tail with 2 or 3 lines of dark pigment Second row of upper labial teeth with
   a broad gap in middle
                          Stripeless Tree Frog, *Hyla meridionalis* (p. 79 and Fig., p. 252)

9. Vent on mid-line   **10**
   Vent opens towards lower edge of tail-fin, on its right side
                          Typical frogs, *Rana*,   **11**

10. Up to about 6.5 cm. Tail rather blunt; beak white with a black edge.
                          Parsley frogs, *Pelodytes* (p. 71 and Fig., p. 252)
    Up to l6cm or even more. Tail sharply pointed; beak black.
                          Spadefoots, *Pelobates* (p. 68 and Fig., p. 252)
    Range is useful in distinguishing the three species. Western Spadefoot has a
    relatively shorter tail than the others.

11. The tadpoles of Typical frogs (*Rana*) are all similar and all rather variable, so they are difficult to identify with certainty. As with other difficult groups, range is sometimes a helpful clue to identity.

American Bullfrog, *Rana catesbeiana* (p. 97, Fig., p. 252). Up to 16cm. A series of ridges across the back. Olive green above, belly whitish. Tail tip clearly pointed.

Water frogs, (p. 89), including Edible frog, *Rana esculenta* (Fig., p. 252) Up to about 8cm but sometimes more. Olive, olive-brown or olive-grey above (often green in later stages), belly white. Tail tip clearly pointed. Teeth 2-3/3 (often 1/3 in Iberian water frog, *Rana perezi*).

Moor Frog, *Rana arvalis* (p. 82). Up to about 4.5cm. Brown above, belly greyish, sometimes large dark spots on upper crest. Tail-tip usually clearly pointed. Teeth usually 2-3/3

Agile Frog, *Rana dalmatina* (p. 83 and Fig., p. 252). Up to 6cm. Pale brown or rufous above with darker spots, belly white with gold spottings. Often some dark spots on upper crest, which is fairly high and tapers abruptly to a very pointed tail tip. Teeth usually 3/4

Italian Agile Frog, *Rana latastei* (p. 84). North Italy etc. Up to about 4.5cm. Rather like tadpole of Agile frog, but crest lower. Brown above, belly whitish, often some dark spots on upper crest.

Common Frog, *Rana temporaria* (p. 80 and Fig., p. 252). Up to about 4.5 cm. Brown to black above, belly grey or black. Tail tip usually blunt. Teeth usually 3/4 (p. 85)

Balkan Stream Frog, *Rana graeca* (p. 82 and 85). Teeth 4–5/4.

Iberian Frog, *Rana iberica* (p. 87). Similar to Common frog, but crest higher and usually blotched. Teeth of Balkan Stream Frog 4–5/4.

Pyrenean Frog, *Rana pyrenaica* (p. 81). Crest low, dark colouring. Tail tip blunt. Teeth 4/4.

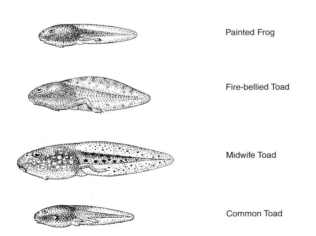

Painted Frog

Fire-bellied Toad

Midwife Toad

Common Toad

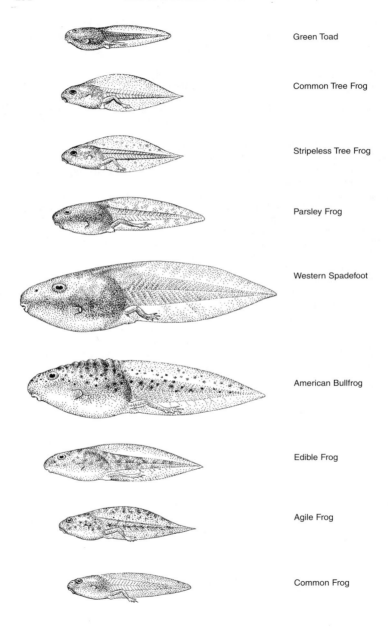

Green Toad

Common Tree Frog

Stripeless Tree Frog

Parsley Frog

Western Spadefoot

American Bullfrog

Edible Frog

Agile Frog

Common Frog

INTERNAL CHARACTERS

In the species texts, identifications have been made using only external characters, but in some difficult groups, differences in skeleton may sometimes be helpful in confirming identities of dead animals.

POND NEWTS (*Triturus*).

Two of the bones on each side of the skull, the frontal and the squamosal, often have projections that may meet to form a structure called the fronto-squamosal arch. Sometimes the arch is completely bony, but often it is interrupted in the middle or is entirely absent. Among European species, the structure usually varies as follows (Fig., below):

Arch completely bony: Palmate and Bosca's newt.
Arch present but interrupted: Common and Italian newts.
Arch completely bony or interrupted: Montandon's newt.
Arch very reduced: Alpine and Marbled newts.
Arch absent: Crested newts.

This feature is most useful in identifying newts apparently intermediate between Common and Palmate newts.
Note The differences listed above are true for adults. Juveniles of all species often lack a complete arch.

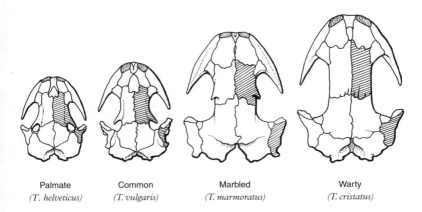

| Palmate | Common | Marbled | Warty |
| *(T. helveticus)* | *(T. vulgaris)* | *(T. marmoratus)* | *(T. cristatus)* |

Newts showing differences in development of fronto-squamosal arch

GREEN LIZARDS (*Lacerta*).

In North Spain, exceptional animals may be found that are not clearly either the Green Lizard or Schreiber's Green lizard. In such cases checking the skull may help. In Schreiber's Green Lizard, the postorbital and postfrontal bones (Fig., right) form a single unit throughout life, whereas in the Green Lizard they are separate, except in some old animals. The bones are best examined from underneath.

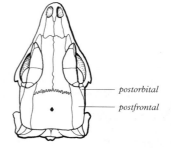

postorbital

postfrontal

Skull of Green Lizard

SMALL LACERTAS (*Lacerta* and *Podarcis*).

European Rock, Meadow and Viviparous lizards (*Lacerta*) can be distinguished from Wall lizards (*Podarcis*) by the shape of the vertebrae in the base of the tail. In both groups, the first 4-6 vertebrae of the tail have a simple projection on each side, but in Wall lizards the next few have two pairs of projections of which the hind one is longer and points backwards. Rock, Meadow and Viviparous lizards may have a single pair of projections on all tail vertebrae, or if any have two pairs, the back one is shorter than the front pair and is parallel to it.

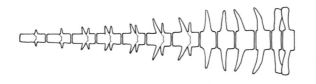

Wall Lizard

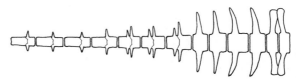

Rock, Meadow and Viviparous Lizards

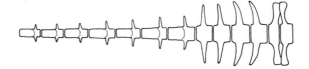

Basal tail vertebrae of Small lacertas

GLOSSARY

Many terms not included in this glossary are explained in the introduction and in the remarks at the beginning of the sections on the main groups of amphibians and reptiles (pp. 28, 54, 98, 111, 195).

Aestivation A period of inactivity during the summer months.

Albino (hence albinism). An animal in which dark pigment fails to develop so that it appears white or pinkish. Albinism occurs as a rare variation in many species.

Amplexus The sexual embrace of amphibians. It occurs in all European frogs and toads, most salamanders and some newts but not in Pond newts. In frogs and toads the male grasps the female just behind the arms in some species, and around the loins in others (Pl. 11).

Anal gland A paired gland situated in the base of the tail of many snakes and opening into the vent. Anal glands often produce a foul-smelling secretion that is ejected when the snake is frightened.

Arthropods Animals with a hard jointed surface that forms an external skeleton, including insects, spiders, centipedes and millipedes, and crustaceans. Arthropods form fuch of the diet of many amphibians and reptiles.

Autotomy The process of shedding a part of the body, either spontaneously or when it is grasped by an attacker. In Europe most lizards and the Golden-striped and Luschan's salamanders (p. 35 and p. 34) are capable of autotomizing the tail if it is held. In lizards the tail usually contains a lot of fat, sometimes as much as half the animal's energy reserves. This often makes it worthwhile for a predator to stop to eat a shed tail rather than continuing to chase the owner. The tail is often made conspicuous by continuing to move for some time after it breaks off and is also sometimes brightly coloured. In lizards, breakage typically takes place at one of the planes of weakness that are present across each vertebra in the tail and, in most cases, a new tail is grown afterwards. The ancestor of agamas and chameleons lost the ability to shed the tail in the very distant past, but some agamas including the Starred Agama (p. 118) have regained this ability, although now breakage takes place between vertebrae, fat deposits are not marked and little regeneration takes place. Tail breakage in salamanders also occurs between vertebrae.

Balkans Used here for the whole of south-east Europe, north to Slovenia, Croatia, Serbia, southern Hungary and Romania.

Bridge The part of the shell of tortoises, terrapins and turtles that joins its upper and lower sections.

Canthus (plural canthi; abbreviation of canthus rostralis). A slight ridge separating the top and side surfaces of the snout in some amphibians and reptiles. Typically it runs from the front corner of the eye to the general region of the nostril.

Carapace Upper section of the shell of tortoises, terrapins, and turtles.

Cervical scale Another name for the nuchal scale at the front of the upper shell of many tortoises, terrapins and turtles (see Fig., p. 99)

Cloaca The cavity into which the contents of the intestine, reproductive organs and kidneys discharge their products in reptiles and amphibians. The cloaca empties through the vent.

Cloacal gland See anal gland.

Cloacal swelling The swollen area around the vent in many newts and salamanders.
Collar A backwardly-directed fold of skin across the lower surface of the neck in some lizards; it is usually covered with enlarged scales that often overlap those behind (see for example. Fig., p. 114).
Constriction Method of holding and sometimes killing prey used by some snakes, in Europe particularly the Sand Boa (p. 201) and Rat Snakes (pp. 211–216). The victim is held very tightly by coils of the body so that breathing and often the action of the heart are prevented.
Costal grooves Vertical grooves on the flanks of some salamanders, newts and their larvae (Fig,. p. 244). They correspond to the position of the ribs. Differences in the number of grooves between the fore and hind limbs can be important in the identification of newt tadpoles, especially in distinguishing those of the Marbled newt and Crested newts (see p. 245).
Crepuscular Active at twilight.
Crest The flat, upright and flexible structure that develops on the tail and often back of male Pond newts in the breeding season. It is believed to be important both as a respiratory structure and a recognition feature that helps females to identify males of their own species. Crests also occur in amphibian larvae, being present above and below the tail and often extending for a variable distance on to the back.
Cytotoxic Used to describe snake venoms that destroy tissue and cells.
Display A conspicuous and usually ritualised pattern of behaviour that is capable of affecting the behaviour of other animals at which the display is directed. Among the obvious ones encountered in reptiles and amphibians are courtship displays in which males 'persuade' females to mate with them. Reptiles especially also often have threat displays in which males particularly may defend their territories and potential mates, often without the need for actual fighting. Other displays are directed towards potential predators.
Diurnal Habitually active by day.
Dorsal To do with the upper surface of the body. In lizards and snakes, dorsal scales usually include those on the sides as well.
Dorsolateral fold A ridge running from the side of the head to the groin in some frogs (see for instance Fig., p. 58).
Dry-stone walls Walls constructed by piling natural, unsmoothed stones without the use of mortar. They contain numerous crevices and hollows used by many reptiles and some amphibians. The abundance of dry-stone walls, and banks constructed in a similar way, has probably increased populations of some reptile species dramatically in many parts of the Mediterranean region.
Emarginate Used here to describe an arrangement where the edge of a scale cuts into the otherwise smooth margin of a larger one. See, for example, Fig., p. 166.
Endemic Restricted to a particular geographic region and, in some but not necessarily originating there.
Fang An enlarged tooth, particularly a hollow or grooved one in snakes, that acts as a means of injecting venom into prey or an attacker.
Femoral pores A series of pores on the underside of the thigh of many lizards (Fig., p. 127), normally better developed in males than females. They exude a waxy secretion that is often used to mark places where the lizard has been, leaving a signal for other members of the species. In some cases, the secretion not only makes it clear that a particular individual has been in a specific place but may give information about its sex and social position. In some cases the deposited secretion also reflects

ultraviolet light, making it visible to many lizards. Femoral pores often extend into the area immediately in front of the vent and, if the ones under the thigh itself are lost as in some geckos, they are then called preanal pores. Pores have been completely lost in some lizards including all chameleons, skinks and *Tarentola* geckos.

Garrigue See scrub.

Gill A breathing organ located in the neck region that functions in water. gills may be external and feathery in salamander and newt larvae and young frog and toad tadpoles, or enclosed, as in older frog and toad tadpoles.

Hemipenis Males of mammals, tortoises, crocodiles and some birds, have a single essentially solid penis with a passage running within it through which sperm is injected into the female. In contrast, lizards and snakes have two organs that serve this purpose, the hemipenes. They are hollow and each is rather like the finger of a glove. When in use they project outwards from the vent but at other times a muscle within each one turns it inside out drawing the organ backwards, so that it lies within the tail base with its free end directed towards the tail tip. Only one hemipenis is used at a time and sperm passes along an often partly enclosed groove on its surface. The two hemipenes are each connected to a separate testicle and sperm reservoir and they are often used alternately. Detailed structure of hemipenes often differs between species and they are sometimes used in identification.

Herpetology The study of reptiles and amphibians.

Hibernation An extended period of rest, immobility or torpor during the winter. Hibernation occurs in most European amphibians and reptiles, including all the more northern ones in which it is often very prolonged. See also p. 16.

Hybrid An individual animal that is the offspring of parents of different species.

Iberia Used here for the Iberian peninsula which is almost entirely made up of Portugal and Spain, although Andorra and the French Pyrenees are often included.

Iris The often brightly coloured and circular area of the eye surrounding the pupil.

Jizz The combination of subtle and usually indefinable features that is often the easiest way of identifying and separating generally similar species. Successful use of jizz depends on having some experience of the animals concerned. It should not be relied on too much until the observer has developed this familiarity.

Karst Limestone formations with characteristic pattern of erosion typified by deep hollows, crevices, caves, and red soil. Karst provides suitable habitats for many reptiles.

Labial To do with the lips. In reptiles, used for a scale that borders these.

Larva A specialised stage in the life history of many animals including most European amphibians, the eggs of which give rise to a free-living aquatic form that finally changes (metamorphoses) to produce the adult stage.

Lateral To do with the sides.

Lek A courtship ritual in which males gather together in an area to display. Females then come to the lek, select partners and mate with them.

Lichenous Used to describe markings which have irregular, often poorly defined edges, rather like lichens.

Lumbar Used for the region of the back just in front of the hind limbs.

Macchia Mediterranean plant communities usually including bushes and small trees. See **Scrub**.

Melanism (hence melanistic, melanic). A condition in which an animal has an unusually large amount of dark pigment so that it appears black or blackish. Melanism occurs in many small-island populations of Wall lizards, but also sporadically in many other species.

Metamorphosis The rapid changes that occur in amphibians when a larva (such as a tadpole) assumes the form of an adult animal. In newts and salamanders the most obvious sign is the disappearance of the feathery gills. In frogs and toads it is the development of the forelegs and the rapid loss of the tail.

Metatarsal tubercle A prominent tubercle on the hind foot of many frogs and toads (see for instance Fig., p. 54).

Montane Living in mountainous country.

Morph See p. 141.

Nasolabial groove A narrow groove running from the nostril to the edge of the upper lip in the Lungless salamanders, including the European Cave salamanders (Fig., p. 28).

Neoteny A condition in which an amphibian larva fails to metamorphose but may grow to a large size and sometimes achieve sexual maturity. Neoteny occurs sporadically in many salamanders and newts. It is common in some populations of the Alpine newt (p. 44) and a universal condition in the Olm (p. 53). Neotonous frog tadpoles are rare and cannot breed.

Neurotoxic Used to describe snake venoms that interfere with nerve conduction and cause paralysis of muscles.

Nuptial pads Rough, and usually dark, areas of skin found on many frogs and toads and some salamanders. Nuptial pads develop to their full extent during the breeding season and help males maintain their grip on females. In frogs and toads they are often situated on the thumb, but other fingers may be involved and some species have rough areas present on the belly, toe-web or forearm.

Ocelli (singular: ocellus). Eye-like spots that often occur in the colouring of reptiles, especially lizards. Ocelli are approximately round with a contrasting border (see for example Pl. 26, Figs 1 and 5).

Ocellated With ocelli.

Paratoid gland Paired glandular swelling on head and neck, situated behind the eye and present in many amphibians, particularly Typical toads and Fire salamanders and their relatives, see for example Fig., p. 54. In these animals the paratoid glands produce a noxious secretion that deters predators.

Pheromone A chemical released into the surrounding air as water that elicits a change of behaviour in another animal, often of the same species. Pheromones may also sometimes be applied directly to the target animal. They are often used in courtship.

Phrygana See Scrub.

Plastron Lower section of the shell of tortoises, terrapins and turtles.

Pleural scales Another name for the costal scales on each side of the upper shell of tortoises, terrapins and turtles (see Fig., p. 99)

Reticulated Used to refer to patterns made up of lines and blotches interconnected in a net-like fashion (see for example Pl. 30, Fig. 3a, and Pl. 32, Fig. 1c).

Salp A soft-bodied, floating marine animal related to sea-squirts. A common food of some marine turtles.

Scree (also known as talus) A slope, usually in a mountain area, composed of rock fragments that have fallen from higher ground **Boulder Scree**. Scree made up of large rocks. Screes contain many openings and crevices and are consequently often good habitats for reptiles and amphibians.

Scrub Used here for a wide variety of plant communities where there is a fairly dense, although often interrupted, low growth. It includes Mediterranean plant communities such as Macchia (the French Maquis) with a mixture of bushes and small trees, and Garriga (the French Garrigue and Greek Phrygana) which is a low, carpet of dense often spiny shrublets.

Serpentine locomotion The main kind of locomotion in snakes in which they body is thrown into horizontal curves which pass backwards, pressing against the ground and vegetation so that the snake moves forwards. It also occurs in many legless lizards and several skink species with very small legs.

Spade Enlarged metatarsal tubercle present on the hind foot of Spadefoot toads and used for digging (Pl. 9).

Species See p. 13.

Spectacle Permanent transparent covering of the eye found in snakes and some lizards (geckos, Snake-eyed Lacertid, Snake-eyed Skink). Lizards usually have openable upper and lower eyelids, but in some kinds there is a transparent window in the lower lid. This enables lizards in hot, dry environments to bask in the sun with their eyes shut, yet still see predators. This ability reduces water loss from the eye which can be considerable in small species. In extreme cases the eyelids fuse together, restricting water loss further, and the window becomes huge to form a spectacle. Once this happens, eyelid separation appears impossible and reptiles with spectacles are condemned to see the world through them for the rest of their evolutionary existence, even if water loss is no longer a problem.

Spermatophore A gelatinous mass containing spermatazoa, produced by male newts and salamanders and taken up by the females.

Spiracle The small opening in the body wall of frog and toad tadpoles that leads from the gill chamber (Fig., p. 248). The enclosed gills are bathed in a stream of water that enters through the mouth and leaves through the spiracle.

Spur Any pointed, projecting structure on the limbs, especially the hind ones. Among European species the word is used for the conical horny scales on the back of the thighs of some tortoises and for the projection on the hind limb of some male Brook salamanders.

Subarticular tubercles Tubercles situated under the joints of the fingers and toes.

Subspecies See p. 14.

Tadpole In its narrow sense the larva of a frog or toad (the word means 'toad head'), but also sometimes used more loosely to include the larvae of newts and salamanders.

Tubercle A small well-defined, sometimes pointed swelling on the skin or an enlarged usually pointed scale.

Tyrrhenian region The islands in and around the Tyrrhenian Sea, particularly Corsica, Sardinia, Monte Cristo, Giglio, and sometimes also Elba and Iles d'Hyères.

Uniform Of a single colour; lacking contrasting markings.

Vent The external entrance/exit of the cloaca.

Ventral To do with the underside. Ventral scales are the belly scales of lizards and snakes. See p. 196 for a more precise definition of snake ventral scales.

Vertebral Used here for the area over the vertebral column (spine) when describing scaling or pattern.

Vocal Sac See p. 17.

Wart Used for any well-defined, raised area on the skin of amphibians.

Web The thin sheet of skin joining the hind toes of many European frogs and toads and some salamanders and newts.

FURTHER READING

General Accounts

W. Böhme and others (eds). 1981–, *Handbuch der Reptilien und Amphibien Europas*. Aula Verlag, Wiesbaden. A multi-volume work designed to cover all the reptiles and amphibians of Europe including the Caucasus area. To date three volumes on lizards, two on snakes, one on chelonians, one on salamanders and one on the Canary islands reptiles have been published and more are in preparation.

Gasc, J.-P.(ed.) 1997. *Atlas of amphibians and Reptiles in Europe*. Societas Europea Herpetologica and Muséum National d'Histoire Naturelle, Paris. 494pp.

Mertens, R. and Wermuth, H. 1960. *Die Amphibien und Reptilien Europas*. Verlag W. Kramer, Frankfurt. 264pp. Lists all subspecies recognised at time of publication.

Corbett, K. 1989. *Conservation of European Reptiles and Amphibians*. Chistopher Helm, London. 274pp.

Amphibians

Nöllert, A and Nöllert, C. 1992. *Die Amphibien Europas*. Franckh-Kosmos Verlags-GmbH, Stuttgart. 382pp.

Griffiths, R. A. 1995. *Newts and Salamanders of Europe*. Poyser, London. 188pp.

Thorn, R. 1968. *Les Salamandres d'Europe, d'Asie et d'Afrique du Nord*. Lechavalier, Paris.

Boulenger, G. A. 1897–98. *The Tailless Batrachians of Europe* (2 vols). Ray Society, London. 376pp. (Some names outdated: *Bombinator = Bombina*; *Rana agilis = R. dalmatina*.)

Savage, R. M. 1961. *The ecology and life history of the Common Frog (Rana temporaria temporaria)*. I. Pitman, London. 221pp.

Reptiles

Brongersma, L. D. 1972. European Atlantic Turtles. *Zoologische Verhandelingen*. E. J. Brill, Leiden. 317pp.

Grüber, U. 1989. Die Schlangen Europas und ums Mittelmeer. Franckh-Kosmos Verlags-GmbH, Stuttgart. 248pp.

Regional accounts

ALBANIA

Bruno, S. 1989. Introduction to a study of the herpetofauna of Albania. *British Journal of Herpetology* **29**: 16-41.

Kopfstein, F. and Wettstein, O. 1920. Reptilien und Amphibien aus Albanien. *Verhandlungen der Zoologisch-Botanischen Gesellschaft in Wien*, 70:387–409.

AUSTRIA

Cabela, A. and Tiedemann, F. 1985. Atlas der Amphibien und Reptilien Osterreichs. *Neue Denkschriften des Naturhistorisches Museum Wien*, 4. 80 pp.

Cabela, A. B. 1982. *Catalogus Faunae Austriae*. 21: Amphibia, Reptilia. Springer-Verlag, Wien.

BELGIUM, NETHERLANDS, LUXEMBURG

Sparreboom, M. (ed.) 1981. *De amphibeën en reptielen van Nederland, België en Luxemburg*. A. A. Balkema, Rotterdam. 284pp.

Sparreboom, M. 1982. *Bibliografie van de Nederlandse Herpetofauna*. Lacerta, Amsterdam. 186pp.

Bergmans, W. and Zuiderwijk, A. 1986. *Atlaas van de Nederlandse Amfibieen en Reptielen en hun Bedreiging*. Lacerta, Amsterdam. 177pp.

Witte, G. F. de. 1948. *Faune de Belgique*. Amphibiens et Reptiles 2nd ed). Musée Royal d'Histoire Naturelle de Belgique, Brussels. 321pp.

BULGARIA

Beskov, V. and Beron, P. 1964. Catalogue et Bibliographie des Amphibiens et des Reptiles en Bulgarie. *Acad. Bulgar. Sci., Sofia.*

Buresch, I. and Zonkov, J. Untersuchungen über die Verbreitung der Reptilien und Amphibien in Bulgarien und auf der Balkanhalbinsel. *Izvestiya na Tsarskite Prirodonauchni Instituti u. Sofia.* (German summaries.). 1 Tortoises and Lizards, 1933: 150–207; 2. Snakes, 1934: 106–188; 3. Salamanders and newts, 1941:171–237; 4. Frogs and toads, 1942: 68–154.

CROATIA, SLOVENIA, SERBIA (INCLUDING MONTENEGRO) AND MACEDONIA

Henle, K. 1985. Biologische, zoogeographische und systematische Bermerkungen zur Herpetofauna Jugoslawiens. *Salamandra* 21: 229–251.

Vogrin, N. 1997. An overview of the herpetofauna of Slovenia. *British Herpetological Society Bulletin*, 58: 26–35.

Radovanovic, M. 1951. *Vodozemci' i gmizavci nase zemlje.* Belgrade. 251pp.

Pozzi, A. 1966. Geonomia e catalogo ragionato degli anfibi e dei rettili della Jugoslavia. *Natura, Milano.* 57: 5–55.

CZECH REPUBLIC AND SLOVAKIA.

Barus, V., Král, B., Oliva, O. et al. 1992. *Obojzivelnici – Amphibia.* Fauna CSFR 25, Praha. 338pp.

Barus, V., Kminiak, M., Král, B. et al. 1992. *Plazi – Reptilia.* Fauna CSFR 26, Praha. 222pp.

Moravec, J. (ed.) 1994. *Atlas of Czech amphibians.* Norodi Muzeu, Praha. 136pp.

Lac, J. 1968. *Obojzivelniky ad Plazy.* Stavovce Slovenska 1, Bratislava.

DENMARK

Bringsøe, H. and Graff, H. 1995. *Bevarelsen af Denmarks padder og krybdyr.* Nordisk Herpetologisk Forening (Aage V. Jensens Fonde).

Shiøtz, A. 1971. Danske Padder. *Natur og Museum*, 15 (1): 1–23.

Soager, O. 1971. Faunaundersogelsen 1970/71. *Nordisk Herpetologisk Forening*, 1971:57–66.

FINLAND

Terhivuo, J. 1981–1993. Provisional atlas and population status of the Finnish amphibian and reptile species with reference to their ranges in northern Europe. *Annales Zoologici Fennici* 18: 139–164; 30: 55–69.

Langerwerf, B. 1975. Reptielen en amfibieën in Finland. *Lacerta,* 34 (1): 4–7. (In Dutch.)

FRANCE

Castanet, J. and Guyetant, R. 1989. *Atlas de Repartition des Amphibiens et Reptiles de France.* Société Herpétologique de France, Paris. 186pp.

Delaguerre, M. and Cheylan, M. 1993. *Batraciens et Reptiles de Corse.* Parc Naturel Regional de Corse. 128pp.

Rollinat, R. 1934. *La Vie des Reptiles de la France Centrale.* Librairie Delagrave, Paris. 343pp.

Angel, F. 1946. *Faune de France. Reptiles et Amphibiens.* Lechavalier, Paris. 204pp.

Fretey, J. 1975. *Guide des Reptiles et Batraciens de France.* Hatier, Paris. 239pp.

GERMANY

Günther, R. (ed.). 1996. *Die Amphibien und Reptilien Deutschlands.* Gustav Fischer Verlag, Jena. 825pp.

GREAT BRITAIN AND IRELAND

Beebee, T. J. C. and Griffiths, R. A. 2000. *Amphibians and Reptiles: A Natural History of the British Herpetofauna.* HarperCollins*Publishers*, London. 270pp.

Arnold, H. R. 1995. Atlas of amphibians and reptiles in Britain. HMSO (ITE Research publication 10).

Lever, C. 1977. *The Naturalised Animals of the British Isles.* Hutchinson, London.

Smith, M. A. 1973. *British Amphibians and Reptiles* (5th ed.). Collins, London. 322pp.

GREECE

Chondropoulos, B. P. 1986. A checklist of Greek reptiles. I.
The lizards. *Amphibia-Reptilia* 7:217-23e.
Chondropoulos, B. P. 1989. A checklist of Greek reptiles.
II. The snakes. *Herpetozoa,Wien* 2: 3-36
Sofianidou, T. S. 1996. Tetrapoda of Greece. In Systematics of the Tetrapoda with appendix of the Greek species. P. 303-370. Giachudis Giapulis, Thessaloniki.
Werner, F. 1938. Die Amphibien und Reptilien Griechenlands. *Zoologica*, 94:1–117, Stuttgart.
Wettstein, O. 1953, 1957. Herpetologia aegaea. *Sitzungsberichte der Osterreichischen Akademie der Wissenschaften. Mathematisch-Naturwissenschaftliche Klasse*, 162: 651–833; 166:123–164.

HUNGARY

Dely, O. G. 1967. Kétéltüek - Amphibia. Magyarország Állatvilága (Fauna Hungariae, 83) 20:1-80.
Dely, O. G. 1978. Hüllöck - Reptilia. Magyarország Állatvilága (Fauna Hungariae, 130) 20:1-120
Fejérváry-Langh, A. M. 1943. Befträge und Berichtigungen zum Amphibien-Teil des ungarischen Faunenkataloges. *Fragmenta Faunistica Hungarica* 6: 42–58. Reptilien-Teil: 81–98.

ITALY

Corti, C. and Lo Cascio, P. 2002. *The Lizards of Italy and Adjacent Area*. Edition Chimaira, Frankfurt am Main.
Bruno, S and Maugeri, S. 1990. *Serpenti d'Italia e d'Europa*. Editoriale Giorgio Mondadori, Milan. 223pp.
Cappocaccia, L. 1968. *Anfibi e Rettili*. Mondadori, Milan. 159pp.
Tortonese, E. and Lanza, B. 1968. *Pesci, Anfibi e Rettili*. Aldo Martello Editore, Milan.
Bruno, S. 1973. Anfibi d'Italia: Caudata. *Natura, Milan*, 64: 209–450.
Bruno, S. 1970. Anfibi e Rettili di Sicilia. *Atti della Accad. Gioenia di Scienze naturli in Catania*. 7th Ser., 2: 1–144.

POLAND

Juszczyk, W. 1987. *Plazy i gady krajowe*. Warsaw. 3 volumes, 838pp.
Berger, L., Jaskowska, J. and Mlynarski, M. 1969. Plazy I gadi. *Katalog Fauny Polski (Catalogus Faunae Poloniae)*, 39:1–73.

ROMANIA

Fuhn, I. 1960. *Fauna Republicii Populare Romîne. Amphibia*. Acad. RPR, Bucurest. 288pp.
Fuhn, I. and Vancea, S. 1961. *Fauna Republicii Populare Romîne*. Reptilia. Acad. RPR, Bucuresti. 352pp.
Cogălniceanu, D., Aioanei, F. and Matei, B. 2000. *Amphibienii din România, Determinator*. Ars Docendi, Bucarest.
Fuhn, I. 1969. *Broaste, serpi, sopîrle*. Editura stiintifica. Bucurest. 257pp.

RUSSIA, UKRAINE, ESTONIA, LATVIA, LITHUANIA, BELARUS, MOLDOVA

Bannikov, A. G., Darevskii, I. S., Inchenko, V. G, Rustamov, A. K. and Shcherbak, N. N. 1977. *Opredelitel Zemnovodnykhe i Presmykayushchikhsya Fauny SSR. Izdatel'stvo* (Guide to the reptiles and amphibians of the fauna of the USSR).
Prosveshenie, Moscow. 414pp.
Kuzmin, S. L. 1999. *The Amphibians of the Former Soviet Union*. Pensoft, Sofia and Moscow. 538pp.
Terentiev, P. V. and Chernov, S. A. 1949 (translated 1965). *Key to Amphibians and Reptiles*. Israel Program for Scientific Translation, Jerusalem. 315pp.

SPAIN AND PORTUGAL

Barbadillo, L. J., Lacomba, J. I., Pérez-Mellado, V., Sancho, V. and Lopez-Jurado, L. F. 1999. *Anfibios y Reptiles de la Península Ibérica, Baleares y Canrias*. Editorial Planeta Barcelona. 419pp.

Salvador, A. (ed.) 1998. Reptiles. *Fauna Iberica*, 10. Museo Nacional de Ciencias Naturales, Madrid. 707pp.

Plegazuelos, J. M. (ed.) 1997. Distribución Biogeografia de los Amfibios y Reptiles en España y Portugal. *Monografias de Herpetología* 3, Granada. 542pp.

Salvador, A. and García París, M. 2001. *Anfibios Españoles*. Esfagnos. 271pp.

Salvador, A. 1986. Guia de Campo de los Anfibios y Reptiles de la Península Ibérica, Islas Baleares y Canarias. Unigraf, Madrid. 255pp.

SWEDEN

Ahlén, I., Andrén, C. and Nilson, G. 1992. *Sveriges grodor, ödlor och ormar*. Författarna och Naturskyddsföreningen. 48pp.

Gislén, T. and Kauri, H. 1959. Zoogeography of the Swedish Amphibians and Reptiles. *Acta Vertebratica*, 1:193–397. Almqvist and Wiksell, Stockholm.

Curry-Lindahl, K. 1975. *Groddjur och kräldjur i färg*. Awe/Gebers, Stockholm.

SWITZERLAND

Grossenbacher, K. 1988. Verbreitungsatlas der Amphibien der Schweiz. Documenta Faunistica Helvetiae, Schweizerischer Bund für Naturschutz, Basel 7:1-207.

Meier, M. 1987 *Amphibien und Reptilien der Schweiz*. Mondo-Verlag. 152pp.

Kramer, E and Stemmler, O. 1986. Schematische Verbreitungskarten der Schweizer Reptilien. *Revue Suisse de Zoologie*, 93: 779–802.

TURKEY

Baran, I and Atatür, M. K. 1998. *Turkish Herpetofauna.*, Turkish Ministry of Environment Ankara. 214pp.

NORTH AFRICA

Bons, J and Geniez, P. 1996. *Amphibiens et Reptiles du Maroc (Sahara Occidental compris). Atlas biogéographique*. Associación Herpetólogica Española. In French and Spanish with a 25-page English summary. 320pp.

Schleich, H. H., Kästle, W and Kabisch, K. 1996. *Amphibians and Reptiles of North Africa: Biology, Ssystematics, Field Guide*. Koeltz Scientific, Koenigstein. 630pp.

Journals

There are a number of specialist journals that deal substantially with European reptiles and amphibians. They include: in English, *Amphibia-Reptilia*, *Herpetological Journal*, and *British Herpetological Society Bulletin*; in Dutch, *Lacerta*; in German, *Salamandra, Mertensiella, Herpetozoa* and *Die Eidechse*; in Danish the journal of the *Nordisk Herpetologisk Forening*; in Spanish, *Revista Española de Herpetología*.

Recordings of frog and toad calls

Anon. 1997. *Lurche Europas*. Von Laar Media, Germany.

Brännström, P. and Mild, K. 1993. *Spelläten hos vara Nordiska groddjur* (Frogs and toads of Northern Europe). Stockholm, Sweden.

Márquez, R. and Matheu, E. 1998. Guía Sonora de las Ranas y *Sapos de España y Portugal*, Alosa, Barcelona

Roche, J.-C. 1997. *Au Pays des Grenouilles*. Sitelle, France.

Roche, J.-C. 1965. *Guide Sonore du Naturaliste*, 2: Batraciens. J.-C. Roche. (LP)

Weissman, C. 1971. *Danske Padder*. Naturhistorisk Museum, Århus, Denmark. (LP)

Further recordings of frog and toad calls can be found on the website of the British Libary National Sound Archive:
http://cadensa.bl.uk.

DISTRIBUTION MAPS

Distribution maps cover the same regions illustrated in the maps showing the area covered by this guide (pp. 12, 22–23). For most species, a general European map is used but, ones confined to the Atlantic islands or to the Aegean area are shown on maps limited to those areas that provide more detail.

The maps show the areas where a particular species can be expected to occur. However, within these areas, it will only be found in suitable habitat, so the relevant section on **Habits** in the text should be checked before deciding whether the species is likely to be encountered at a specific locality. Additional information on locality may be found in the appropriate section on **Range**. In the maps, question marks indicate uncertain areas of distribution.

The accuracy of the maps varies. This is due to a number of causes. For many countries relatively little up to date information on distribution is available, but for others it may be scant or very diffuse and hard to locate. In some places, known localities for a given species are widely separated and it may be difficult to tell whether this represents a broken range, or whether insufficient observations have been made in the area. Another problem concerns retreating species: several forms are no longer found in parts of their original range, but the old records for these areas may never have been refuted. Account is taken in the maps of large, well established introductions but small-scale ones and those of short duration are not shown.

On the whole, omissions in the maps are probably fairly minor and, if a species is apparently encountered very far outside its marked range, this will most probably be due to misidentification. In such cases, the animal in question should be checked carefully against both the relevant key and description.

1. Fire Salamander
Salamandra salamandra Pl. 1, p. 31

2. Corsican Salamander
Salamandra corsica Pl. 1, p. 33

3. Alpine Salamander
Salamandra atra Pl. 1, p. 33

Lanza's Alpine Salamander
alamandra lanzai Pl. 1, p. 34

5. Luschan's Salamander
Mertensiella luschani Pl. 3, p. 34

6. Spectacled Salamander
Salamandrina terdigitata Pl. 2, p. 35

. Golden-striped Salamander
Chioglossa lusitanica Pl. 2, p. 35

8. Sharp-ribbed Newt
Pleurodeles waltl Pl. 3, p. 36

9. Pyrenean Brook Newt
Euproctus asper Pl. 3, p. 37

0. Corsican Brook Newt
Euproctus montanus Pl. 3, p. 38

11. Sardinian Brook Newt
Euproctus platycephalus Pl. 3, p. 39

12. Marbled Newt
Triturus marmoratus Pl. 4, p. 42

13. Southern Marbled Newt
Triturus pygmaeus, p. 42

14. Northern Crested Newt
Triturus cristatus Pl. 4, p. 43

15. Italian Crested Newt
Triturus carnifex Pl. 4, p. 44

16. Balkan Crested Newt
Triturus karelinii Pl. 4, p. 44

17. Danube Crested Newt
Triturus dobrogicus Pl. 4, p. 44

18. Alpine Newt
Triturus alpestris Pl. 6, p. 44

19. Montandon's Newt
Triturus montandoni Pl. 6, p. 45

20. Common Newt
Triturus vulgaris Pl. 5, p. 46

21. Palmate Newt
Triturus helveticus Pl. 5, p. 48

. Bosca's Newt
turus boscai Pl. 5, p. 48

23. Italian Newt
Triturus italicus Pl. 6, p. 49

24. Ambrosi's Cave Salamander
Speleomantes ambrosii Pl. 2, p. 51

. Italian Cave Salamander
peleomantes italicus Pl. 2, p. 51

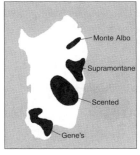

26. Sardinian Cave Salamanders
Speleomantes spp. Pl. 2, p. 52

27. Olm
Proteus anguinus Pl. 3, p. 53

. Yellow-bellied Toad
ombina variegata Pl. 11, p. 60

29. Fire-bellied Toad
Bombina bombina Pl. 11, p. 61

30. Painted Frog
Discoglossus pictus Pl. 7, p. 62

31. West Iberian Painted Frog
Discoglossus galganoi Pl. 7, p. 62

32. East Iberian Painted Frog
Discoglossus jeanneae, p. 63

33. Tyrrhenian Painted Frog
Discoglossus sardus Pl. 7, p. 63

34. Corsican Painted Frog
Discoglossus montalentii Pl. 7, p. 64

35. Common Midwife Toad
Alytes obstetricans Pl. 8, p. 64

36. Iberian Midwife Toad
Alytes cisternasii Pl. 8, p. 66

37. Southern Midwife Toad
Alytes dickhilleni Pl. 8, p. 67

38. Majorcan Midwife Toad
Alytes muletensis Pl. 8, p. 67

39. Western Spadefoot
Pelobates cultripes Pl. 9, p. 68

. Common Spadefoot
elobates fuscus Pl. 9, p. 69

41. Eastern Spadefoot
Pelobates syriacus Pl. 9, p. 69

42. Parsley Frog
Pelodytes punctatus Pl. 9, p. 71

. Iberian Parsley Frog
elodytes ibericus, Pl. 8, p. 72

44. Common Toad
Bufo bufo Pl.10, p. 73

45. Natterjack
Bufo calamita Pl.10, p. 74

. Green Toad
ufo viridis Pl.10, p. 76

47. Common Tree Frog
Hyla arborea Pl. 11, p. 77

48. Italian Tree Frog
Hyla intermedia, p. 78

49. Tyrrhenian Tree Frog
Hyla sarda Pl. 11, p. 78

50. Stripeless Tree Frog
Hyla meridionalis Pl. 11, p. 79
(Also Canary Islands, Madeira)

51. Common Frog
Rana temporaria Pl. 12, p. 80

52. Pyrenean Frog
Rana pyrenaica Pl. 12, p. 81

53. Moor Frog
Rana arvalis Pl. 12, p. 82

54. Agile Frog
Rana dalmatina Pl. 13, p. 83

55. Italian Agile Frog
Rana latastei Pl. 13, p. 84

56. Balkan Stream Frog
Rana graeca Pl. 13, p. 85

57. Italian Stream Frog
Rana italica, p. 86

58. Iberian Frog
Rana iberica Pl. 13, p. 87

59. Marsh Frog
Rana ridibunda Pl. 14, p. 89
(Widely introduced in France)

60. Pool Frog
Rana lessonae Pl. 14, p. 90

61. Edible Frog
Rana kl. *esculenta* Pl. 14, p. 92

62. Iberian Water Frog
Rana perezi, p. 93
(Also Canary Islands, Madeira)

63. Graf's Hybrid Frog
Rana kl. *grafi*, p. 93

64. Italian Pool Frog
Rana bergeri, p. 94

65. Italian Hybrid Frog
Rana kl. *hispanica*, p. 94

66. Albanian Pool Frog
Rana shqiperica

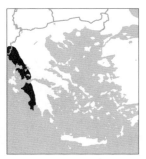

67. Epirus Water Frog
Rana epeirotica, p. 95

68. Greek Marsh Frog
Rana balcanica, p. 96

69. Karpathos Water Frog
Rana cerigensis, p. 96

70. Cretan Water Frog
Rana cretensis, p. 96

71. Levant Water Frog
Rana bedriagae, p. 97

72. American Bullfrog
Rana catesbeiana Pl. 14, p. 97

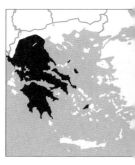

73. Hermann's Tortoise
Testudo hermanni Pl. 15, p. 100

74. Spur-thighed Tortoise
Testudo graeca Pl. 15, p. 101
(Introductions elsewhere)

75. Marginated Tortoise
Testudo marginata Pl. 15, p. 102
(Also Sardinia)

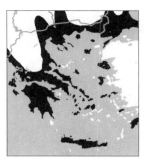

76. European Pond Terrapin
Emys orbicularis Pl. 16, p. 103
(Reintroduced into C. Europe)

77. Spanish Terrapin
Mauremys leprosa Pl. 16, p. 104

78. Balkan Terrapin
Mauremys rivulata Pl. 16, p. 105

79. Starred Agama
Laudakia stellio Pl. 19, p. 118
(Also Corfu, Malta)

80. Mediterranean Chameleon
Chamaeleo chamaeleon Pl. 19,
p. 119
(Introduced elsewhere)

81. African Chameleon
Chamaeleo africanus Pl. 19, p. 120

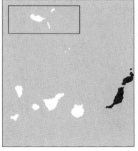

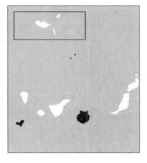

82. Moorish Gecko
Tarentola mauritanica Pl. 20, p. 121
(Also Madeira)

83. East Canary Gecko
Tarentola angustimentalis, p. 122

84. Gran Canaria Gecko
Tarentola boettgeri, p. 122

274 DISTRIBUTION MAPS

85. Tenerife Gecko
Tarentola delalandii, p. 123

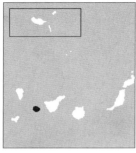

86. La Gomera Gecko
Tarentola gomerensis, p. 123

87. Turkish Gecko
Hemidactylus turcicus Pl. 20,
p. 124
(Also Canary Islands)

88. European Leaf-toed Gecko
Euleptes europaea, Pl. 20,
p. 124

89. Kotschy's Gecko
Cyrtopodion kotschyi Pl. 20,
p. 125

90. Large Psammodromus
Psammodromus algirus Pl. 21,
p. 127

91. Spanish Psammodromus
Psammodromus hispanicus
Pl. 21, p. 128

92. Spiny-Footed Lizard
Acanthodactylus erythrurus
Pl. 21, p. 129

93. Steppe Runner
Eremias arguta Pl. 21, p. 130

. Snake-eyed Lacertid
phisops elegans Pl. 21, p. 130

95. Dalmatian Algyroides
Algyroides nigropunctatus Pl. 22,
p. 131

96. Greek Algyroides
Algyroides moreoticus Pl.22,
p. 132

7. Pygmy Algyroides
Algyroides fitzingeri Pl. 22, p. 132

98. Spanish Algyroides
Algyroides marchi Pl. 22, p. 133

99. Ocellated Lizard
Lacerta lepida Pls 23 and 26,
p. 136

00. Schreiber's Green Lizard
Lacerta schreiberi Pls 23 and 26,
. 137

101. Green Lizard
Lacerta viridis Pls 24 and 26,
p. 138

102. Balkan Green Lizard
Lacerta trilineata Pls 21 and 22,
p. 139

103. Sand Lizard
Lacerta agilis Pls 25 and 26,
p. 140

104. Viviparous Lizard
Lacerta vivipara Pl. 27, p. 144

105. Common Wall Lizard
Podarcis muralis Pl. 27, p. 145

106. Iberian Rock Lizard
Lacerta monticola Pl. 28, p. 148

107. Pyrenean Rock Lizard
Lacerta bonnali Pl. 29, p. 150

108. Aurelio's Rock Lizard
Lacerta aurelioi Pl. 29, p. 151

109. Aran Rock Lizard
Lacerta aranica, p. 151

110. Iberian Wall Lizard
Podarcis hispanica Pl. 28, p. 151

111. Bocage's Wall Lizard
Podarcis bocagei Pl. 28, p. 152

112. Carbonell's Wall Lizard
Podarcis carbonelli Pl. 29, p. 153

113. Columbretes Wall Lizard
Podarcis atra Pl. 29, p. 154

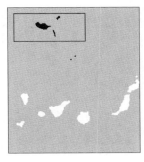

114. Madeira Lizard
Lacerta dugesii Pl. 29, p. 154

115. Moroccan Rock Lizard
Lacerta perspicillata Pl. 30, p. 156

116. Lilford's Wall Lizard
Podarcis lilfordi Pl. 30, p. 157

117. Ibiza Wall Lizard
Podarcis pityusensis Pl. 30, p. 158

118. Bedriaga's Rock Lizard
Lacerta bedriagae Pl. 31, p. 159

119. Tyrrhenian Wall Lizard
Podarcis tiliguerta Pl. 31, p. 160

120. Italian Wall Lizard
Podarcis sicula Pl. 32, p. 162

121. Sicilian Wall Lizard
Podarcis wagleriana Pl. 32, p. 163

122. Aeolian Wall Lizard
Podarcis raffonei Pl. 32, p. 164

123. Maltese Wall Lizard
Podarcis filfolensis Pl. 31, p. 165

124. Horvath's Rock Lizard
Lacerta horvathi Pl. 33, p. 167

125. Mosor Rock Lizard
Lacerta mosorensis Pl. 33, p. 168

126. Sharp-snouted Rock Lizard
Lacerta oxycephala Pl. 33, p. 168

127. Dalmatian Wall Lizard
Podarcis melisellensis Pl. 33, p. 169

128. Meadow Lizard
Lacerta praticola Pl. 34, p. 173

129. Greek Rock Lizard
Lacerta graeca Pl. 35, p. 174

130. Balkan Wall Lizard
Podarcis taurica Pl. 34, p. 175

131. Erhard's Wall Lizard
Podarcis erhardii Pl. 36, p. 176

132. Peloponnese Wall Lizard
Podarcis peloponnesiaca Pl. 35,
p. 177

133. Milos Wall Lizard
Podarcis milensis Pl. 36, p. 178

134. Skyros Wall Lizard
Podarcis gaigeae, p. 178

135. Caucasian Rock Lizard
Lacerta saxicola Pl. 34, p. 179

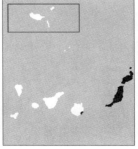

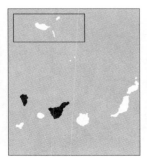

136. Anatolian Rock Lizard
Lacerta anatolica Pl. 35, p. 179

137. Atlantic Lizard
Gallotia atlantica Pl. 37, p. 181

138. Tenerife Lizard
Gallotia galloti Pl. 37, p. 182

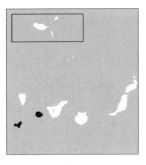

139. Boettger's Lizard
Gallotia caesaris Pl. 37, p. 182

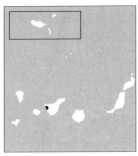

140. Tenerife Speckled Lizard
Gallotia intermedia Pl. 38, p. 183

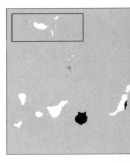

141. Gran Canaria Giant Lizard
Gallotia stehlini Pl. 38, p. 184

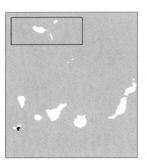

142. El Hierro Giant Lizard
Gallotia simonyi Pl. 38, p. 184

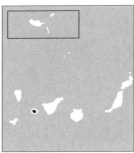

143. La Gomera Giant Lizard
Gallotia gomerana Pl. 38, p. 185

144. Snake-Eyed Skink
Ablepharus kitaibellii Pl. 39, p. 18

145. Ocellated Skink
Chalcides ocellatus Pl. 39, p. 187

146. Bedriaga's Skink
Chalcides bedriagai Pl. 39, p. 188

147. Western Three-toed Skink
Chalcides striatus Pl. 39, p. 188

8. Italian Three-toed Skink
Chalcides chalcides Pl. 39, p. 189

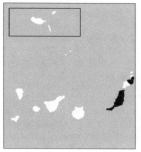

149. East Canary Skink
Chalcides simonyi Pl. 40, p. 189

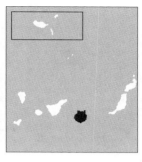

150. Gran Canaria Skink
Chalcides sexlineatus Pl. 40,
p. 190

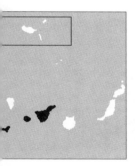

51. West Canary Skink
Chalcides viridianus Pl.40, p. 190

152. Levant Skink
Mabuya aurata Pl. 40, p. 191

153. Limbless Skink
Ophiomorus punctatissimus
Pl. 41, p. 191

54. Slow Worm
Anguis fragilis Pl. 41, p. 192

155. Peloponnese Slow Worm
Anguis cephallonica, p. 193

156. European Glass Lizard
Ophisaurus apodus Pl. 41, p. 193

157. Iberian Worm Lizard
Blanus cinereus Pl. 41, p. 194

158. Anatolian Worm Lizard
Blanus strauchi Pl. 41, p. 194

159. Worm Snake
Typhlops vermicularis Pl. 42, p. 20
(Also N to S Croatia)

160. Sand Boa
Eryx jaculus Pl. 42, p. 201
(Also N Bulgaria, SE Romania)

161. Montpellier Snake
Malpolon monspessulanus Pl. 42,
p. 202

162. Horseshoe Whip Snake
Coluber hippocrepis Pl. 42, p. 20

163. Algerian Whip Snake
Coluber algirus Pl. 43, p. 205

164. Dahl's Whip Snake
Coluber najadum Pl. 44, p. 205

165. Reddish Whip Snake
Coluber collaris Pl. 44, p. 206

66. Western Whip Snake
Coluber viridiflavus Pl. 43, p. 207

167. Balkan Whip Snake
Coluber gemonenesis Pl. 43,
p. 208

168. Gyaros Whip Snake
Coluber gyarosensis Pl. 43, p. 208

69. Large Whip Snake
Coluber caspius Pl. 43, p. 209

170. Coin-Marked Snake
Coluber nummifer Pl. 43, p. 210

171. Dwarf Snake
Eirenis modestus Pl. 44, p. 210

72. Leopard Snake
Elaphe situla Pl. 44, p. 211

173. Four-lined Snake
Elaphe quatuorlineata Pl. 45, p. 211

174. Steppe Snake
Elaphe dione Pl. 44, p. 213

175. Aesculapian Snake
Elaphe longissima Pl. 45, p. 214

176. Italian Aesculapian Snake
Elaphe lineata Pl. 45, p. 215

177. Ladder Snake
Elaphe scalaris Pl. 45, p. 215

178. Grass Snake
Natrix natrix Pl. 46, p. 216

179. Viperine Snake
Natrix maura Pl. 46, p. 218

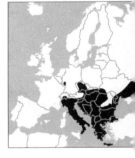

180. Dice Snake
Natrix tessellata Pl.46, p. 220

181. Smooth Snake
Coronella austriaca Pl. 47, p. 221

182. Southern Smooth Snake
Coronella girondica Pl. 47, p. 222

183. False Smooth Snake
Macroprotodon cucullatus Pl. 47
p. 223

184. Cat Snake
Telescopus fallax Pl. 47, p. 225

185. Orsini's Viper
Vipera ursinii Pl. 48, p. 228

186. Adder
Vipera berus Pl. 48, p. 230

187. Seoane's Viper
Vipera seoanei Pl. 48, p. 231

188. Asp Viper
Vipera aspis Pls. 48 and 49,
p. 232

189. Lataste's Viper
Vipera latasti Pl. 49, p. 234

190. Nose-horned Viper
Vipera ammodytes Pl. 49, p. 235

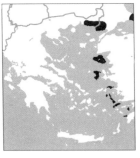

191. Ottoman Viper
Vipera xanthina Pl. 49, p. 235

192. Milos Viper
Vipera schweizeri Pl. 49, p. 237

INDEX

In this index, bold numbers refer to the colour plates. Subspecies are not included, but can be found in the text of the species concerned.

Ablepharus kitaibelii, **39**, 186
Acanthodactylus erythrurus, **21**, 129
Adder, **48**, 230
Agama, Starred, **19**, 118
Agamas,**118**
Agama stellio, **19**, 118
Agamidae, **118**
Agkistrodon, 2/3
Algyroides, Dalmatian, **22**, 131
  Greek, **22**, 132
  Pygmy, **22**, 132
  Spanish, **22**, 133
Algyroides fitzingeri, **22**, 132
  hidalgoi, **22**, 133
  marchi, **22**, 133
  moreoticus, **22**, 133
  nigropunctatus, **22**, 131
Alytes cisternasii, **8**, 66
  dickhelleni, **8**, 67
  muletensis, **8**, 67
  obstetricans, **8**, 64
Amphisbaena alba, 194
Amphisbaenidae, 194
Anguidae, 191
Anguis apodus, 193
  cephallonica, 193
  fragilis, **41**, 192
  peloponnesiaca, 193
Bataguridae, 103
Blanus cinereus, **41**, 194
  strauchi, **41**, 194
Boa, Sand, **42**, 201
Bombina bombina, **11**, 61
  variegata, **11**, 60
Boidae, 201
Bufo bufo, **10**, 73
  calamita, **10**, 74
  mauritanicus, **10**, 74
  viridis, **10**, 76
Bufonidae, 73
Bullfrog, American, 97
Caretta caretta, **17**, 107
Chalcides bedriagai, **39**, 188
  chalcides, **39**, 189
  ocellatus, **39**, 187
  pistaceae, **188**
  sexlineatus, **40**, 190
  simonyi, **40**, 189
  striatus, **39**, 188
  viridianus, **40**, 190
Chamaeleo africanus, **19**, 120
  chamaeleon, **19**, 119
Chamaeleonidae, 119
Chameleon, African, 120
  Mediterranean, 119

Chameleons, 119
  Chelonia mydas, **18**, 108
Cheloniidae, 106
Chelydra serpentina, 26
Chioglossa lusitanica, **2**, 35
Chrysemys picta, **26**, 66
Coluber algirus, **43**, 205
  caspius, **43**, 209
  collaris, **44**, 206
  gemonensis, **43**, 208
  gyarosensis, **43**, 208
  hippocrepis, **42**, 204
  jugularis, **43**, 209
  najadum, **44**, 205
  nummifer, **43**, 210
  rubriceps, **44**, 206
  viridiflavus, **43**, 207
Colubridae, 202
Coronella austriaca, **47**, 221
  girondica, **47**, 222
Cyrtodactylus kotschyi, 126
Cyrtopodion kotschyi, **20**, 125
Dermochelyidae, **106**
Dermochelys coriacea, **18**, 110
Discoglossidae, 59
Discoglossus galganoi, 62
  jeanneae, 63
  montalentii, **7**, 64
  nigriventer 64
  pictus, **7**, 62
  sardus, **7**, 63
Eirenis modestus, **44**, 210
Elaphe dione, **44**, 213
  lineata, **45**, 215
  longissima, **45**, 214
  quatuorlineata, **45**, 211
  rechingeri, 212
  sauromates, **45**, 212
  scalaris, **45**, 215
  situla, **44**, 211
Emydidae, 103
Emys orbicularis, 16
Eremias arguta, **21**, 130
Eretmochelys imbricata, **18**, 109
Eryx jaculus, **42**, 201
Euleptes europaea, **20**, 124
Euproctus asper, **3**, 37
  montanus, **3**, 38
  platycephalus, **3**, 39
Frog, Agile, **13**, 83
  Albanian Pool, 95
  American Bull, **14**, 97
  Balkan Stream, **13**, 85
  Common, **12**, 80
  Common Tree, **11**, 77
  Corsican Painted, **7**, 64

  Cretan Water, 96
  East Iberian painted, 63
  Edible, **14**, 92
  Epirus Water, 95
  Graf's Hybrid, 93
  Grass, 81
  Greek Marsh, 96
  Iberian, **13**, 87
  Iberian Parsley, **8**, 72
  Iberian Water, 93
  Italian Agile, **13**, 84
  Italian Hybrid, 94
  Italian Pool, 94
  Italian Stream, 86
  Italian Tree, 78
  Karpathos Water, 96
  Lake, 90
  Levant Painted, 64
  Levant Water, 97
  Marsh, **14**, 89
  Moor, **12**, 82
  Painted, **7**, 62
  Parsley, 71
  Pool, **14**, 90
  Pyrenean, **12**, 81
  Stripeless Tree, **11**, 79
  Tyrrhenian Painted, **7**, 63
  Tyrrhenian Tree, **11**, 70
  West Iberian Painted, 62
Frogs, Brown
  Parsley, 71
  Painted, 59
  Tree, 77
  Water, 88
Gallotia, 180
Gallotia atlantica, **37**, 181
  caesaris, **37**, 182
  galloti, **37**, 182
  goliath, 185
  gomerana, **38**, 185
  intermedia, **38**, 183
  simonyi, **38**, 184
  stehlini, **38**, 184
Gecko, East Canary, 122
  European leaf-toed, **20**, 124
  Gran Canaria, 122
  Kotschy's, **20**, 125
  La Gomera, 123
  Moorish, **20**, 121
  Tenerife, 123
  Turkish, **20**, 124
Geckos, 120
Gekkota, 120
Hardun, 118
Hemidactylus turcicus, **20**, 124
Hydromantes, 50

Hyla arborea, **11**, 77
　intermedia, **11**, 78
　italica, 78
　maculata, 78
　meridionalis, **11**, 79
　sarda, **11**, 78
　variegata, 78
Hylidae, 77
Lacerta, 134, 142
Lacerta agilis, 25, 26, 140
　anatolica, **36**, 179
　aranica, 151
　aurelioi, **29**, 151
　bedriagae, **31**, 159
　bilineata, 138
　bonnali, **29**, 150
　cyreni, 150
　dugesii, **29**, 154
　graeca, **35**, 174
　horvathi, **33**, 167
　lepida, **23**, **26**, 136
　monticola, **28**, 148
　mosorensis, **33**, 168
　nevadensis, 136
　oertzeni, 186
　oxycephala, **33**, 168
　perspicillata, **30**, 156
　praticola, **34**, 173
　saxicola, **34**, 179
　schreiberi, **23**, 137
　trilineata, **24**, 139
　viridis, 25, 138
　vivipara, **27**, 144
Lacertas, Small, 142
Lacertidae, 126
Laudakia stellio, **19**, 118
Lepidochelys kempii, **17**, 107
　olivacea, 108
Lizard, Aeolian Wall, **32**, 164
　Anatolian Rock, **36**, 179
　Aran Rock, 151
　Atlantic, **37**, 181
　Aurelio's Rock, **29**, 151
　Balkan Green, **24**, **26**, 139
　Balkan Wall, **34**, 175
　Bedriaga's Rock, **31**, 159
　Bocage's Wall, **28**, 152
　Boettger's, **37**, 182
　Carbonell's Wall, **29**, 153
　Caucasian Rock, **34**, 179
　Columbretes Wall, **29**, 154
　Common, 145
　Common Wall, **27**, 145
　Dalmatian Wall, **33**, 169
　El Hierro Giant, **38**, 184
　Erhard's Wall, **36**, 170
　European glass, **41**, 193
　Gran Canaria Giant, **38**, 184
　Greek Rock, **35**, 174
　Green, **24**, **26**, 138
　Horvath's Rock, **33**, 167
　Iberian Rock, **28**, 148
　Iberian Wall, **28**, 151

Ibiza Wall, **30**, 158
Italian Wall, **32**, 162
La Gomera Giant, **38**, 185
Lilford's Wall, **30**, 157
Madeira, **29**, 154
Maltese Wall, **31**, 165
Meadow, **34**, 173
Milos Wall, **36**, 178
Moroccan Rock, **30**, 156
Mosor Rock, **33**, 168
Ocellated, **23**, **26**, 136
Peloponnese Wall, **35**, 177
Pyrenean Rock, **29**, 150
Sand, 25, **26**, 140
Schreiber's Green, **23**, **26**, 137
Sharp-snouted Rock, **33**, 168
Sicilian Wall, **32**, 163
Skyros Wall, **34**, 178
Snake-eyed, **21**, 130
Spanish Worm, 194
Spiny-footed, **21**, 129
Tenerife, **37**, 182
Tenerife Giant, 185
Tenerife Speckled, **38**, 183
Tyrrhenian Wall, **31**, 160
Viviparous, **27**, 144
Lizards, Green, 134
　Canary Island, 180
　Lacertid, 126
　Rock, 147
　Wall, 142
　Worm, 194
Mabuya aurata, **40**, 191
Macroprotodon cucullatus, **47**, 233
Macrovipera Schweizeri, 237
Malpolon monspessulanus, **42**, 202
Mauremys caspica, 105
　leprosa, **16**, 104
　rivulata, **16**, 105
Mertensiella luschani, **3**, 34
Natrix maura, **46**, 218
　megalocephala, **46**, 218
　natrix, **46**, 216
　tessellata, **46**, 220
Natterjack, **10**, 74
Newt, Alpine, **6**, 44
　Balkan Crested, 44
　Bosca's, **5**, 48
　Common, **5**, 46
　Corsican Brook, **3**, 38
　Danube Crested, **4**, 44
　Italian, **6**, 49
　Italian Crested, **4**, 44
　Marbled, **4**, 42
　Montandon's, **6**, 45
　Northern Crested, **4**, 43
　Palmate, **5**, 48
　Pyrenean Brook, **3**, 37
　Sardinian Brook, **3**, 39
　Sharp-ribbed, **3**, 36
　Smooth, 47
　Southern Marbled, 42
　Warty, 43

Newts, Pond, 39
Olm, **3**, 53
Ophiomorus punctatissimus, **41**, 191
Ophisaurus apodus, **41**, 193
Ophisops elegans, **21**, 130
Pelobates cultripes, **9**, 63
　fuscus, **9**, 69
　syriacus, **9**, 70
Pelobatidae,
Pelodiscus sinensis, 26
Pelodytidae, 71
Pelodytes ibericus, **8**, 72
　punctatus, **8**, 71
Phyllodactylus europaeus, 125
Plethodontidae, 50
Pleurodeles waltl, **3**, 36
Podarcis, 142
Podarcis ater, 154
　atra, **29**, 154
　bocagei, **28**, 152
　carbonelli, **29**, 153
　dugesii, **29**, 155
　erhardii, **36**, 176
　filfolensis, **31**, 165
　gaigeae, **34**, 172
　hispanica, **28**, 151
　hispanicus, 152
　lilfordi, **30**, 157
　melisellensis, **33**, 169
　milensis, **36**, 178
　muralis, **27**, 145
　peloponnesiaca, **35**, 177
　peloponnesiacus, 178
　perspicillata, **30**, 157
　pityusensis, **30**, 158
　raffonei, **32**, 164
　sicula, **32**, 162
　siculus, 163
　taurica, **34**, 175
　tauricus, 175
　tiliguerta, **31**, 160
　wagleriana, **32**, 163
　waglerianus, 164
Proteidae, 53
Proteus anguinus, **3**, 53
Psammodromus, Large, **21**, 127
　Spanish, **21**, 128
Psammodromus algirus, **21**, 127
　hispanicus, **21**, 128
Pseudopus apodus, 193
Rana arvalis, **12**, 82
　balcanica, 96
　bedriagae, 97
　bergeri, 94
　catesbeiana, **14**, 97
　cerigensis, 96
　cretensis, 96
　dalmatina, **13**, 83
　epeirotica, 95
　esculenta, **14**, 92
　graeca, **13**, 85
　grafi, 93
　hispanica, 94

iberica, 13, 87
italica, 86
kl. *esculenta*, 92
kl. *grafi*, 93
kl. *hispanica*, 94
kurtmuelleri, 96
latastei, 13, 84
lessonae, 14, 90
levantina, 97
maritima, 94
perezi, 93
pyrenaica, 12, 81
ridibunda, 14, 89
shqiperica, 95
temporaria, 12, 80
Ranidae, 80
Ridley, Kemp's, 17, 107
    Pacific, 108
Salamander, Alpine, 1, 33
    Ambrosi's cave, 51
    Corsican, 1, 33
    Fire, 1, 31
    Gené's Cave, 2, 521
    Golden-striped, 2, 351
    Italian Cave, 2, 51
    Lanza's Alpine, 1, 34
    Luschan's, 3, 34
    Monte Albo Cave, 2, 52
    Scented Cave, 2, 52
    Spectacled, 2, 35
    Supramontane Cave, 2, 52
Salamanders, Cave, 50
Salamandra atra, 1, 33
    corsica, 1, 33
    lanzai, 1, 34
    luschani, 34
    salamandra, 1, 31
Salamandridae, 30
Salamandrina terdigitata, 2, 35
Sand Boa, 42, 201
Scheltopusik, 193
Scincidae, 185
Skink, Bedriaga's, 188
    East Canary, 40, 189
    Gran Canaria, 40, 190
    Italian Three-toed, 39, 189
    Levant, 40, 191
    Limbless, 41, 191
    Ocellated, 39, 187
    Snake-eyed, 39, 186
    West Canary, 40, 190
    Western Three-toed, 39, 189
Skinks, 185
Slow Worm, 41, 192
    Peloponnese, 193
    Slow worms, 191
Snake, Aesculapian, 45, 214
    Algerian Whip, 43, 205
    Balkan Whip, 43, 208
    Blotched, 45, 212
    Cat, 47, 225
    Coin-marked, 43, 210
    Dahl's Whip, 44, 205

Dice, 46, 220
Dwarf, 44, 210
Dwarf Four-lined, 212
False Smooth, 47, 223
Four-lined, 45, 211
Gyaros Whip, 43, 208
Grass, 46, 216
Horseshoe Whip, 42, 204
Italian Aesulapian, 45, 215
Ladder, 45, 215
Large Whip, 43, 209
Leopard, 44, 211
Montpellier, 42, 203
Rechinger's, 212
Reddish Whip, 44, 206
Smooth, 47, 221
Southern Smooth, 47, 222
Steppe, 44, 213
Viperine, 46, 218
Western Four-lined, 212
Western Whip, 43, 207
Worm, 42, 200
Snake-eyed Lacertid, 21, 130
Spadefoots, 68
Spadefoot, Common, 9, 69
    Eastern, 9, 70
    Western, 9, 68
Spadefoots, 68
Speleomantes ambrosii, 2, 51
    flavus, 2, 52
    genei, 2, 52
    imperialis, 2, 52
    italicus, 2, 51
    strinatii, 51
    supramontis, 52
Stellio stellio, 118
Steppe runner, 130
Tarentola geckos, 122
Tarentola angustimentalis, 122
    boettgeri, 122
    delalandii, 123
    gomerensis, 123
    mauritanica, 20, 121
Teira dugesii, 155
    perspicillata, 157
Telescopus fallax, 47, 225
Tenuidactylus kotschyi, 126
Terrapin, Balkan, 16, 105
    Caspian, 105
    European Pond, 16, 103
    Painted, 106
    Red-eared, 16, 105
    Spanish, 16, 104
Testudinidae, 106
Testudo graeca, 15, 101
    hermanni, 15, 100
    marginata, 15, 102
    terrestris, 102
    weissingeri, 102
Timon lepidus, 137
Toad, Clawed, 7, 25
    Common, 10, 73
    Common Midwife, 8, 64

Fire-bellied, 11, 61
Green, 10, 76
Iberian Midwife, 8, 66
Majorcan Midwife, 67
Mauritanian, 74
Natterjack, 10, 74
Southern Midwife, 10, 67
Spadefoot, 9, 68
Yellow-bellied, 11, 60
Tortoise, Greek 102
    Hermann's, 15, 100
    Marginated, 15, 102
    Moorish, 102
    Spur-thighed, 15, 101
    Weissinger's 102
Trachemys scripta, 16, 105
Triturus alpestris, 6, 44
    blasii, 42
    boscai, 5, 48
    carnifex, 4, 44
    cristatus, 4, 43
    dobrogicus, 4, 44
    helveticus, 5, 48
    italicus, 6, 49
    karelinii, 4, 44
    marmoratus, 4, 42
    montandoni, 6, 45
    pygmaeus, 42
    vulgaris, 5, 46
Turtle, Chinese Soft-shelled 26
    Green, 18, 108
    Hawksbill, 18, 109
    Leathery, 18, 110
    Loggerhead, 17, 107
    Painted, 26, 106
    Snapping, 26
Typhlopidae, 200
Typhlops vermicularis, 42, 200
Viper, Asp, 48, 49, 232
    Blunt-nosed, 237
    Common, 231
    Lataste's, 49, 234
    Meadow, 230
    Milos, 49, 237
    Northern, 231
    Nose-horned, 49, 235
    Orsini's, 48, 228
    Ottoman, 49, 236
    Seoane's, 48, 231
Viperidae, 226
Vipers, Pit, 226
Vipera 226
Vipera, ammodytes, 49, 235
    aspis, 48, 49, 232
    berus, 48, 230
    latasti, 49, 234
    lebetina, 237
    schweizeri, 49, 237
    seoanei, 48, 231
    ursinii, 48, 228
    xanthina, 49, 236
Xenopus laevis, 7, 25
Zootoca vivipara, 145